Hamlet's Mill

God creating the stars, with the planetary spheres shown inside, according to the Ptolemaic order. Each sphere is marked by a star, the fourth sphere, that of the sun, being indicated by a half-visible circle.

Hamlet's Mill

An essay on myth and the frame of time

GIORGIO de SANTILLANA

&

HERTHA von DECHEND

A NONPAREIL BOOK

David R. Godine · Publisher · Boston

Much of the research for this book was supported by
a grant from the Twentieth Century Fund.

Lines from "As I Walked Out One Evening," by W. H. Auden.
Copyright © 1940 and renewed 1968 by W. H. Auden. Reprinted
from *Collected Shorter Poems 1927–1957*, by W. H. Auden,
by permission of Random House, Inc.

Line from Ulysses by James Joyce, reprinted by permission of
Random House, Inc. Copyright 1914, 1918 by Margaret Caroline
Anderson. Copyright 1934 by The Modern Library.
Copyright 1942, 1946 by Nora Joseph Joyce.
New edition, corrected and reset, 1961.

First paperback edition published in 1977 by
DAVID R. GODINE, PUBLISHER, INC.
Box 450
Jaffrey, New Hampshire 03452
www.godine.com

Library of Congress Catalog Card Number: 69-13267
ISBN: 978-0-87923-215-3

Tenth printing, 2015
Printed in the United States of America

Preface

As the senior, if least deserving, of the authors, I shall open the narrative.

Over many years I have searched for the point where myth and science join. It was clear to me for a long time that the origins of science had their deep roots in a particular myth, that of *invariance*. The Greeks, as early as the 7th century b.c., spoke of the quest of their first sages as the Problem of the One and the Many, sometimes describing the wild fecundity of nature as the way in which the Many could be deduced from the One, sometimes seeing the Many as unsubstantial variations being played on the One. The oracular sayings of Heraclitus the Obscure do nothing but illustrate with shimmering paradoxes the illusory quality of "things" in flux as they were wrung from the central intuition of unity. Before him Anaximander had announced, also oracularly, that the cause of things being born and perishing is their mutual injustice to each other in the order of time, "as is meet," he said, for they are bound to atone forever for their mutual injustice. This was enough to make of Anaximander the acknowledged father of physical science, for the accent is on the real "Many." But it was true science after a fashion.

Soon after, Pythagoras taught, no less oracularly, that "things are numbers." Thus mathematics was born. The problem of the origin of mathematics has remained with us to this day. In his high old age, Bertrand Russell has been driven to avow: "I have wished to know how the stars shine. I have tried to apprehend the Pythagorean power by which number holds sway above the flux. A little of this, but not much, I have achieved." The answers that he found, very great answers, concern the nature of logical clarity, but not of philosophy proper. The problem of number remains to perplex us, and from it all of metaphysics was born. As a historian, I went on investigating the "gray origins" of science, far into its pre-Greek beginnings, and how philosophy was born of it, to go on puzzling us. I condensed it into a small book, *The Origins of Scientific Thought*. For both philosophy and science came from that fountainhead; and it is clear that both were children of the same myth.[1] In a number of studies, I continued to pursue it under the name of "scientific rationalism"; and I tried to show that through all the immense developments, the "Mirror of Being" is always the object of true science, a metaphor which still attempts to reduce the Many to the One. We now make many clear distinctions, and have come to separate science from philosophy utterly, but what remains at the core is still the old myth of eternal invariance, ever more remotely and subtly articulated, and what lies beyond it is a multitude of procedures and technologies, great enough to have changed the face of the world and to have posed terrible questions. But they have not answered a single philosophical question, which is what myth once used to do.

If we come to think of it, we have been living in the age of Astronomical Myth until yesterday. The careful and rigorous edifice of Ptolemy's *Almagest* is only window dressing for Plato's theology, disguised as elaborate science. The heavenly bodies are moving in "cycle and epicycle, orb in orb" of a mysterious motion according to the divine decree that circular motions ever more intricate would account for the universe. And Newton himself, once he had accounted for it, simply replaced the orbs with the understandable force of gravitation, for which he "would feign no hypotheses." The hand of God was still the true motive force; God's will and God's own mathematics went on, another name for Aristotle's Prime Mover. And shall we deny that Einstein's spacetime is nothing other than a pure pan-mathematical myth, openly acknowledged at last as such?

I was at this point, lost between science and myth, when, on the occasion of a meeting in Frankfurt in 1959, I met Dr. von Dechend,

---

[1] The Pythagorean problem is at the core of my *Origins*. My efforts came eventually to fruition in my *Prologue to Parmenides* of 1964 (reprinted in *Reflections on Men and Ideas* [1968], p. 80).

one of the last pupils of the great Frobenius, whom I had known; and with her I recalled his favorite saying: "What the hell should I care for my silly notions of yesterday?" We were friends from the start. She was then Assistant to the Chair of the History of Science, but she had pursued her lonely way into cultural ethnology, starting in West Africa on the tracks of her "Chef," which were being opened up again at the time by that splendid French ethnologist, the late Marcel Griaule. She too had a sense that the essence of myth should be sought somewhere in Plato rather than in psychology, but as yet she had no clue.

By the time of our meeting she had shifted her attention to Polynesia, and soon she hit pay dirt. As she looked into the archaeological remains on many islands, a clue was given to her. The moment of grace came when, on looking (on a map) at two little islands, mere flyspecks on the waters of the Pacific, she found that a strange accumulation of *maraes* or cult places could be explained only one way: they, and only they, were both exactly sited on two neat celestial coordinates: the Tropics of Cancer and of Capricorn.

Now let Dechend take over the narrative:

"To start from sheer opposition to ruling opinions is not likely to lead to sensible insight, at least so we think. But anyhow, I did not start from there, although there is no denying that my growing wrath about the current interpretations (based upon discouraging translations) was a helpful spur now and then. In fact, there was nothing that could be called a 'start,' least of all the intention to explore the astronomical nature of myth. To the contrary, on my side, having come from ethnology to the history of science, there existed 'in the beginning' only the firm decision never to become involved in astronomical matters, under any condition. In order to keep safely away from this frightening field, my subject of inquiry was meant to be the mythical figure of the craftsman god, the Demiurge in his many aspects (Hephaistos, Tvashtri, Wayland the Smith, Goibniu, Ilmarinen, Ptah, Khnum, Kothar-wa-Hasis, Enki/Ea, Tane, Viracocha, etc.). Not even a whiff of suspicion came to me during the investigation of Mesopotamian myths—of all cultures!—everything looked so very terrestrial, though slightly peculiar. It was after having spent more than a year over at least

10,000 pages of Polynesian myths—collected in the 19th century (there are many more pages available than these)—that the annihilating recognition of our complete ignorance came down upon me like a sledge hammer: there was no single sentence that could be understood. But then, if anybody was entitled to be taken seriously, it had to be the Polynesians guiding their ships securely over the largest ocean of our globe, navigators to whom our much praised discoverers from Magellan to Captain Cook confided the steering of their ships more than once. Thus, the fault had to rest with us, not with Polynesian myth. Still, I did not then 'try astronomy for a change'—there *was* a strict determination on my part to avoid this field. I looked into the archaeological remains of the many islands, and there a clue was given to me (to call it being struck by lightning would be more correct) which I duly followed up, and then there was no salvation anymore: astronomy could not be escaped. First it was still 'simple' geometry—the orbit of the sun, the Tropics, the seasons—and the adventures of gods and heroes did not make much more sense even then. Maybe one should *count*, for a change? What could it mean, when a hero was on his way slightly more than two years, 'returning' at intervals, 'falling into space,' coming off the 'right' route? There remained, indeed, not many possible solutions: it had to be *planets* (in the particular case of Aukele-nui-a-iku, Mars). If so, planets had to be constitutive members of every mythical personnel; the Polynesians did not invent this trait by themselves."

This text of Professor von Dechend, in its intellectual freedom and audacity, bears the stamp of her inheritance from the heroic and innocent and cosmopolitan age of German science around the eighteen-thirties. Its heroes, Justus von Liebig and Friedrich Woehler, were the objects of her work done before 1953. Another of those virtues, scornful indignation, will come to the fore in the appendices, which are so largely the product of her efforts.

Now I resume:

Years before, I had once looked at Dupuis' *L'Origine de tous les cultes*, lost in the stacks of Widener Library, never again consulted. It was a book in the 18th-century style, dated "An III de la République." The title was enough to make one distrustful—one of those "enthusiastic" titles which abounded in the 18th century and promised far too much. How could it explain the Egyptian system, I thought, since hieroglyphics had not yet been deciphered? (Athanasius Kircher was later to show us how it was done out of Coptic tradition.) I had dropped the forbidding tome, only jotting down a sentence: "Le mythe est né de la science; la science seule l'expliquera." I had the answer there, but I was not ready to understand.

This time I was able to grasp the idea at a glance, because I was ready for it. Many, many years before, I had questioned myself, in a note, about the meaning of *fact* in the crude empirical sense, as applied to the ancients. It represents, I thought, not the intellectual surprise, not the direct wonder and astonishment, but first of all an immense, steady, minute attention to the seasons. What is a solstice or an equinox? It stands for the capacity of coherence, deduction, imaginative intention and reconstruction with which we could hardly credit our forefathers. And yet there it was. I *saw*.

Mathematics was moving up to me from the depth of centuries; not after myth, but before it. Not armed with Greek rigor, but with the imagination of astrological power, with the understanding of astronomy. Number gave the key. Way back in time, before writing was even invented, it was *measures* and *counting* that provided the armature, the frame on which the rich texture of real myth was to grow.

Thus we had returned to the true beginnings, in the Neolithic Revolution. We agreed that revolution was essentially technological. The earliest social scientist, Democritus of Abdera, put it in one striking sentence: men's progress was the work not of the mind but of the *hand*. His late successors have taken him too literally, and concentrated on artifacts. They have been unaware of the enormous intellectual effort involved, from metallurgy to the arts, but especially in astronomy. The effort of sorting out and identifying the only presences which totally eluded the action of our hands led to those pure objects of contemplation, the stars in their courses. The Greeks would not have misapprehended that effort: they called astronomy the Royal Science. The effort at organizing the cosmos took shape from the supernal presences, those alone which thought might put in control of reality, those from which all arts took their meaning.

But nothing is so easy to ignore as something that does not yield freely to understanding. Our science of the past flowered in the fullness of time into philology and archaeology, as learned volumes on ancient philosophy have continued to pour forth, to little avail. A few masters of our own time have rediscovered these "preliterate" accomplishments. Now Dupuis, Kircher and Boll are gone like those archaic figures, and are equally forgotten. That is the devouring way of time. The iniquity of oblivion blindly scattereth her poppies.

It is well known how many images of the gods have to do with the making of fire, and an American engineer, J. D. McGuire, discovered that also certain Egyptian images, until then unsuspected, presented deities handling a fire drill. Simple enough: fire itself was the link between what the gods *did* and what man could do. But from there, the mind had once been able to move on to prodigious feats of intellect. That world of the mind was fully worthy of those Newtons and Einsteins long forgotten—those masters, as d'Alembert put it, of whom we know nothing, and to whom we owe everything.

We had the idea. It was simple and clear. But we realized that we would run into formidable difficulties, both from the point of view of modern, current scholarship and from the no less unfamiliar approach needed for method. I called it playfully, for short, "the cat on the keyboard," for reasons that will appear presently. For how can one catch time on the wing? And yet the flow of time, the time of music, was of the essence, inescapable, baffling to the systematic mind. I searched at length for an inductive way of presentation. It was like piling Pelion upon Ossa. And yet this was the least of our difficulties. For we also had to face a wall, a veritable Berlin Wall, made of indifference, ignorance, and hostility. Humboldt, that wise master, said it long ago: First, people will deny a thing; then they will belittle it; then they will decide that it had been known long ago. Could we embark upon an enormous task of detailed scholarship on the basis of this more than dubious prospect? But our own task was set: to rescue those intellects of the past, distant and recent, from oblivion. "Thus saith the Lord God: 'Come from the four winds, O breath, and breathe upon these slain, that they may live.' " Such poor scattered bones, *ossa vehementer sicca*, we had to revive.

This book reflects the gradually deepening conviction that, first of all, respect is due these fathers of ours. The early chapters will make, I think, for easy reading. Gradually, as we move above timberline, the reader will find himself beset by difficulties which are not of our making. They are the inherent difficulties of a science which was fundamentally reserved, beyond our conception. Most frustrating, we could not use our good old simple catenary logic, in which principles come first and deduction follows. This was not the way of the archaic thinkers. They thought rather in terms of what we might call a fugue, in which all notes cannot be constrained into a single melodic scale, in which one is plunged directly into the midst of things and must follow the temporal order created by their thoughts. It is, after all, in the nature of music that the notes cannot all be played at once. The order and sequence, the very meaning, of the composition will reveal themselves—with patience—in due time. The reader, I suggest, will have to place himself in the ancient "Order of Time."

Troilus expressed the same idea in a different image: "He that will have a cake out of the wheat must needs tarry the grinding."

GIORGIO DE SANTILLANA

ACKNOWLEDGMENTS

Illustration from a postcard. Copyright 1962 by Verlag Karl Alber. Reprinted by permission of Verlag Karl Alber, Freiburg im Breisgau.

Figures 1 and 2 on page 21 in "Études sur la cosmologie des Dogon et des Bambara du Soudan Français" by D. Zahan and S. de Ganay, *Africa*, vol. 21, 1951. Copyright 1951 by D. Zahan, S. de Ganay, and the International African Institute. Reprinted by permission of the International African Institute, London.

Figure 6 on page 21 in *The Carta Marina of Olaus Magnus: Venice 1539 and Rome 1572* by Edward Lynam, Tall Tree Library Publication 12, 1941. Reprinted by permission of Tall Tree Library, Jenkintown, Pa.

Illustrations on page 26 and facing page 48 in *Gesammelte. Werke*, vol. 8, by Johannes Kepler, ed. Franz Hammer, 1963. Reprinted by permission of C. H. Beck'sche Verlagsbuchhandlung, Munich.

Illustration on page 179 in *Gesammelte Werke*, vol. 1, by Johannes Kepler, ed. Max Caspar, 1938. Reprinted by permission of C. H. Beck'sche Verlagsbuchhandlung, Munich.

Figures 63 and 64 on pages 44 and 45 in *The Flammarion Book of Astronomy*, ed. Camille Flammarion, Simon and Schuster, 1964. Copyright 1955 by Librairie Flammarion. Reprinted by permission of Flammarion Publishers, Paris.

Figure 177 on page 742 in "Primitive Methods of Drilling" by J. D. McGuire, Annual Report of the U.S. National Museum, 1894. Reprinted by permission of the Smithsonian Institution Press, Washington, D.C.

Figures 724 and 970 on pages 663 and 748 in *Gesammelte Abhandlungen*, vol. 4, by E. Seler, 1960. Copyright 1960 by Akademische Druck- und Verlagsanstalt. Reprinted by permission of Akademische Druck- und Verlagsanstalt, Graz.

xv · *Acknowledgments*

Illustration on page 96 of the *Codices Selecti Phototypice Impressi*, vol. 8, ed. Dr. F. Anders, 1967. Copyright 1952 by Akademische Druck- und Verlagsanstalt. Reprinted by permission of Akademische Druck- und Verlagsanstalt, Graz.

Illustration on page 434 in *L'Uranographie Chinoise* by G. Schlegel, 1875. Reprinted by permission of Martinus Nijhoff, The Hague.

Figure 104 on page 377 in *Science and Civilisation in China*, vol. 3, by J. Needham, 1959. Copyright 1959 by Cambridge University Press. Reprinted by permission of Cambridge University Press, New York.

Plate II facing page 22 in *Die Geschichte der Sternkunde* by E. Zinner, 1931. Reprinted by permission of Springer Verlag, Berlin.

Figures 36, 70, 71, and 75 on pages 91, 117, and 120 in *Die Erscheinungen am Sternenhimmel* by H. v. Baravalle, 1962. Copyright 1962 by Verlag Freies Geistesleben. Reprinted by permission of Verlag Freies Geistesleben, Stuttgart.

Illustration on page 289 in *The Dawn of Astronomy* by J. N. Lockyer, 1894. Reprinted by permission of the Massachusetts Institute of Technology Press, Cambridge, Mass.

Figure 15 on plate IV in *Catalogue of Engraved Gems, Greek, Etruscan and Roman* by G. M. A. Richter, 1956. Reprinted by permission of the Metropolitan Museum of Art, Fletcher Fund, 31.11.14, New York.

Figure 70 on page 540 in "Animal Figures on Prehistoric Pottery from Mimbres Valley, New Mexico" by J. W. Fewkes. Reprinted by permission of the American Anthropological Association from the *American Anthropologist*, vol. 18, 1916, Washington, D.C.

Figures 1427 and 1444 on plates 107 and 109 in *La Glyptique Mesopotamienne Archaïque* by P. Amiet, 1961. Reprinted by permission of the Centre National de la Recherche Scientifique, Paris.

Figures 7 and 9 on pages 66 and 69 in *Anfänge der Astronomie* by B. L. van der Waerden, 1965. Copyright 1965 by N. V. Erben P. Noordhoff's Uitgeverszaak, renewed 1968 by Birkhäuser Verlag. Reprinted by permission of Birkhäuser Verlag, Basel.

Figure 17 on page 99 in *L'Arbre Cosmique dans la Pensée populaire et dans la Vie quotidienne du Nord-Ouest Africain* by Viviana Pâques, 1964. Copyright 1965 by Institut d'Ethnologie. Reprinted by permission of Institut d'Ethnologie, Paris.

Illustrations on pages 67 and 68 in "Ein zweites Goldland Salomos" by J. Dahse, *Zeitschrift für Ethnologie*, vol. 43, 1911. Reprinted by permission of Dr. Günther Hartmann.

Drawing of the Precession of the Equinoxes by Stefan Fuchs. Reprinted by permission of Stefan Fuchs, University of Frankfurt, Germany.

We are indebted to Mrs. Katharina Lommel, Staatliches Museum für Völkerkunde, München, for obtaining most of the illustrations used in our book.

CONTENTS

ILLUSTRATIONS

ABBREVIATIONS

ABAW	Abhandlungen der Bayerischen Akademie der Wissenschaften
AEG. WB.	Wörterbuch der Aegyptischen Sprache
AEG. Z.	Zeitschrift für Aegyptische Sprache und Altertumskunde
AFO	Archiv für Orientforschung
AJSL	American Journal of Semitic Languages and Literature
ANET	Ancient Near Eastern Texts relating to the Old Testament
AN. OR.	Analecta Orientalia (Roma)
AOTAT	Altorientalische Texte zum Alten Testament
APAW	Abhandlungen der Preussischen Akademie der Wissenschaften
AR	Annual Report
ARBAE	Annual Report of the Bureau of American Ethnology (Washington)
ARW	Archiv für Religionswissenschaft
ATAO	A. Jeremias: Das Alte Testament im Lichte des Alten Orients
AV	Atharva Veda
BA	Baessler Archiv (Berlin)
BAE	Bureau of American Ethnology
BASOR	Bulletin of the American Schools of Oriental Research
BIFAO	Bulletin de l'Institut Français d'Archéologie Orientale (Cairo)
BPB MUS.	Bernice Pauahi Bishop Museum (Honolulu)
BVSGW	Berichte über die Verhandlungen der Sächsischen Gesellschaft der Wissenschaften (Leipzig)
BT	Bibliotheca Teubneriana
EE	Enuma elish, the Babylonian Creation Epic
ERE	Encyclopaedia of Religion and Ethics (ed. James Hastings)

FFC	Folklore Fellows Communications (Helsinki)
FUF	Finnisch-Ugrische Forschungen
GE	Gilgamesh Epic
HAOG	A. Jeremias: Handbuch der Altorientalischen Geisteskultur
HUCA	Hebrew Union College Annual (Cincinnati)
IAFE	Internationales Archiv für Ethnographie (Leiden)
JAOS	Journal of the American Oriental Society
JCS	Journal of Cuneiform Studies
JNES	Journal of Near Eastern Studies
JRAS	Journal of the Royal Asiatic Society
JSA	Journal de la Société des Africanistes
LCL	Loeb Classical Library
MAGW	Mitteilungen der Anthropologischen Gesellschaft Wien
MAR	Mythology of All Races (Boston)
MBH.	Mahabharata
MVAG	Mitteilungen der Vorderasiatischen Gesellschaft
OLZ	Orientalistische Literaturzeitung
OR.	Orientalia, New Series (Roma)
PB	A. Deimel: Pantheon Babylonicum
RA	Revue d'Assyriologie et d'Archéologie Orientale
RC	Revue Celtique
RE	Realencyclopaedie der Klassischen Altertumswissenschaften (ed. Pauly-Wissowa)
RH. MUS.	Rheinisches Museum für Philologie
RLA	Reallexikon der Assyriologie
ROSCHER	Ausführliches Lexikon der griechischen und römischen Mythologie
RV	Rigveda
SBAW	Sitzungsberichte der Bayerischen Akademie der Wissenschaften
SBE	Sacred Books of the East

SHAW	Sitzungsberichte der Heidelberger Akademie der Wissenschaften
SOAW	Sitzungsberichte der Oesterreichischen Akademie der Wissenschaften
SPAW	Sitzungsberichte der Preussischen Akademie der Wissenschaften
TM	J. Grimm: Teutonic Mythology
WB. MYTH.	Wörterbuch der Mythologie
WZKM	Wiener Zeitschrift für die Kunde des Morgenlandes
ZA	Zeitschrift für Assyriologie und vorderasiatische Archaeologie
ZDMG	Zeitschrift der Deutschen Morgenländischen Gesellschaft
ZFE	Zeitschrift für Ethnologie
ZVV	Zeitschrift des Vereins für Volkskunde

Introduction

The unbreakable fetters which bound down the Great Wolf Fenrir had been cunningly forged by Loki from these: the footfall of a cat, the roots of a rock, the beard of a woman, the breath of a fish, the spittle of a bird.

The Edda

Toute vue des choses qui n'est pas étrange est fausse.

Valéry

THIS IS meant to be only an essay. It is a first reconnaissance of a realm well-nigh unexplored and uncharted. From whichever way one enters it, one is caught in the same bewildering circular complexity, as in a labyrinth, for it has no deductive order in the abstract sense, but instead resembles an organism tightly closed in itself, or even better, a monumental "Art of the Fugue."

The figure of Hamlet as a favorable starting point came by chance. Many other avenues offered themselves, rich in strange symbols and beckoning with great images, but the choice went to Hamlet because he led the mind on a truly inductive quest through a familiar landscape—and one which has the merit of its literary setting. Here is a character deeply present to our awareness, in whom ambiguities and uncertainties, tormented self-questioning and dispassionate insight give a presentiment of the modern mind. His personal drama was that he had to be a hero, but still try to avoid the role Destiny assigned him. His lucid intellect remained above the conflict of motives—in other words, his was and is a truly contemporary consciousness. And yet this character whom the poet made one of us, the first unhappy intellectual, concealed a past as a legendary being, his features predetermined, preshaped by long-standing myth. There was a numinous aura around him, and many clues led up to him. But it was a surprise to find behind the mask an ancient and all-embracing cosmic power—the original master of the dreamed-of first age of the world.

Yet in all his guises he remained strangely himself. The original Amlodhi,* as his name was in Icelandic legend, shows the same characteristics of melancholy and high intellect. He, too, is a son dedicated to avenge his father, a speaker of cryptic but inescapable truths, an elusive carrier of Fate who must yield once his mission is accomplished and sink once more into concealment in the depths of time to which he belongs: Lord of the Golden Age, the Once and Future King.

This essay will follow the figure farther and farther afield, from the Northland to Rome, from there to Finland, Iran, and India; he will appear again unmistakably in Polynesian legend. Many other Dominations and Powers will materialize to frame him within the proper order.

Amlodhi was identified, in the crude and vivid imagery of the Norse, by the ownership of a fabled mill which, in his own time, ground out peace and plenty. Later, in decaying times, it ground out salt; and now finally, having landed at the bottom of the sea, it is grinding rock and sand, creating a vast whirlpool, the Maelstrom (i.e., the grinding stream, from the verb *mala*, "to grind"), which is supposed to be a way to the land of the dead. This imagery stands, as the evidence develops, for an astronomical process, the secular shifting of the sun through the signs of the zodiac which determines world-ages, each numbering thousands of years. Each age brings a World Era, a Twilight of the Gods. Great structures collapse; pillars topple which supported the great fabric; floods and cataclysms herald the shaping of a new world.

The image of the mill and its owner yielded elsewhere to more

* The indulgence of specialists is asked for the form of certain transliterations throughout the text; for example, Amlodhi instead of Amlodi, Grotte instead of Grotti, etc. (Ed.)

sophisticated ones, more adherent to celestial events. In Plato's powerful mind, the figure stood out as the Craftsman God, the Demiurge, who shaped the heavens; but even Plato did not escape the idea he had inherited, of catastrophes and the periodic rebuilding of the world.

Tradition will show that the measures of a new world had to be procured from the depths of the celestial ocean and tuned with the measures from above, dictated by the "Seven Sages," as they are often cryptically mentioned in India and elsewhere. They turn out to be the Seven Stars of Ursa, which are normative in all cosmological alignments on the starry sphere. These dominant stars of the Far North are peculiarly but systematically linked with those which are considered the operative powers of the cosmos, that is, the planets as they move in different placements and configurations along the zodiac. The ancient Pythagoreans, in their conventional language, called the two Bears the Hands of Rhea (the Lady of Turning Heaven), and called the planets the Hounds of Persephone, Queen of the Underworld. Far away to the south, the mysterious ship *Argo* with its Pilot star held the depths of the past; and the Galaxy was the Bridge out of Time. These notions appear to have been common doctrine in the age before history—all over the belt of high civilizations around our globe. They also seem to have been born of the great intellectual and technological revolution of the late Neolithic period.

The intensity and richness, the coincidence of details, in this cumulative thought have led to the conclusion that it all had its origin in the Near East. It is evident that this indicates a diffusion of ideas to an extent hardly countenanced by current anthropology. But this science, although it has dug up a marvelous wealth of details, has been led by its modern evolutionary and psychological bent to forget about the main source of myth, which was astronomy—the Royal Science. This obliviousness is itself a recent turn of events—barely a century old. Today expert philologists tell us that Saturn and Jupiter are names of vague deities, subterranean or atmospheric, superimposed on the planets at a "late" period; they neatly sort out folk origins and "late" derivations, all unaware that planetary periods, sidereal and synodic, were known and rehearsed in numerous ways by celebrations already traditional in archaic times. If a scholar has never known those periods even from elementary science, he is not in the best position to recognize them when they come up in his material.

Ancient historians would have been aghast had they been told that obvious things were to become unnoticeable. Aristotle was proud to state it as known that the gods were originally stars, even if popular fantasy had later obscured this truth. Little as he believed in progress, he felt this much had been secured for the future. He could not guess that W. D. Ross, his modern editor, would condescendingly annotate: "This is historically untrue." Yet we know that Saturday and Sabbath had to do with Saturn, just as Wednesday and Mercredi had to do with Mercury. Such names are as old as time; as old, certainly, as the planetary heptagram of the Harranians. They go back far before Professor Ross' Greek philology. The inquiries of great and meticulous scholars such as Ideler, Lepsius, Chwolson, Boll and, to go farther back, of Athanasius Kircher and Petavius, had they only been read carefully, and noted, would have taught several relevant lessons to the historians of culture, but interest shifted to other goals, as can be seen from current anthropology, which has built up its own idea of the "primitive" and what came after.

One still reads in that most unscientific of records, the Bible, that God disposed all things by number, weight and measure; ancient Chinese texts say that "the calendar and the pitch pipes have such a close fit, that you could not slip a hair between them." People read it, and think nothing of it. Yet such hints might reveal a world of vast and firmly established complexity, infinitely different from ours. But the experts now are benighted by the current folk fantasy, which is the belief that they are beyond all this—critics without nonsense and extremely wise.

In 1959 I wrote:

The dust of centuries had settled upon the remains of this great world-wide archaic construction when the Greeks came upon the scene. Yet something of it survived in traditional rites, in myths and fairy tales no longer understood. Taken verbally, it matured the bloody cults intended to procure fertility, based on the belief in a dark universal force of an ambivalent nature, which seems now to monopolize our interest. Yet its original themes could flash out again, preserved almost intact, in the later thought of the Pythagoreans and of Plato.

But they are tantalizing fragments of a lost whole. They make one think of those "mist landscapes" of which Chinese painters are masters, which show here a rock, here a gable, there the tip of a tree, and leave the rest to imagination. Even when the code shall have yielded, when the techniques shall be known, we cannot expect to gauge the thought of those remote ancestors of ours, wrapped as it is in its symbols.

Their words are no more heard again
Through lapse of many ages . . .

We think we have now broken part of that code. The thought behind these constructions of the high and far-off times is also lofty, even if its forms are strange. The theory about "how the world began" seems to involve the breaking asunder of a harmony, a kind of cosmogonic "original sin" whereby the circle of the ecliptic (with the zodiac) was tilted up at an angle with respect to the equator, and the cycles of change came into being.

This is not to suggest that this archaic cosmology will show any great physical discoveries, although it required prodigious feats of concentration and computing. What it did was to mark out the unity of the universe, and of man's mind, reaching out to its farthest limits. Truly, man is doing the same today.

Einstein said: "What is inconceivable about the universe, is that it should be at all conceivable." Man is not giving up. When he discovers remote galaxies by the million, and then those quasi-stellar radio sources billions of light-years away which confound his speculation, he is happy that he can reach out to those depths. But he pays a terrible price for his achievement. The science of astrophysics reaches out on a grander and grander scale without losing its footing. Man as man cannot do this. In the depths of space he loses himself and all notion of his significance. He is unable to fit himself into the concepts of today's astrophysics short of schizophrenia. Modern man is facing the nonconceivable. Archaic man, however, kept a firm grip on the conceivable by framing within his cosmos an order of time and an eschatology that made sense to him and reserved a fate for his soul. Yet it was a prodigiously vast theory, with no concessions to merely human sentiments. It, too, dilated the mind beyond the bearable, although without destroying man's role in the cosmos. It was a ruthless metaphysics.

Not a forgiving universe, not a world of mercy. That surely not. Inexorable as the stars in their courses, *miserationis parcissimae*, the Romans used to say. Yet it was a world somehow not unmindful of man, one in which there was an accepted place for everything, rightfully and not only statistically, where no sparrow could fall unnoted, and where even what was rejected through its own error would not go down to eternal perdition; for the order of Number and Time was a total order preserving all, of which all were members, gods and men and animals, trees and crystals and even absurd errant stars, all subject to law and measure.

This is what Plato knew, who could still speak the language of archaic myth. He made myth consonant with his thought, as he built the first modern philosophy. We have trusted his clues as landmarks even on occasions when he professes to speak "not quite seriously." He gave us a first rule of thumb; he knew what he was talking about.

Behind Plato there stands the imposing body of doctrine attributed to Pythagoras, some of its formulation uncouth, but rich with the prodigious content of early mathematics, pregnant with a science and a metaphysics that were to flower in Plato's time. From it come such words as "theorem," "theory," and "philosophy." This in its turn rests on what might be called a proto-Pythagorean phase, spread all over the East but with a focus in Susa. And then there was something else again, the stark numerical computing of Babylon. From it all came that strange principle: "Things are numbers."

Once having grasped a thread going back in time, then the test of later doctrines with their own historical developments lies in their congruence with tradition preserved intact even if half understood. For there are seeds which propagate themselves along the jetstream of time.

And universality is in itself a test when coupled with a firm design. When something found, say, in China turns up also in Babylonian astrological texts, then it must be assumed to be relevant, for it reveals a complex of uncommon images which nobody could claim had risen independently by spontaneous generation.

Take the origin of music. Orpheus and his harrowing death may be a poetic creation born in more than one instance in diverse places. But when characters who do not play the lyre but blow pipes get themselves flayed alive for various absurd reasons, and their identical end is rehearsed on several continents, then we feel we have got hold of something, for such stories cannot be linked by internal sequence. And when the Pied Piper turns up both in the medieval German myth of Hamelin and in Mexico long before Columbus, and is linked in both places with certain attributes like the color red, it can hardly be a coincidence. Generally, there is little that finds its way into music by chance.

Again, when one finds numbers like 108, or 9×13, reappearing under several multiples in the *Vedas*, in the temples of Angkor, in Babylon, in Heraclitus' dark utterances, and also in the Norse Valhalla, it is not accident.

There is one way of checking signals thus scattered in early data, in lore, fables and sacred texts. What we have used for sources may seem strange and disparate, but the sifting was considered, and it had its reasons. Those reasons will be given later in the chapter on method. I might call it comparative morphology. The reservoir of myth and fable is great, but there are morphological "markers" for what is not mere storytelling of the kind that comes naturally. There is also wonderfully preserved archaic material in "secondary" primitives, like American Indians and West Africans. Then there are courtly stories and annals of dynasties which look like novels: the *Feng Shen Yen I*, the Japanese *Nihongi*, the Hawaiian *Kumulipo*. These are not merely fantasy-ridden fables.

In hard and perilous ages, what information should a well-born man entrust to his eldest son? Lines of descent surely, but what else? The memory of an ancient nobility is the means of preserving the *arcana imperii,* the *arcana legis* and the *arcana mundi,* just as it was in ancient Rome. This is the wisdom of a ruling class. The Polynesian chants taught in the severely restricted *Whare-wānanga* were mostly astronomy. That is what a liberal education meant then.

Sacred texts are another great source. In our age of print one is tempted to dismiss these as religious excursions into homiletics, but originally they represented a great concentration of attention on material which had been distilled for relevancy through a long period of time and which was considered worthy of being committed to memory generation after generation. The tradition of Celtic Druidism was delivered not only in songs, but also in tree-lore which was much like a code. And in the East, out of complicated games based on astronomy, there developed a kind of shorthand which became the alphabet.

As we follow the clues—stars, numbers, colors, plants, forms, verse, music, structures—a huge framework of connections is revealed at many levels. One is inside an echoing manifold where everything responds and everything has a place and a time assigned to it. This is a true edifice, something like a mathematical matrix, a World-Image that fits the many levels, and all of it kept in order by strict measure. It is measure that provides the countercheck, for there is much that can be identified and redisposed from rules like the old Chinese saying about the pitch pipes and the calendar. When we speak of measures, it is always some form of Time that provides them, starting from two basic ones, the solar year and the octave, and going down from there in many periods and intervals, to actual weights and sizes. What modern man attempted in the merely conventional metric system has archaic precedents of great complexity. Down the centuries there comes an echo of Al-Biruni's wondering a thousand years ago, when that prince of scientists discovered that the Indians, by then miserable astronomers, calculated aspects and events by means of stars—and were not able to show him any one star that he asked for. Stars had become items for them, as they were to become again for Leverrier and Adams, who never troubled to look at Neptune in their life although they had computed and discovered it in 1847. The Mayas and the Aztecs in their unending calculations seem to have had similar attitudes. The connections were what counted. Ultimately so it was in the archaic universe, where all things were signs and signatures of each other, inscribed in the hologram, to be divined subtly. And Number dominated them all (appendix #1).

This ancient world moves a little closer if one recalls two great transitional figures who were simultaneously archaic and modern in their habits of thought. The first is Johannes Kepler, who was of the old order in his unremitting calculations and his passionate devotion to the dream of rediscovering the "Harmony of the Spheres." But he was a man of his own time, and also of ours, when this dream began to prefigure the polyphony that led up to Bach. In somewhat the same way, our strictly scientific world view has its counterpart in what John Hollander, the historian of music, has described as "The Untuning of the Sky." The second transitional figure is no less a man than Sir Isaac Newton, the very inceptor of the rigorously scientific view. There is no real paradox in mentioning Newton in this connection. John Maynard Keynes, who knew Newton as well as many of our time, said of him:

> Newton was not the first of the Age of Reason. He was the last of the magicians, the last of the Babylonians and Sumerians, the last great mind which looked out on the visible and intellectual world with the same eyes as those who began to build our intellectual world rather less than 10,000 years ago ... Why do I call him a magician? Because he looked on the whole universe and all that is in it *as a riddle*, as a secret which could be read by applying pure thought to certain evidence, certain mystic clues which God had laid about the world to allow a sort of philosopher's treasure hunt to the esoteric brotherhood. He believed that these clues were to be found partly in the evidence of the heavens and in the constitution of elements (and that is what gives the false suggestion of his being an experimental natural philosopher), but also partly in certain papers and traditions handed down by the brethren in an unbroken chain back to the original cryptic revelation in Babylonia. He regarded the universe as a cryptogram set by the Almighty—just as he himself wrapt the discovery of the calculus in a cryptogram when he communicated with Leibniz. By pure thought, by concentration of mind, the riddle, he believed, would be revealed to the initiate.[1]

[1] "Newton the Man," in *The Royal Society*. Newton Tercentenary Celebrations (1947), p. 29.

Lord Keynes' appraisal, written ca. 1942, remains both unconventional and profound. He knew, we all know, that Newton failed. Newton was led astray by his dour sectarian preconceptions. But his undertaking was truly in the archaic spirit, as it begins to appear now after two centuries of scholarly search into many cultures of which he could have had no idea. To the few clues he found with rigorous method, a vast number have been added. Still, the wonder remains, the same that was expressed by his great predecessor Galileo:

> But of all other stupendous inventions, what sublimity of mind must have been his who conceived how to communicate his most secret thoughts to any other person, though very far distant either in time or place, speaking with those who are in the Indies, speaking to those who are not yet born, nor shall be this thousand or ten thousand years? And with no greater difficulty than the various arrangement of two dozen little signs upon paper? Let this be the seal of all the admirable inventions of man.

Way back in the 6th century A.D., Grégoire de Tours was writing: "The mind has lost its cutting edge, we hardly understand the Ancients." So much more today, despite our wallowing in mathematics for the million and in sophisticated technology.

It is undeniable that, notwithstanding our Classics Departments' labors, the wilting away of classical studies, the abandonment of any living familiarity with Greek and Latin has cut the *omphaloessa*, the umbilical cord which connected our culture—at least at its top level—with Greece, in the same manner in which men of the Pythagorean and Orphic tradition were tied up through Plato and a few others with the most ancient Near East. It is beginning to appear that this destruction is leading into a very up-to-date Middle Ages, much worse than the first. People will sneer: "Stop the World, I want to get off." It cannot be changed, however; this is the way it goes when someone or other tampers with the reserved knowledge that science is, and was meant to represent.

But, as Goethe said at the very onset of the Progressive Age, "Noch ist es Tag, da rühre sich der Mann! Die Nacht tritt ein, wo niemand wirken kann." ("It is still day, let men get up and going—the night creeps in, when there is nothing doing.") There might come once more some kind of "Renaissance" out of the hopelessly condemned and trampled past, when certain ideas come to life again, and we should not deprive our grandchildren of a last chance at the heritage of the highest and farthest-off times. And if, as looks infinitely probable, even that last chance is passed up in the turmoil of progress, why then one can still think with Poliziano, who was himself a master humanist, that there will be men whose minds find a refuge in poetry and art and the holy tradition "which alone make men free from death and turn them to eternity, so long as the stars will go on, still shining over a world made forever silent." Right now, there is still left some daylight in which to undertake this first quick reconnaissance. It will necessarily leave out great and significant areas of material, but even so, it will investigate many unexpected byways and crannies of the past.

CHAPTER I

The Chronicler's Tale

> . . . you of changeful counsel,
> undefiled Titan of exceeding
> strength, you who consume all
> and increase it again, you who
> hold the indestructible bond by
> the unlimited order of the Aeon,
> wily-minded, originator of gen-
> eration, you of crooked coun-
> sel . . .
>
> *From the Orphic Hymns*

THE PROPER GATE through which to enter the realm of pre-Shakespearean Hamlet is the artless account given by Saxo Grammaticus (c. 1150–c. 1216) in books III and IV of his *Gesta Danorum*. What follows is the relevant part of book III in Elton's translation, only slightly shortened.

The story begins with the feats of Orvendel, Amlethus' father—especially his victory over King Koll of Norway—which drove Orvendel's brother Fengö, "stung with jealousy," to murder him (appendix #2). "Then he took the wife of the brother he had butchered, capping unnatural murder with incest." (So Saxo qualifies it.)

Amleth beheld all this, but feared lest too shrewd a behaviour might make his uncle suspect him. So he chose to feign dullness, and pretend an utter lack of wits. This cunning course not only concealed his intelligence, but ensured his safety. Every day he remained in his mother's house utterly listless and unclean, flinging himself on the ground and bespattering his person with foul and filthy dirt. His discoloured face and visage smutched with slime denoted foolish and grotesque madness. All he said was of a piece with these follies; all he did savoured of utter lethargy . . .

He used at times to sit over the fire, and, raking up the embers with his hands, to fashion wooden crooks, and harden them in the fire, shaping at their tips certain barbs, to make them hold more tightly to their fastenings. When asked what he was about, he said that he was preparing sharp javelins to avenge his father. This answer was not a little scoffed at, all men deriding his idle and ridiculous pursuit; but the thing helped his purpose afterwards. Now it was his craft in this matter that first awakened in the deeper observers a suspicion of his cunning. For his skill in a trifling art betokened the hidden talent of a craftsman . . . Lastly, he always watched with the most punctual care over his pile of stakes that he had pointed in the fire. Some people, therefore, declared that his mind was quick enough, and fancied that he only played the simpleton . . . His wiliness (said these) would be most readily detected, if a fair woman were put in his way in some secluded place, who should provoke his mind to the temptations of love . . . , if his lethargy were feigned, he would seize the opportunity, and yield straightway to violent delights.

So men were commissioned to draw the young man in his rides into a remote part of the forest, and there assail him with a temptation of this nature. Among these chanced to be a foster-brother of Amleth, who had not ceased to have regard to their common nurture . . . He attended Amleth among his appointed train . . . and was persuaded that he would suffer the worst if he showed the slightest glimpse of sound reason, and above all if he did the act of love openly. This was also plain enough to Amleth himself. For when he was bidden mount his horse, he deliberately set himself in such a fashion that he turned his back to the neck and faced about, fronting the tail; which he proceeded to encompass with the reins, just as if on that side he would check the horse in its furious pace . . . The reinless steed galloping on, with the rider directing its tail, was ludicrous enough to behold.

Amleth went on, and a wolf crossed his path amid the thicket; when his companions told him that a young colt had met him, he retorted that in Fengö's stud there were too few of that kind fighting. This was a gentle but witty fashion of invoking a curse upon his uncle's riches. When they averred that he had given a cunning answer, he answered that he had spoken deliberately: for he was loth to be thought prone to lying about any matter, and wished to be held a stranger to falsehood; and accordingly he mingled craft and candour in such wise that, though his words did lack truth, yet there was nothing to betoken the truth and betray how far his keenness went.

Again, as he passed along the beach, his companions found the rudder[1] of a ship which had been wrecked, and said they had discovered a huge knife. "This," said he, "was the right thing to carve such a huge ham"; by which he really meant the sea, to whose infinitude, he thought, this enormous rudder matched.

Also, as they passed the sandhills, and bade him look at the meal, meaning the sand, he replied that it had been ground small by the hoary tempests of the ocean. His companions praising his answer, he said that he had spoken wittingly. Then they purposely left him, that he might pluck up more courage to practice wantonness.

The woman whom his uncle had dispatched met him in a dark spot, as though she had crossed him by chance; and he took her and would have ravished her, had not his foster-brother, by a secret device, given him an inkling of the trap . . . Alarmed, and fain to possess his desire in greater safety, he caught up the woman in his arms and dragged her off to a distant and impenetrable fen. Moreover, when they had lain together, he conjured her earnestly to disclose the matter to none, and the promise of silence was accorded as heartily as it was asked. For both of them had been under the same fostering in their childhood; and this early rearing in common had brought Amleth and the girl into great intimacy.

So, when he had returned home, they all jeeringly asked him whether he had given way to love, and he avowed that he had ravished the maid. When he was next asked where he did it, and what had been his pillow, he said that he had rested upon the hoof of a beast of burden, upon a cockscomb, and also upon a ceiling. For, when he was starting into temptation, he had gathered fragments of all these things, in order to avoid lying . . .

The maiden, too, when questioned on the matter, declared that he had done no such thing; and her denial was the more readily credited when it was found that the escort had not witnessed the deed.

But a friend of Fengö, gifted more with assurance than judgment, declared that the unfathomable cunning of such a mind could not be detected by a vulgar plot, for the man's obstinacy was so great that it ought not to be assailed with any mild measures . . . Accordingly, said he, his own profounder acuteness had hit on a more delicate way, which was well fitted to be put in practice, and would effectually discover what they desired to know. Fengö was purposely to absent himself, pretending affairs of great import. Amleth should be closeted alone with his mother in her chamber; but a man should first be commissioned to place himself in a concealed part of the room and listen heedfully to what they talked about . . . The speaker, loth to seem

[1] Saxo, however, wrote *gubernaculum*, i.e., steering oar (3.6.10; *Gesta Danorum*, C. Knabe and P. Herrmann, eds. [1931], p. 79).

readier to devise than to carry out the plot, zealously proffered himself as the agent of the eavesdropping. Fengö rejoiced of the scheme, and departed on pretence of a long journey. Now he who had given this counsel repaired privily to the room where Amleth was shut up with his mother, and lay down skulking in the straw. But Amleth had his antidote for the treachery.

Afraid of being overheard by some eavesdropper, he at first resorted to his usual imbecile ways, and crowed like a noisy cock, beating his arms together to mimic the flapping of wings. Then he mounted the straw and began to swing his body and jump again and again, wishing to try if aught lurked there in hiding. Feeling a lump beneath his feet, he drove his sword into the spot, and impaled him who lay hid. Then he dragged him from his concealment and slew him. Then, cutting his body into morsels, he seethed it in boiling water, and flung it through the mouth of an open sewer for the swine to eat, bestrewing the stinking mire with his hapless limbs. Having in this wise eluded the snare, he went back to the room. Then his mother set up a great wailing and began to lament her son's folly to his face but he said: "Most infamous of women! dost thou seek with such lying lamentations to hide thy most heavy guilt? Wantoning like a harlot, thou hast entered a wicked and abominable state of wedlock, embracing with incestuous bosom thy husband's slayer . . ." With such reproaches he rent the heart of his mother and redeemed her to walk in the ways of virtue.

When Fengö returned, nowhere could he find the man who had suggested the treacherous espial . . . Amleth, among others, was asked in jest if he had come on any trace of him, and replied that the man had gone to the sewer, but had fallen through its bottom and been stifled by the floods of filth, and that he had then been devoured by the swine that came up all about that place. This speech was flouted by those who heard; for it seemed senseless, though really it expressly avowed the truth.

Fengö now suspected that his stepson was certainly full of guile, and desired to make away with him, but durst not do the deed for fear of the displeasure, not only of Amleth's grandsire Rorik, but also of his own wife. So he thought that the King of Britain should be employed to slay him, so that another could do the deed, and he be able to feign innocence . . .

Amleth, on departing, gave secret orders to his mother to hang the hall with knotted tapestry, and to perform pretended obsequies for him a year from thence; promising that he would then return.

Two retainers of Fengö then accompanied him, bearing a letter graven in wood . . . ; this letter enjoined the King of the Britons to put to death the youth who was sent over to him. While they were reposing, Amleth searched their coffers, found the letter, and read the instructions therein. Whereupon he erased all the writing on the surface, substituted fresh characters, and so, changing the purport of the instructions, shifted his own doom upon his companions. Nor was he satisfied with removing from himself the sentence of death and passing the peril on to others, but added an entreaty that the King of Britain would grant his daughter in marriage to a youth of great judgment whom he was sending to him. Under this was falsely marked the signature of Fengö.

Now when they had reached Britain, the envoys went to the king and proffered him the letter which they supposed was an implement of destruction to another, but which really betokened death to themselves. The king dissembled the truth, and entreated them hospitably and kindly. Then Amleth scouted all the splendour of the royal banquet like vulgar viands, and abstaining very strangely, rejected that plenteous feast, refraining from the drink even as from the banquet. All marvelled that a youth and a foreigner should disdain the carefully cooked dainties of the royal board and the luxurious banquet provided, as if it were some peasant's relish. So, when the revel broke up, and the king was dismissing his friends to rest, he had a man sent into the sleeping room to listen secretly, in order that he might hear the midnight conversation of his guests. Now, when Amleth's companions asked him why he had refrained from the feast of yestereve, as if it were poison, he answered that the bread was flecked with blood and tainted; that there was a tang of iron in the liquor; while the meats of the feast reeked the stench of a human carcase, and were infected by a kind of smack of the odour of the charnel. He further said that the king had the eyes of a slave, and that the queen had in three ways shown the behaviour of a bondmaid. Thus he reviled with insulting invective not so much the feast as its givers. And presently his companions, taunting him with his old defect of wits, began to flout him with many saucy jeers . . .

All this the king heard from his retainer; and declared that he who could say such things had either more than mortal wisdom or more than mortal folly . . . Then he summoned his steward and asked him whence he had procured the bread . . . The king asked where the corn had grown of which it was made, and whether any sign was to be found there of human carnage? The other answered, that not far off was a field, covered with the ancient bones of slaughtered men, and still bearing plainly all the signs of ancient carnage . . . The king . . . took the pains to learn also what had been the source of the lard. The other declared that his hogs had, through negligence, strayed from keeping, and battened on the rotten carcase of a robber, and that perchance their pork had thus come to have something of a corrupt smack. The king, finding that Amleth's judgment was right in this thing also, asked of what liquor the steward had mixed the drink? Hearing that it had been brewed of water and meal, he had the spot of the spring pointed out to him, and set to digging deep down; and there he found rusted away, several swords, the tang whereof it was thought had tainted the waters. Others relate that Amleth blamed the drink because, while quaffing it, he had detected some bees that had fed in the paunch of a dead man; and that the taint, which had formerly been imparted to the combs, had reappeared in the taste. The king . . . had a secret interview with his mother, and asked her who his father had really been. She said she had submitted to no man but the king. But when he threatened that he would have the truth out of her by a trial, he was told that he was the offspring of a slave . . . Abashed as he was with shame for his low estate, he was so ravished with the young man's cleverness that he asked him why he had aspersed the queen with the reproach that she had demeaned herself like a slave? But while resenting that the courtliness of his wife had been accused in the midnight gossip of a guest, he found that her mother had been a bondmaid . . .

Then the king adored the wisdom of Amleth as though it were inspired, and gave him his daughter to wife; accepting his bare word as though it were a witness from the skies.

Moreover, in order to fulfill the bidding of his friend, he hanged Amleth's companions on the morrow. Amleth, feigning offence, treated this piece of kindness as a grievance, and received from the king, as compensation, some gold which he afterwards melted in the fire, and secretly caused to be poured into some hollowed sticks.

When he had passed a whole year with the king he obtained leave to make a journey, and returned to his own land, carrying away of all his princely wealth and state only the sticks which held the gold. On reaching Jutland, he exchanged his present attire for his ancient demeanour, which he had adopted for righteous ends . . .

Covered with filth, he entered the banquet-room where his own obsequies were being held, and struck all men utterly aghast, rumour having falsely noised abroad his death. At last terror melted into mirth, and the guests jeered and taunted one another, that he, whose last rites they were celebrating as though he were dead, should appear in the flesh. When he was asked concerning his comrades, he pointed to the sticks he was carrying, and said, "Here is both the one and the other." This he observed with equal truth and pleasantry . . . for it pointed at the weregild of the slain as though it were themselves.

Thereon, wishing to bring the company into a gayer mood, he joined the cupbearers, and diligently did the office of plying the drink. Then, to prevent his loose dress hampering his walk, he girded his sword upon his side, and purposely drawing it several times, pricked his fingers with its point. The bystanders accordingly had both sword and scabbard riveted across with an iron nail. Then, to smooth the way more safely to his plot, he went to the lords and plied them heavily with draught upon draught, and drenched them all so deep in wine, that their feet were made feeble with drunkenness, and they turned to rest within the palace, making their bed where they had revelled . . .

So he took out of his bosom the stakes he had long ago prepared, and went into the building, where the ground lay covered with the bodies of the nobles wheezing off their sleep and their debauch. Then, cutting away its supports, he brought down the hanging his mother had knitted, which covered the inner as well as the outer walls of the hall. This he flung upon the snorers, and then applying the crooked stakes, he knotted and bound them in such insoluble intricacy, that not one of the men beneath, however hard he might struggle, could contrive to rise. After this he set fire to the palace. The flames spread, scattering the conflagration far and wide. It enveloped the whole dwelling, destroyed the palace, and burnt them all while they were either buried in deep sleep or vainly striving to arise.

Then he went to the chamber of Fengö, who had before this been conducted by his train into his pavilion; plucked up a sword that chanced to be hanging to the bed, and planted his own in its place. Then, awakening his uncle, he told him that his nobles were perishing in the flames, and that Amleth was here, armed with his old crooks to help him, and thirsting to exact the vengeance, now long overdue, for his father's murder. Fengö, on hearing this, leapt from his couch, but was cut down while, deprived of his own sword, he strove in vain to draw the strange one . . . O valiant Amleth, and worthy of immortal fame, who being shrewdly armed with a feint of folly, covered a wisdom too high for human wit under a marvellous disguise of silliness! and not only found in his subtlety means to protect his own safety, but also by its guidance found opportunity to avenge his father. By this skillful defence of himself, and strenuous revenge for his parent, he has left it doubtful whether we are to think more of his wit or his bravery.

It is a far cry from Saxo's tale and its uncouth setting to the Renaissance refinements of Shakespeare. This is nowhere more obvious than in the scene in the Queen's hall, with its heaped straw on the floor, its simmering caldrons, its open sewer, and the crude manner of disposing of "Polonius," all befitting the rude Middle Ages. The whole sad, somber story of the lonely orphan prince is turned by Saxo into a *Narrenspiel*, yet a strong tradition permeates the artless narrative. Hamlet is the avenging power whose superior intellect confounds evildoers, but his intellect also brings light and strength to the helpless and ill-begotten who are made to recognize their misery. There is nothing pleasant in the revelation brought home to the English king, yet he humbles himself before the ruthless insight and "adores" Hamlet's wisdom as "though it were inspired." More clearly than in Shakespeare, Hamlet is the ambivalent power dispensing good and evil. It is clear also that certain episodes, like the exchange of swords with Fengö, are crude and pointless devices going counter to the heroic theme. These are set dramatically right only when handled by Shakespeare, but they seem to indicate an original rigid pattern based on the *Ruse of Reason*, as Hegel would say. Evil is never attacked frontally, even when convention would require it. It is made to defeat itself. Hamlet must not be conceived as a heroic misfit, but as a distributor of justice. Shakespeare has focused exactly right. He has avoided restoring the brutal, heroic element required by the saga, and made the drama instead wholly one of the mind. In the light of a higher clarity, who can 'scape whipping?

It would be pointless to compare all over again the several versions of the Hamlet scheme in the north and west of Europe, and in ancient Rome. This has been done very effectually.[2] Thus, it is possible to rely on the "identity" of the shadowy Icelandic Amlodhi (in a so-called fairy tale his name is Brjam), who is first mentioned in the 10th century, and appears anew in Iceland as a Danish reimport in the "Ambales Saga," written in the 16th or 17th century. Parallels to Amlethus' behavior and career have been found

---

[2] Besides F. Y. Powell's introduction and appendix to Elton's translation of Saxo Grammaticus' *The First Nine Books of the Danish History of Saxo Grammaticus* (1894), already cited at the opening of the chapter, see the following: P. Herrmann, *Die Heldensagen des Saxo Grammaticus* (1922); I. Gollancz, *Hamlet in Iceland* (1898); R. Zenker, *Boeve-Amlethus* (1905); E. N. Setälä, "Kullervo-Hamlet," FUF (1903, 1907, 1910).

in the Sagas of Hrolf Kraki, of Havelok the Dane, as well as in several Celtic myths.[3]

In the version reported by Saxo, Hamlet goes on to reign successfully. The sequel of his adventures is taken up in book IV of the Chronicle, but this narrative shows a very different hand. It is an inept job, made of several commonplaces from the repertory of ruse and fable, badly stitched together. When Hamlet, in addition to the English King's daughter, is made to marry the Queen of Scotland, and bring his two wives home to live together in harmony, we can suspect an incompetent attempt to establish a dynastic claim of the House of Denmark to the realm of Britain. Hamlet eventually falls in battle, but there is not much in the feats recounted to justify Saxo's dithyrambic conclusion that if he had lived longer he might have been another Hercules. The true personage has been overlaid beyond recognition, although there still clings to him a numinous aura. Curiously enough, the misconstruing of Hamlet's story in the direction of success continues today. In the recent Russian film version of Shakespeare's play, Hamlet is shown as a purposeful, devious and ruthless character, bent only on carrying off a coup d'état. Yet, in Saxo's first part, the tragic meaning is clearly adumbrated when Hamlet's return is timed to coincide with his own obsequies. The logic requires that he perish together with the tyrant.

The name Amleth, Amlodhi, Middle English Amlaghe, Irish Amlaidhe, stands always for "simpleton," "stupid," "like unto a dumb animal." It also remained in use as an adjective. Gollancz has pointed out that in "The Wars of Alexander," an alliterative poem from the north of England largely translated from the *Historia de Preliis*, Alexander is twice thus mentioned contemptuously by his enemies:

Thou Alexander, thou ape, thou amlaghe out of Greece
Thou little thefe, thou losangere (1), thou lurkare in cities . . .

[3] See, for Hrolfssaga Kraki, scil., the youth of Helgi and Hroar, and the related story of Harald and Haldan (told in Saxo's seventh book): Zenker, *Boeve-Amlethus*, pp. 121–26; Herrmann, *Die Heldensagen*, pp. 271f.; Setälä, "Kullervo-Hamlet," FUF *3* (1903), pp. 74f.

Darius, inquiring about Alexander's appearance, is shown by his courtiers a caricature graphically described:

> *And thai in parchment him payntid, his person him shewid,*
> *Ane amlaghe, ane asaleny (2), ane ape of all othire,*
> *A wirling (3), a wayryngle (4), a waril-eghid (5) shrewe,*
> *The caitifeste creatour, that cried (6) was evire*[4]

This image of the "caitiffest creature" goes insistently with certain great figures of myth. With the figure of Hamlet there goes, too, the "dog" simile. This is true in Saxo's Amlethus, in Ambales, in the Hrolfssaga Kraki, where the endangered ones, the two princes Helgi and Hroar (and in Saxo's seventh book Harald and Haldan), are labeled dogs, and are called by the dog-names "Hopp and Ho."

Next comes what looks at first like the prototype of them all, the famous Roman story of Lucius Junius Brutus, the slayer of King Tarquin, as told first by Titus Livius. (The nickname Brutus again connotes the likeness to dumb brutes.) Gollancz says of it:

> The merest outline of the plot cannot fail to show the striking likeness between the tales of Hamlet and Lucius Iunius Brutus. Apart from general resemblance (the usurping uncle; the persecuted nephew, who escapes by feigning madness; the journey; the oracular utterances; the outwitting of the comrades; the well-matured plans for vengeance), there are certain points in the former story which must have been borrowed directly from the latter. This is especially true of Hamlet's device of hiding the gold inside the sticks. This could not be due to mere coincidence; and moreover, the evidence seems to show that Saxo himself borrowed this incident from the account of Brutus in Valerius Maximus; one phrase at least from the passage in the *Memorabilia* was transferred from Brutus to Hamlet (Saxo says of Hamlet "obtusi cordis esse," Valerius "obtusi se cordis esse simulavit"). Saxo must have also read the Brutus story as told by Livy, and by later historians, whose versions were ultimately based on Dionysius of Halicarnassus.[5]

To juxtapose the twin brothers Hamlet and Brutus, here is the earlier portion of the tale of Brutus as told by Livy (*1.56*). The subsequent events connected with the rape of Lucrece are too well known to need repeating.

---

[4] (1) liar; (2) little ass; (3) dwarf; (4) little villain; (5) wall-eyed; (6) created.
[5] Gollancz, pp. xxi–xxiv.

While Tarquin was thus employed (on certain defensive measures), a dreadful prodigy appeared to him; a snake sliding out of a wooden pillar, terrified the beholders, and made them fly into the palace; and not only struck the king himself with sudden terror, but filled his breast with anxious apprehensions: so that, whereas in the case of public prodigies the Etrurian soothsayers only were applied to, being thoroughly frightened at this domestic apparition, as it were, he resolved to send to Delphi, the most celebrated oracle in the world; and judging it unsafe to entrust the answers of the oracle to any other person, he sent his two sons into Greece, through lands unknown at that time, and seas still more unknown. Titus and Aruns set out, and, as a companion, there was sent with them Junius Brutus, son to Tarquinia, the king's sister, a young man of a capacity widely different from the assumed appearance he had put on. Having heard that the principal men in the state, and among the rest his brother, had been put to death by his uncle, he resolved that the king should find nothing in his capacity which he need dread, nor in his fortune which he need covet; and he determined to find security in contempt since in justice there was no protection. He took care, therefore, to fashion his behaviour to the resemblance of foolishness, and submitted himself and his portion to the king's rapacity. Nor did he show any dislike of the surname Brutus, content that, under the cover of that appellation, the genius which was to be the deliverer of the Roman people should lie concealed, and wait the proper season for exertion . . . He was, at this time, carried to Delphi by the Tarquinii, rather as a subject of sport than as a companion; and is said to have brought, as an offering to Apollo, a golden wand inclosed in a staff of cornel wood, hollowed for the purpose, an emblem figurative of the state of his own capacity. When they were there, and had executed their father's commission, the young men felt a wish to enquire to which of them the kingdom of Rome was to come; and we are told that these words were uttered from the bottom of the cave.—"Young men, whichever of you shall first kiss your mother, he shall possess the sovereign power at Rome" . . . Brutus judged that the expression of Apollo had another meaning, and as if he had accidentally stumbled and fallen, he touched the earth with his lips, considering that she was the common mother of all mankind.

For most conventional-minded philologists, Brutus was the answer to a prayer, even to the gold enclosed in a stick. They had the sound classical source, from which it is reassuring to derive developments in the outlying provinces. They felt their task to be at an end. With a few trimmings of seasonal cults and fertility rites, the whole

Amlethus package was wrapped, sealed and delivered, to join the growing pile of settled issues.

Yet even the Roman version was not without its disturbing peculiarities. Livy reports only the answer to the private question of the two princes. But if Tarquin had sent them to Delphi, it was to get an answer to his own fears. And the answer is to be found in Zonaras' compendium of the early section of Dio Cassius' lost Roman history. Delphic Apollo said that the king would lose his reign "when a dog would speak with human voice."[6] There is no evidence that Saxo read Zonaras.

There is also a strange variant to Tarquin's prophetic nightmare reported by Livy. It does not lack authority, for it is mentioned in Cicero's *De divinatione* (*1.22*) and taken from a lost tragedy on Brutus by Accius, an early Roman poet. Says Tarquin: "My dream was that shepherds drove up a herd and offered me two beautiful rams issued of the same mother. I sacrificed the best of the two, but the other charged me with its horns. As I was lying on the ground, gravely wounded, and looked up at heaven, I saw a great portent: the flaming orb of the sun coming from the right, took a new course and melted." Well may the Etruscan soothsayers have been exercised about the rams and the changed course of the sun in the same image, for they were concerned with astronomy. This problem will be dealt with later. An interesting variant of this dream is found in the Ambales Saga, and it can hardly have come from Cicero.[7]

However all that may be, there is more than enough to suspect that the story goes back even farther than the Roman kings. Accordingly scholars undertook to investigate the link with the Persian legend of Kyros, which turned out not to be rewarding. But Saxo himself, even if he read Valerius Maximus, contains features which are certainly outside the classical tradition, and he shows another way.

From the *Narrenspiel* the account of Hamlet's ride along the shore is worth a second look: He notices an old steering oar (*guber-*

[6] Zenker, pp. 149f.
[7] Gollancz, p. 105.

naculum) left over from a shipwreck, and he asks what it might be. "Why," they say, "it is a big knife." Then he remarks, "This is the right thing to carve such a huge ham"—by which he really means the sea. Then, Saxo goes on, "as they passed the sandhills and bade him look at the meal, meaning the sand, he replied that it had been ground small by the hoary tempests of the ocean. His companions praising his answer, he said he had spoken wittingly."

It is clear that Saxo at this point does not know what to do with the remarks, for he has always pointed out that Amlethus' answers were meaningful. "For he was loth to be thought prone to lying about any matter, and wished to be held a stranger to falsehood, although he would never betray how far his keenness went." This being the systematic theme of Hamlet's adventure, a theme worked out and contrived to show him as a Sherlock Holmes in disguise, the two remarks quoted are the only ones left to look pointlessly silly. They do not fit.

In fact, they come from a vastly different story. Snorri Sturluson, the learned poet of Iceland (1178–1241), in his *Skaldskaparmal* ("The Language of the Bards") explains many kenningar of famous bards of the past. He quotes a verse from Snaebjörn, an Icelandic skald who had lived long before. This kenning has been the despair of translators, as is the case in any very ancient, partly lost poetic language. There are no less than three terms in the nine lines that can be considered *hapax legomena*, i.e., terms which occur only once. The most authoritative translation is that of Gollancz and here it is:

> 'Tis said, sang Snaebjörn, that far out, off yonder ness, the Nine Maids of the Island Mill stir amain the host-cruel skerry-quern—they who in ages past ground Hamlet's meal. The good chieftain furrows the hull's lair with his ship's beaked prow. Here the sea is called Amlodhi's mill.[8]

That is enough. Whatever the obscurities and ambiguities, one thing is clear: goodbye to Junius Brutus and the safe playgrounds of classical derivations.

[8] Gollancz, p. xi.

This deals with the gray, stormy ocean of the North, its huge breakers grinding forever the granite skerries, and Amlodhi is its king. The quern has not vanished from our language. It is still the surf mill. Even the *British Island Pilot*, in its factual prose, conveys something of the power of the Nine Maids, whose very name is echoed in the Merry Men of Mey on Pentland Firth:

> When an ordinary gale has been blowing for many days, the whole force of the Atlantic is beating against the shores of the Orkneys; rocks of many tons in weight are lifted from their beds, and the roar of the surge may be heard for twenty miles; the breakers rise to the height of 60 feet . . .

As the storm heightens, "all distinction between air and water is lost, everything seems enveloped in a thick smoke." Pytheas, the first explorer of the North, called it the "sea lung," and concluded this must be the end of the earth, where sky and water rejoin each other in the original chaos.

This introduces a much more ancient and certainly independent tradition, whose sources are in early Norse myth—or at least run through it from a still more ancient lineage.

CHAPTER II

The Figure in Finland

Now THE DISCUSSION leaps, without apologies, over the impassable fence erected by modern philologists to protect the linguistic family of Indo-European languages from any improper dealings with strange outsiders. It is known that Finland, Esthonia and Lapland are a cultural island, ethnically related to the Hungarians and to other faraway Asian peoples: Siryenians, Votyaks, Cheremissians, Mordvinians, Voguls, Ostyaks. They speak languages which belong to the Ugro-Finnish family, as totally unrelated to Germanic as Basque would be. These languages are described as "agglutinative" and often characterized by vowel harmonization, such as is found in Turkish. These cultural traditions until quite recently were segregated from the Scandinavian environment. Even if Western culture—and Christianity with it—seeped through among the literati from the Middle Ages on, their great epic, the *Kalevala*, remained intact, entrusted as it was to oral transmission going back in unchanged form to very early times. It shows arrestingly primitive features, so primitive that they discourage any attempt at a classical derivation. It was collected in writing only in the 19th century by Dr. Elias Lönnrot. But even in this segregated tradition, startling parallels were found with Norse and Celtic myth, which must go back to times before their respective recorded histories. The main line of the poem will be dealt with later. Here, it is important to look at the story of Kullervo Kalevanpoika ("the son of Kaleva"), which has been carefully analyzed by E. N. Setälä in his masterly inquiry "Kullervo-Hamlet."[1] His material is

---

[1] FUF 3 (1903), pp. 61–97, 188–255; 7 (1907), pp. 188–224; 10 (1910), pp. 44–127.

necessary, as well as that collected by Kaarle Krohn,[2] in order to take into account many variants (which Lönnrot has not incorporated into the runes *31–36* of the official *Kalevala*) dealing with Kullervo.

The first event is the birth of Kullervo's father and uncle, who are, according to rune *31*, swans (or chickens), driven from one another by a hawk. Usually it is told that a poor man, a plowman, made furrows around a tree trunk (or on a small hill) which split open, and out of it were born two boys. One of them, Kalervo, grew up in Carelia, the other, Untamo, in Suomi-Finland. The hate between the brothers arises usually in the following manner: Kalervo sows oats before the door of Untamo, Untamo's sheep eat them, Kalervo's dog kills the sheep; or there is a quarrel about the fishing grounds (rune *31*.19ff.). Untamo then produces the war. In fact, he makes the war out of his fingers, the army out of his toes, soldiers of the sinews of his heel. But there are versions where Untamo arms trees and uses them as his army. He kills Kalervo and all his family, except Kalervo's wife, who is brought to Untamo's home and there gives birth to our hero, Kullervo. The little one is rocked in the cradle for three days,

> *when the boy began his kicking,*
> *and he kicked and pushed about him,*
> *tore his swaddling clothes to pieces,*
> *freed himself from all his clothing,*
> *then he broke the lime-wood cradle.*[3]

At the age of three months,

> *when a boy no more than knee-high,*
> *he began to speak in this wise:*
> *"Presently when I am bigger,*
> *And my body shall be stronger,*
> *I'll avenge my father's slaughter,*

---

[2] *Kalevalastudien 6. Kullervo* (1928).

[3] Translated by W. F. Kirby (Everyman's Library). The original rough meter has been made to sound like a poor man's *Hiawatha*, but it was the original metric model for Longfellow.

And my mother's tears atone for."
This was heard by Untamoinen,
And he spoke the words which follow:
"He will bring my race to ruin,
Kalervo reborn is in him."

And the old crones all considered,
how to bring the boy to ruin,
so that death might come upon him.

Untamo tries hard to kill the child, with fire, with water, by hanging. A large pyre is built, Kullervo is thrown into it. When the servants of Untamo come after three days to look,

knee-deep sat the boy in ashes,
in the embers to his elbows,
in his hands he held a coal-rake,
and was stirring up the fire.

Setälä reports a version where the child, sitting in the midst of the fire, the (golden) hook in his hand, and stirring the fire, says to Untamo's servants that he is going to avenge the death of his father.[4] Kullervo is thrown into the sea; after three days they find him sitting in a golden boat, with a golden oar, or, according to another version, he is sitting in the sea, on the back of a wave, measuring the waters

Which perchance might fill two ladles,
Or if more exactly measured,
Partly was a third filled also.

Next, they hang the child on a tree, or a gallows is erected—again with frustrating results:

Kullervo not yet has perished,
Nor has he died on the gallows.
Pictures on the tree he's carving,

[4] "Kullervo-Hamlet," FUF 7, p. 192.

In his hands he holds a graver.
All the tree is filled with pictures,
All the oak-tree filled with carvings.

One tradition says that he is carving the names of his parents with a golden stylus. After this the sequence of events is difficult to establish. There are variants, where Kullervo performs his revenge very soon—he merely goes to a smithy and procures the arms. Or he is at once sent out of the country to the smith to serve as cowherd and shepherd. But in rune *31*, he is first given smaller commissions: to guard and rock a child—he blinds and kills it. Then he is sent to clear a forest, and to fell the slender birch trees.

Five large trees at length had fallen,
Eight in all he felled before him.[5]

He sits down afterwards and speaks (*31*.273ff.).

"Lempo [the Devil] *may the work accomplish,*
Hiisi now may shape the timber!"
In a stump he struck his axe-blade,
And began to shout full loudly,
And he piped, and then he whistled,
And he said the words which follow:
"Let the woods be felled around me,
Overthrown the slender birch-trees,
Far as sounds my voice resounding,
Far as I can send my whistle.
Let no sapling here be growing,
Let no blade of grass be standing,
Never while the earth endureth,

[5] There is a strange Dindsencha (this word applies to the explanations of place-names which occur repeatedly in Irish tradition; see W. Stokes, "The Prose Tales in the Rennes Dindsenchas," RC *16*, pp. 278f.) about the felling of five giant trees —three ash trees, one oak, one yew. "The oak fell to the south, over Mag n-Ailte, as far as the Pillar of the Living Tree. 900 bushels was its crop of acorns, and three crops it bore every year . . . apples, nuts, and acorns. The ash of Tortu fell to the South-east, that from Usnach to the North. The yew north-east, as far as Druinn Bairr it fell. The ash of Belach Dahli fell upwards as far as Carn Uachtair Bile."

Or the golden moon is shining,
Here in Kalervo's son's forest,
Here upon the good man's clearing."[6]

In the *Kalevala,* Untamo next orders Kullervo to build a fence, and so he does, out of whole pines, firs, ash trees. But he made no gateway into it, and announced:

He who cannot raise him birdlike,
Nor upon two wings can hover,
Never may he pass across it,
Over Kalervo's son's fencing!

Untamo is taken aback:

Here's a fence without an opening . . .
Up to heaven the fence is builded,
To the very clouds uprising.[7]

[6] The Esthonian Kalevipoeg (= son of Kaleva, the same as Finnish Kalevan-poika) makes the soil barren wherever he has plowed with his wooden plow (Setälä, FUF 7, p. 215), but he, too, fells trees with noise—as far as the stroke of his axe is heard, the trees fall down (p. 203). As for Celtic tradition, one of the Rennes Dindsenchas tells that arable land is changed into woodland because brother had killed brother, "so that a wood and stunted bushes overspread Guaire's country, because of the parricide which he committed" (Stokes, RC *16*, p. 35). Whereas J. Loth (*Les Mabinogion du Livre Rouge de Hergest,* vol. *1*, p. 272, n. 6) gives the names of three heroes who make a country sterile: "Morgan Mwynvawr, Run, son of Beli, and Llew Llaw Gyffes, who turn the ground red. Nothing grew for a year, herb or plant, where they passed: Arthur was more 'rudvawc' than they. Where Arthur had passed, for seven years nothing would grow." *Rudvawc* means "red ravager," as we learn from Rachel Bromwich (*Trioedd Ynys Prydein: The Welsh Triads* [1961], p. 35). Seven years was the cycle of the German Wild Hunter; Arthur was a Wild Hunter, too. The "Waste Land" is, moreover, a standard motif of the legends spun around the Grail and the Fisher King. All this will make sense eventually.

[7] This might originally have been the same story as the one about Romulus drawing a furrow around the new city and killing Remus for jumping over it. In the Roman tradition, the murder makes no sense. Without following up this key phenomenon here, we would like to say that in Finland the stone labyrinth (the English "Troy town") is called Giant's Fence, and also St. Peter's Game, Ruins of Jerusalem, Giant's Street, and Stone Fence (see W. H. Matthews, *Mazes and Labyrinths,* p. 150). Whereas Al-Biruni (*India 1,* p. 306), when dealing with Lanka (Ceylon)—i.e., Ravana's labyrinth that was conquered by Rama and Hanuman—remarks that in Muslim countries this "labyrinthic fortress is called Yavana-Koti, which has been frequently explained as Rome."

Kullervo does some more mischief, threshing the grain to mere chaff, ripping a boat asunder, feeding the cow and breaking its horn, heating the bath hut and burning it down—these are the usual feats of the "Strong Boy" (the "Starke Hans" of German tales, who with us became Paul Bunyan). So, finally, he is sent out of the country, to the house of Ilmarinen the divine smith, as a cowherd. There is, however, a remarkable variant where it is said that he was "sent to Esthonia to bark under the fence; he barked one year, another one, a little from the third; three years he barked at the smith as his uncle, at the wife [or servant] of the smith as his daughter-in-law." This sounds strange indeed, and the translator himself added question marks. There is a still stranger parallel in the great Irish hero Cuchulainn, a central figure of Celtic myth, whose name means "Dog of the Smith Culan." This persistent dog-gishness will bear investigation at another point and so will Smith Ilmarinen himself.

The wife of Ilmarinen (often called Elina, Helena) makes Kullervo her herdsman, and maliciously bakes a stone into his lunch bread so that he breaks his knife, the only heirloom left from his father. A crow then advises Kullervo to drive the cattle into the marshes and to assemble all the wolves and bears and change them into cattle. Kullervo said:

> *"Wait thou, wait thou, whore of Hiisi,*
> *For my father's knife I'm weeping,*
> *Soon wilt thou thyself be weeping."* (33.125ff.)

He acts on the crow's advice, takes a whip of juniper, drives the cattle into the marshes, and the oxen into the thicket.

> *Half of these the wolves devoured,*
> *To the bears he gave the others,*
> *And he sang the wolves to cattle,*
> *And he changed the bears to oxen.*

Kullervo carefully instructs the wolves and the bears on what they are expected to do, and (33.153ff.)

Then he made a pipe of cow-bone,
And a whistle made of ox-horn,
From Tuomikki's leg a cow-horn,
And a flute from heel of Kirjo,
Then upon the horn blew loudly,
And upon his pipe made music.
Thrice upon the hill he blew it,
Six times at the pathway's opening.

He drives the "cattle" home, Helena goes to the stables to milk, and is torn by wolf and bear.

This fierce retaliation gives point to an event that is only a feeble joke in Saxo's version. A wolf crosses Hamlet's path, and he is told it is a horse. "Why," he remarks, "in Fengö's stud there are too few of that kind fighting." Saxo tries to explain: "This was a gentle but witty fashion of invoking a curse upon his uncle's riches." It makes little sense. One suspects here instead an echo of the theme revealed by Kullervo, who drives home wolves and bears in place of cattle. The hero's mastery of wild beasts evokes memories of classical myth. This has not escaped Karl Kerényi,[8] whose comment is useful, although not his line of psychological speculation: "It is impossible to try to derive Finnish mythology from the Greek, or conversely. Yet it is also impossible not to notice that Kullervo, who is the Miraculous Child and the Strong Servant in one, shows himself at last to be Hermes and Dionysos. He appears as Hermes in the making of musical instruments tied up with the destruction of cattle . . . He shows himself as Dionysos in what he does with wild beasts and with his enemy. It is Dionysos-like behavior—if we see it through the categories of Greek myth—to make wolves and bears appear by magic as tame animals, and it is again Dionysos-like to use them for revenge against his enemy. We recognize with awe the tragic-ironic tone of Euripides' Bacchae, when we read the dramatic scene of the milking of wild beasts. An even closer analogy is given by the fate of the Etruscan pirates, Dionysos' enemies, who are chastised by the intervention of wild animals . . ."

[8] K. Kerényi, "Zum Urkind-Mythologen," *Paideuma 1* (1940), p. 255.

In rune *35*, Lönnrot makes Kullervo return to his parents and brothers and sisters. This is unexpected inasmuch as they have been killed a number of runes earlier, although the crux of the many rune songs is that the names of the heroes are far from stable and, as has already been said, the original order of things is impossible to reconstruct. But one event stands out. A sister is not at home. On one occasion the hero meets a maiden in the woods, gathering berries. They lie together and realize later in conversing that they are brother and sister. The maiden drowns herself, but Kullervo's mother dissuades him from suicide. So he goes to war, and in so doing he fulfills his revenge. First he asks the great god Ukko for the gift of a sword (36.242ff.).

> *Then the sword he asked was granted,*
> *And a sword of all most splendid,*
> *And he slaughtered all the people,*
> *Untamo's whole tribe was slaughtered,*
> *Burned the houses all to ashes,*
> *And with flame completely burned them,*
> *Leaving nothing but the hearthstones,*
> *Nought but in each yard the rowan.*

Returning home, Kullervo finds no living soul; all have died. When he weeps over his mother's grave, she awakes,

> *And beneath the mould made answer:*
> *"Still there lives the black dog, Musti,*
> *Go with him into the forest,*
> *At thy side let him attend thee."*

There in the thicket reside the blue forest-maidens, and the mother advises him to try to win their favor. Kullervo takes the black dog and goes into the forest, but when he comes upon the spot where he had dishonored his sister, despair overcomes him, and he throws himself upon his own sword.

Here at last a point is made explicitly which in other stories remains a dark hint. There is a sin that Hamlet has to atone for. The knowledge that Kullervo and his sister killed themselves for unwitting incest calls to mind the fact that in Saxo the adolescent Prince is initiated to love by a girl who does not betray him "because she happened to be his foster-sister and playmate since childhood." This seems contrived, as if Saxo had found there a theme he does not grasp. The theme becomes manifest in King Arthur. It is ambiguous and elusive, but all the more inexorable in Shakespeare. Hamlet must renounce his true love, as he has to renounce himself in his predicament:

> "Get thee to a nunnery. Why wouldst thou be a breeder of sinners? . . . What should such fellows as I do nawling between earth and heaven? We are arrant knaves all; believe none of us. Go thy ways to a nunnery."

In the play-within-a-play, the Prince feels free to step out of character:

> *Lady, shall I lie in your lap?*
> *No, my lord.*
> *I mean, my head upon your lap?*
> *Ay, my lord.*
> *Do you think I meant country matters?*
> *I think nothing, my lord.*
> *That is a fair thought to lie between maid's legs.*
> *What is, my lord?*
> *Nothing.*

But the die is cast. Ophelia's suicide by drowning, like Kullervo's sister's, brings about the death of her lover—and of her brother too. The two aspects join in the final silence. At least Hamlet, ever conscious, has had a chance to describe in despair the insoluble knot of his guilt:

> "I loved Ophelia; forty thousand brothers
> could not with all their quantity of love
> make up my sum. What wilt thou do for her?"

And now Kullervo. Setälä's analysis of the whole parallel goes as follows:

As concerns generalities: brother kills brother; one son survives, who sets his mind on revenge from earliest childhood; the uncle tries to kill him, but he succeeds in achieving his revenge. As concerns details: Setälä wants to identify the stakes and hooks, which the hero in all northern versions shapes or carves, sitting at the hearth—Brjam does it in a smithy—with the golden hook or rake that little Kullervo, sitting in the middle of the fire, holds in his hands, stirring the flames. Each hero (including Kullervo in one of the versions found by Setälä) makes it clear that he means to avenge his father.

With some puzzlement Setälä brings out one other point which will turn out to be crucial later on. In every northern version there is some dark utterance about the sea. The words are weird. Hamlet wants to "cut the big ham" with the steering oar; the child Kullervo is found measuring the depth of the sea with an oar or with a ladle. Kalevipoeg, the Esthonian counterpart of Kullervo Kalevanpoika, measures the depth of lakes with his height. Amlodhi-Ambales, sitting by a bottomless mountain lake, says only: "Into water wind has come, into water wind will go."

CHAPTER III

The Iranian Parallel

For from today new feasts and
customs date
Because tonight is born Shah Kai
Khusrau

Shahnama

T HE HAMLET THEME moves now to Persia. Firdausi's *Shahnama*,
the Book of Kings, is the national epic of Iran,[1] and Firdausi
(ca. A.D. 1010) is still today the national poet. At the time Firdausi
wrote, his protector, Sultan Mahmud of Ghazna, had shifted the
center of his power to India, and the Iranian empire had long been
only a memory. With prodigious scholarship, Firdausi, like Homer
before him, undertook to organize and record the Zendic tradition,
which extended back from historic times into the purely mythical.
The first section on the Pishdadian and Kaianian dynasties must be
considered mythical throughout, although it does reach into historic times and encompasses four of the nine volumes of the Book
of Kings in the English translation. Khusrau (Chosroes in Greek) is
also the name of a line of historical rulers, one of whom, Khusrau
Anushirvan, gave sanctuary to the last philosophers of Greece, the
members of the Platonic Academy driven out by Justinian in
A.D. 529. But Firdausi's Kai Khusrau is the towering figure of his
own mythical age. Almost one-fifth of the whole work is allotted
to him. He is actually the Haosravah of the *Zend Avesta*, and also

[1] We cite here the English translation of Arthur and Edward Warner (1905–
1909).

the Rigvedic Sushravah, an identity which raises again the much discussed question of a common Indo-European "Urzeit," the time of origins.

The common features of Saxo's Amlethus and Kai Khusrau are so striking that Jiriczek, and after him Zenker, undertook detailed comparative studies.[2] But they concluded that the Greek saga of Bellerophon might provide a common origin, and that was the end of their quest. Classical antiquity has a magnetic quality for the scholarly mind. It acts upon it like the Great Lodestone Mountain in Sindbad. The frail philological bark comes apart as soon as Greece looms over the horizon. Bellerophon's somber tale would provide a parallel too, but does that have to be the end of the trail? As Herodotus ruefully remarks, his own Hellenic antiquity goes back in recorded memory but a few centuries; beyond that, it blends with the Indo-European patrimony of legends.

In the vast flow of the *Shahnama*, one prominent feature is the perpetual war between "Untamo" and "Kalervo," here the two rival peoples of Turan and Iran. Because the vicissitudes of the Kaianian dynasty of Iran are spread over a narrative twice as long as both epics of Milton combined, it is necessary here to concentrate on one essential aspect.

The Iranian plot shows some "displacement" in that Afrasiyab the Turanian kills, instead of his brother, his nephew Siyawush who is also his son-in-law, so that the "avenger" of this crime is bound to come forth as the common grandson of the hostile Turanian Shah Afrasiyab and his brother, the noble Iranian Shah Kai Ka'us (the same one who plays no small role in the *Rigveda* as Kavya Ushanas, and in the *Avesta* as Kavi Usan). Siyawush, as commander of his father's army, offers peace to the Turanian Afrasiyab, who accepts the offer because he has had a catastrophic dream.[3] This dream resembles those of Tarquin and Ambales. Kai Ka'us does not trust Afrasiyab and declines peace. Siyawush, not wishing to break his own treaty with Turan, goes to live with Afrasiyab.

---

[2] O. L. Jiriczek, "Hamlet in Iran," ZVV *10* (1900), pp. 353–64; R. Zenker, *Boeve-Amlethus* (1905), pp. 207–82.

[3] Firdausi, Warner trans., vol. *2*, pp. 232f.

Afrasiyab honors the young man in every way, and gives him a large province which he rules excellently, i.e., in the "Golden Age" style of his father Kai Ka'us. Siyawush marries first a daughter of the Turanian Piran, then Shah Afrasiyab gives him his own daughter Farangis. But there is a serpent in that garden. Afrasiyab's jealous brother Garsiwas, an early Polonius, plots so successfully against Siyawush that Afrasiyab finally sends an army against the blameless young ruler. Siyawush is captured and killed. The widowed Farangis escapes, accompanied by Piran (Siyawush's first father-in-law) to Piran's home where she gives birth to a boy of great beauty, Kai Khusrau, Afrasiyab's and Kai Ka'us' common grandson:

> *One dark and moonless night, while birds, wild beasts*
> *And cattle slept, Piran in dream beheld*
> *A splendour that outshone the sun itself,*
> *While Siyawush, enthroned and sword in hand,*
> *Called loudly to him saying: "Rest no more!*
> *Throw off sweet sleep and think of times to come.*
> *For from today new feasts and customs date,*
> *Because to-night is born Shah Kai Khusrau!"*
> *The chieftain roused him from his sweet repose:*
> *Gulshahr the sunny-faced woke. Piran*
> *Said unto her: "Arise, Betake thyself*
> *To minister to Farangis, for I*
> *Saw Siyawush in sleep a moment since,*
> *Surpassing both the sun and moon in lustre,*
> *And crying: 'Sleep no more, but join the feast*
> *Of Kai Khusrau, the monarch of the world!'"*
> *Gulshahr came hasting to the Moon and saw*
> *The prince already born; she went with cries*
> *Of joy that made the palace ring again*
> *Back to Piran the chief. "Thou wouldest say,"*
> *She cried, "that king and Moon are fairly matched!"*[4]

---

[4] Firdausi, Warner trans., vol. 2, pp. 325f.

With this prophetic dream of a great new age begins a long time of trials for the predestined hero. The boy grows up among the shepherds; he becomes a great hunter with a crude bow and arrows that he makes for himself without arrowheads or feathers, like Hamlet whittling his stakes. Grandfather Afrasiyab, being afraid of the boy, orders the prince brought to him so that he can convince himself his victim is harmless. Although Afrasiyab has sworn solemnly not to hurt Khusrau, Piran urges the boy to play the village idiot for his own safety. When the tyrant questions him with feigned benevolence, Kai Khusrau answers in the very same style as Amlethus did, in riddles which sound senseless and indicate that young Khusrau likens himself to a dog. The usurper feels relieved: "The fellow is a fool!"

Now, the tale of vengeance, unduly abbreviated by Saxo's report and in other versions, is told by Firdausi with an appropriately majestic setting and on a grand scale. The anger of Iran and the world, stemming from the death of Siyawush, is orchestrated apocalyptically into a cosmic tumult:

> *The world was all revenge and thou hadst said:*
> *"It is a seething sea!" Earth had no room*
> *For walking, air was ambushed by the spears;*
> *The stars began to fray, and time and earth*
> *Washed hands in mischief . . .*[5]

Still, the two archcriminals manage to escape and hide with inexhaustible resourcefulness. Afrasiyab even plays Proteus in the waters of a deep salt lake, constantly assuming new shapes to evade capture. Finally, two volumes and a multitude of events later, Afrasiyab and the evil counsellor are caught with a lasso or a net and both perish.

Only by going back to the Avestan tradition can one make sense of the many vicissitudes to which the Yashts or hymns of the *Avesta* allude repeatedly.[6] The Shahs Kai Khusrau and Afrasiyab

[5] Firdausi, Warner trans., vol. *2*, p. 342.
[6] Yasht 5.41–49; *19*.56–64, 74.

were contending in a quest for the enigmatic *Hvarna*, rendered as the "Glory," or the Charisma of Fortune. To obtain it the Shahs kept sacrificing a hundred horses, a thousand oxen, ten thousand lambs to the goddess Anahita, who is a kind of Ishtar-Artemis. Now this Glory "that belongs to the Aryan nations, born and unborn, and to the holy Zarathustra" was in Lake Vurukasha. Afrasiyab, Shah of the non-Aryan Turanians, was not entitled to it. But leaving his hiding place in an underground palace of iron "a thousand times the height of man" and illuminated by artificial sun, moon and stars, he tried three times to capture the Hvarna, plunging into Lake Vurukasha. However, "the Glory escaped, the Glory fled away, the Glory changed its seat." There will be more discussion of Afrasiyab's attempts and his "horrible utterances," in the chapter "Of Time and the Rivers." The Glory was, instead, allotted to Kai Khusrau, and it was bestowed upon him without much ado. At this point it is fair to say that Hvarna stands for Legitimacy, or Heavenly Mandate, which is granted to rulers, but is also easily withdrawn. Yima (Jamshyd), the earliest "world ruler," lost it three times.

The story of diving Afrasiyab has had many offshoots in Eurasian folklore. There the Turanian Shah is spelled "Devil," and God causes him to dive to the bottom of the sea, so that in the meantime one of the archangels, or St. Elias, can steal a valuable object which is the legal property of the Devil. Sometimes the object is the sun, sometimes the "divine power," or thunder and lightning, or even a treaty between God and Devil which had turned out to be unprofitable for God.

There remains the essential dénouement. During those eventful years, Kai Ka'us held joint rulership with his grandson, secure in the Glory. Shortly after the victory over the upstart, Kai Ka'us dies and Kai Khusrau ascends the Ivory Throne. For sixty years, says the poem, "the whole world was obedient to his sway." It is striking that there is no word of any event after Kai Ka'us' death. Maybe it is because all has been achieved. Happy reigns have no history. But it is told that Kai Khusrau falls into deep melancholy and soul-searching.[7] He fears he may "grow arrogant in soul, corrupt in thought" like his predecessors Yima (Jamshyd) and, among others, Kai Ka'us himself, who had tried to get himself carried to heaven by eagles like the Babylonian Etana. So he makes the supreme decision:

> *"And now I deem it better to depart*
> *To God in all my glory . . .*
> *Because this Kaian crown and throne will pass."*

The great Shah, then, who had once stated (at his first joint enthronement):

> *"The whole world is my kingdom, all is mine*
> *From Pisces downward to the Bull's head,"*[8]

prepares his departure, takes leave of his paladins, waving aside their supplications and those of his whole army:

> *A cry rose from the army of Iran:*
> *The sun hath wandered from its way in heaven!*

The dream of Tarquin finds here an early echo. The Shah appoints as his successor Luhrasp and wanders off to a mountaintop, accompanied by five of his paladins, to whom he announces in the evening, before they sit down for the last time to talk of the great past they have lived together:

> *"What time the radiant sun shall raise its flag,*
> *And turn the darksome earth to liquid gold,*
> *Then is the time when I shall pass away*
> *And haply with Surush[9] for company."*

Toward dawn he addresses his friends once more:

> *"Farewell for ever! When the sky shall bring*
> *The sun again ye shall not look on me*

[7] Firdausi, Warner trans., vol. 4, pp. 272ff.
[8] Firdausi, Warner trans., vol. 2, p. 407.
[9] Surush = Avestic Sraosha, the "angel" of Ahura Mazdah.

Henceforth save in your dreams. Moreover be not
Here on the morrow on these arid sands,
Although the clouds rain musk, for from the Mountains
Will rise a furious blast and snap the boughs
And leafage of the trees, a storm of snow
Will shower down from heaven's louring rack,
And towards Iran ye will not find the track."

The chieftains' heads were heavy at the news.
The warriors slept in pain, and when the sun
Rose over the hills the Shah had disappeared.

The five paladins are lost and buried in the snowstorm.[10]

---

[10] This theme of sleep in the "hour of Gethsemane" will occur more than once, e.g., in Gilgamesh. The myth of Quetzalcouatl is even more circumstantial. The exiled ruler is escorted by the dwarves and hunchbacks, who are also lost in the snow along what is now the Cortez Pass, while their ruler goes on to the sea and departs. But here at least he promises to come back and judge the living and the dead.

CHAPTER IV

History, Myth and Reality

"Let us try, then, to set forth in
our statement what things these
are, and of what kind, and how
one should learn them . . . It is,
indeed, a rather strange thing to
hear; but the name that we, at
any rate, give it—one that people
would never suppose, from inex-
perience in the matter—is astronomy; people are ignorant that he
who is truly an astronomer must
be wisest, not he who is an astronomer in the sense understood
by Hesiod . . . ; but the man who
has studied the seven out of the
eight orbits, each travelling over
its own circuit in such a manner
as could not ever be easily observed by any ordinary nature
that did not partake of a marvellous nature."

Epinomis 989 E–990 B

THE STRANGE END of the Iranian tale, which concludes with an
ascent to heaven like that of Elias, leaves the reader wondering. If
this is the national epos (almost one half of it in content), where is
the epic and the tragic element? In fact, there is a full measure of
the Homeric narrative in Firdausi that had to be left aside, there are
great battles as on the windy plains of Troy, challenges and duels,
the incredible feats of heroes like Rustam and Zal, abductions and
intrigues, infinite subplots to the tale, enough for a bard to entertain

his patrons for weeks and to ensure him a durable supply of haunches of venison. But the intervention of the gods in the tale is not so humanized as in the *Iliad,* although it shows through repeatedly in complicated symbolism and bizarre fairy tales. The conflict of will and fate is not to the measure of man. What has been traced above is a confusing story of dynastic succession under a shadowy Glory, a Glory without high events, keyed to a Hamlet situation and an unexplained melancholy. The essence is an unsubstantial pageant of ambiguous abstractions, an elusive ballet of wildly symbolic actions tied to ritual magic and religious doctrines, with motivations which bear no parallel to normal ones. The whole thing is a puzzle to be interpreted through hymns—very much as in the *Rigveda.*

But here at last there is given *apertis verbis* one key to the imagery: Khusrau's crowning words:

> *"The whole world is my kingdom: all is mine*
> *From Pisces downward to the Bull's head."*

If a hero of the western hemisphere were to proclaim: "All of this continent is mine, from Hatteras to Eastport," he would be considered afflicted with a one-dimensional fancy. Does that stretch of coast stand in his mind for a whole continent? Yet here the words make perfect sense because Kai Khusrau does not refer to the earth. He designates that section of the zodiac comprised between Pisces and Aldebaran, the thirty degrees which cover the constellation Aries. It means that his reign is not only of heaven, it is essentially of Time. The dimension of heaven is Time. Kai Khusrau comes in as a function of time, preordained by events in the zodiac.

> *"For from today new feasts and customs date . . ."*

Why Aries, and what it all imports, is not relevant at this point. It turns out that "ruler of Aries" was the established title of supreme power in Iran,[1] and it may have meant as much or as little as "Holy

[1] Persia "belongs" to Aries according to Paulus Alexandrinus. See Boll's *Sphaera,* pp. 296f., where it is stated that this was the oldest scheme. It is still to be found in the *Apocalypse.* Moses' ram's horns stand for the same world-age.

Roman Emperor" in the West. What counts is that Rome is a *place* on earth, whose prestige is connected with a certain historical period, whereas Aries is a zone of heaven, or rather, since heaven keeps moving, a certain *time* determined by heavenly motion in connection with that constellation. Rome is a historic fact, even "Eternal Rome," which was once and then is left only to memory. Aries is a labeled time, and is bound to come back within certain cycles.

Even if Kai Khusrau is conceived as a worldly ruler in an epos which prefaces history, it is clear that no modern historical or naturalistic imagination can provide the key to such minds as those of the Iranian bards out of whose rhapsodies the learned Firdausi organized the story. No basis in history can be found, no fertility or seasonal symbolism can be traced into it, and even the psychoanalysts have given up trying. This type of thought can be defined in one way: it is essentially cosmological.

This is not to make things uselessly difficult, but to outline the real frame of mythical thought, such as is actually quite familiar and yet by now hardly recognized. It even appears in the mode of lyrical meditation, at least in the English of Fitzgerald:

> *Iram indeed is gone with all its Rose*
> *And Jamshyd's Sev'n-ringed Cup where no one knows*
> > *But still the Vine her ancient Ruby yields*
> *And still a Garden by the Water blows.*
>
> *And look—a thousand Blossoms with the Day*
> *Woke—and a thousand scatter'd into Clay*
> > *And the first Summer month that brings the Rose*
> *Shall take Jamshyd and Kai Kubad away.*
>
> *But come with old Khayyam, and leave the lot*
> *Of Kai Kubad and Kai Khosrau forgot . . .*

Omar Khayyam may speak as a weary skeptic or as mystical Sufi, but all he speaks of is understood as real. The heroes of the past are as real as the friends for whom he is writing, as the vine and the roses and the waters, as his own direct experience of flux and impermanence in life. When he makes his earthenware pots to feel and think, it is no literary trope; it is the knowledge that all transient things are caught in the same transmutation, that all substance is one: the stuff that pots and men and dreams are made of.

This is what could be called living reality, and it is singularly different from ordinary or objective reality. When the poet thinks that this brick here may be the clay that was once Kai Khusrau, he rejoins Hamlet musing in the graveyard: "To what base uses we return, Horatio! Why may not imagination trace the noble dust of Alexander till'a find it stopping a bung-hole?" Here are already four characters, two of them unreal, two lost in the haze of time, yet all equally present in *our* game, whereas most concrete characters, say the Director of Internal Revenue, are not, however they may affect us otherwise. In that realm of "true existence" we shall find stars and vines and roses and water, the eternal forms, and it will include also the ideas of mathematics, another form of direct experience. The world of history is outside it as a whole. Khayyam does not, any more than Firdausi a generation later, mention the glories of Cyrus and Artaxerxes, but only mythical heroes, just as our own Middle Ages ignored history and spoke of Arthur and Gawain. It had been all "once upon a time," and if Dante brings back myth so powerfully to life, it is because his own contemporaries believed themselves truly descended from Dardanus and Troy, and wondered whether the Lord Ulysses might not still be alive; whereas Kaiser Barbarossa, asleep in his Kyffhäuser mountain—that must surely be a fable like Snow White. Or is it? Fairy tales are easily dismissed for their familiar sound. But it might turn out that such great imperial figures turned into legend have a hidden life of their own, that they follow the laws of myth laid down long before them. Even as King Arthur did not really die but lives on in the depth of the mystic lake, according to Merlin's prophecy, so Godfrey of Viterbo (c. 1190), who had been in Barbarossa's service, alone brings the "true" version. It is the orthodox one in strangely preserved archaic language: the Emperor sleeps on in the depth of the Watery Abyss (cf. chapter XI and appendix #33) where the retired rulers of the world are.

Voire, ou sont de Constantinople
L'empereurs aux poings dorez . . .

A distinction begins to appear between myth and fable. Hamlet is showing himself in the aspect of a true myth, a universal one. He is still that now. And Khayyam was the greatest mathematician of his time, the author of a planned calendar reform which turned out to be even more precise than the one that was adopted later as the Gregorian calendar, an intellect in whom trenchant skepticism could coexist with profound Sufi intellection. He knew full well that Jamshyd's seven-ringed cup is not lost, since it stands for the seven planetary circles of which Jamshyd is the ruler, just as Jamshyd's magic mirror goes on reflecting the whole world, as it is the sky itself. But it is natural to let them retain their iridescent mystery, since they belong to the living reality, like Plato's whorls and his Spindle of Necessity. Or like Hamlet himself.

What then were Jamshyd or Kai Khusrau? To the simple, a magic image, a fable. To those who understood, a reflection of Time itself, obviously one of its major aspects. They could be recognized under many names in many places, even conflicting allusions. It was always the same myth, and that was enough. It expressed the laws of the universe, in that specific language, the language of Time. This was the way to talk about the cosmos.

All that is living reality, *sub specie transeuntis*, has a tale, as it appears in awesome, or appalling, or comforting aspects, in the "fearful symmetry" of tigers or theorems, or stars in their courses, but always alive to the soul. It is a play of transmutations which include us, ruled by Time, framed in the eternal forms. A thought ruled by Time can be expressed only in myth. When mythical languages were universal and self-explanatory, thought was also self-sufficient. It could seek no explanation of itself in other terms, for it was reality expressed as living. As Goethe said, "Alles Vergängliche—ist nur ein Gleichnis."

Men today are trained to think in spatial terms, to localize objects. After childhood, the first question is "where and when did it happen?" As science and history invade the whole landscape of thought, the events of myth recede into mere fable. They appear as escape fantasies: unlocated, hardly serious, their space ubiquitous, their time circular.

Yet some of those stories are so strong that they have lived on vividly. These are true myths. These personages are unmistakably identified, yet elusively fluid in outline. They tell of gigantic figures and superhuman events which seem to occupy the whole living space between heaven and earth. Those figures often lend their names to historical persons in passing and then vanish. Any attempt to tie them down to history, even to the tradition of great and catastrophic events, is invariably a sure way to a false trail. Historical happenings will never "explain" mythical events. Plutarch already knew as much. Instead, mythical figures have invaded history under counterfeit presentments, and subtly shaped it to their own ends. This is a working rule which was established long ago, and it has proved constantly valid if one is dealing with true myth and not with ordinary legends. To be sure, mythical figures are born and pass on, but not quite like mortals. There have to be characteristic styles for them like *The Once and Future King*. Were they once? Then they have been before, or will be again, in other names, under other aspects, even as the sky brings back forever its configurations. Surely, if one tried to pinpoint them as persons and things, they would melt before his eyes, like the products of sick fantasy. But if one respects their true nature, they will reveal that nature as *functions*.

Functions of what? Of the general order of things as it could be conceived. These figures express the behavior of that vast complex of variables once called the cosmos. They combine in themselves variety, eternity, and recurrence, for such is the nature of the cosmos itself. That the cosmos might be infinite seems to have remained beyond the threshold of awareness of humankind up to the time of Lucretius, of Bruno and Galileo. And Galileo himself, who had serious doubts on the matter, agreed with all his predecessors that surely the universe is eternal, and that hence all its changes come under the law of periodicity and recurrence. "What is eternal," Aristotle said, "is circular, and what is circular is eternal."

That was the mature conclusion of human thought over millennia. It was, as has been said, an obsession with circularity. There is nothing new under the sun, but all things come back in ever-varying recurrence. Even the hateful word "revolution" referred once only to those of the celestial orbs. The cosmos was one vast system full of gears within gears, enormously intricate in its connections, which could be likened to a many-dialed clock. Its functions appeared and disappeared all over the system, like strange cuckoos in the clock, and wonderful tales were woven around them to describe their behavior; but just as in an engine, one cannot understand each part until one has understood the way all the parts interconnect in the system.

Similarly, Rudyard Kipling in a droll allegory, "The Ship That Found Herself," once explained what happens on a new ship in her shakedown voyage. All the parts spring into clamorous being as each plays its role for the first time, the plunging pistons, the groaning cylinders, the robust propeller shaft, the straining bulkheads, the chattering rivets, each feeling at the center of the stage, each telling the steam about its own unique and incomparable feats, until at last they subside into silence as a new deep voice is heard, that of the ship, who has found her identity at last.

This is exactly what happens with the great array of myths. All the myths presented tales, some of them weird, incoherent or outlandish, and some epic and tragic. At last it is possible to understand them as partial representations of a system, as functions of a whole. The vastness and complexity of the system is only beginning to take shape, as the parts fall into place. The only thing to do is proceed inductively, step by step, avoiding preconceptions and letting the argument lead toward its own conclusions.

In the simple story of Kai Khusrau, the Hamlet-like features are curiously preordained, although it is not clear to what end. The King's power is explicitly linked, in time and space, with the moving configurations of the heavens. It is common knowledge that heaven in its motion does provide coordinates for time and place on earth. The navigator's business is to operate on this connection between above and below. But in the early centuries, the connection was infinitely richer in meaning. No historical monarch, however convinced of his charisma, could have said: "The whole world is my kingdom, all is mine from Pisces to Aldebaran." Earthly concepts seem to have been transferred to heaven, and inversely. In fact, this world of myth imbricates uranography and geography into a whole which is really one cosmography, and the "geographical" features referred to can be mystifying, as they may imply either of these domains or both.

For instance, when the "rivers" Okeanos or Eridanos are mentioned, are they not conceived as being first in heaven and then eventually on earth, too? It is as if any region beyond ancient man's direct ken were to be found simply "upwards." True events, even in an official epic like the *Shahnama*, are not "earth-directed." They tend to move "upwards." This is the original form of astrology, which is both vaster and less defined than the later classic form which Ptolemy set forth. Even as the cosmos is one, so cosmography is made up of inextricably intertwined data. To say that events on earth reflect those in heaven is a misleading simplification to begin with. In Aristotelian language, form is said to be metaphysically prior to matter, but both go together. It is still necessary to discover which is the focus of "true" events in heaven.

To recapitulate for clarity, whatever is true myth has no historical basis, however tempting the reduction, however massive and well armed the impact of a good deal of modern criticism on that belief. The attempt to reduce myth to history is the so-called "euhemerist" trend, from the name of Euhemeros, the first debunker. It was a wave of fashion which is now receding, for it was too simpleminded to last. Myth is essentially cosmological. As heaven in the cosmos is so vastly more important than our earth, it should not be surprising to find the main functions deriving from heaven. To identify them under a variety of appearances is a matter of mythological judgment, of the capacity to recognize essential forms through patient sifting of the immense amount of material.

Hamlet "is" here Kullervo, there Brutus or Kai Khusrau, but always recognizably the same. Jamshyd reappears as Yama among the Indo-Aryans, as Huang-ti, the Yellow Emperor, in China, and under many other names. There was always the tacit understanding, for those who spoke the archaic language, who were involved in the archaic cosmos, that he is everywhere the same function. And who is the Demiurge? He has many names indeed. Plato does not care to explain in our terms. Is this personage a semi-scientific fiction, the manufacturer of a planetarium, just as the Lost Continent of Atlantis is a semi-historical fiction? The author himself says only that such stories are "not quite serious." Yet they are surely not a spoof. Plato, who shaped what is called philosophy and its language, who was the master of its penetrating distinctions, reverts to the language of myth when he feels he has to; and he uses that ancient language as if to the manner born.[2]

In this accounting for past myths, the heart of the problem remains elusive. Kipling was a writer still marvelously attuned to the juvenile mind that lives in most of us. But the fact is that myth itself, as a whole, is a lost world. The last forms—or rehearsals—of a true myth took place in medieval culture: the *Romance of Alexander*, and the Arthurian myth as it is found in Malory.[*]

There are other stories—we call them history—of man's conquest over nature, the telling of the great adventure of mankind as a whole. But here it is only faceless social man who is winning man's victories. It is not the history of technology; it is, if anything, science fiction that can bring in the adventures of the future. Science fiction, when it is good, is a wholly valid attempt at restoring a mythical element, with its adventures and tragedies, its meditations on man's errors and man's fate. For true tragedy is an essential component or outcome of myth. Possibly, history can be given a minute

[2] In his Seventh Letter (341C–344D) he denies strongly that scientific "names" and "sentences" (*onomata, rēmata*) could assist in obtaining essential insight. Cf. also Clemens Alexandrinus, *Stromata* 5.9.58.

[*] Still, there have been modern attempts deserving the name of myth. One, of course, is Sir Thomas More's *Utopia*, which has taken on so much meaning through the centuries. We realize today that it, too, was partly oracular. And we should not forget *Alice in Wonderland*, the perfect nonsense myth, as significant and as nonsensical as the *Kalevala* itself. This parallel will appear relevant at the end of the appendices. Today, there is Austin Wright's *Islandia*, which appeared in 1942, and its present sequel, *The Islar*, by Mark Saxton, to be published in the autumn of 1969.

of timeliness and then dismissed with its load of interpretations and apprehensions that last as long as the reading—but the real present, the only thing that counts, is the eternal Sphinx.

Today's children, that impassive posterity to whom all reverence is due, know where to look for myths: in animal life, in the *Jungle Books,* in the stories of Lassie and Flipper, where innocence is unassailable, in Western adventures suitably arranged by grownups for the protection of law and order. Much of the rest sedulously built up by mass media is modern prejudice and delusion, like the glamor of royalty, or the perfection of super-detergents and cosmetics: *super-stitio,* leftovers. So one might feel tempted to say: actually, however, no particle of myth today is left over, and we have to do only with a deliberate lie about the human condition. Tolkien's efforts at reviving the genre, whatever the talent employed, carry as much conviction as the traditional three-dollar bill.

The assumed curious child would have been pleased only if he had been told the "story" of the engine just as Kipling tells it, which is hardly the style of a mechanical engineer. But suppose now the child had been confronted with the "story" of a planet as it emerges from the textbooks of celestial mechanics, and had been asked to calculate its orbits and perturbations. This would be a task for a joyless grownup, and a professional one at that. Who else could face the pages bristling with partial differential equations, with long series of approximations, with integrals contrived from pointless quadratures? Truly a world of reserved knowledge. But if, on the other hand, a person living several thousand years ago had been confronted with cunningly built tales of Saturn's reign, and of his exorbitant building and modeling activities—after he had "separated Heaven and Earth" by means of that fateful sickle, that is, after he had established the obliquity of the ecliptic . . . If he had heard of Jupiter's ways of command and his innumerable escapades, populating the earth with gentle nymphs forever crossed in their quest for happiness, escapades that were invariably successful in spite of the constant watchfulness of his jealous "ox-eyed" or sometimes "dog-eyed" spouse . . . If this person also learned of the fierce adventures of Mars, and the complex mutual involvement of gods and heroes expressing themselves in terms of action and unvarying numbers, he would have been a participant in the process of mythical knowledge. This knowledge would have been transmitted by his elders, confirmed by holy commands, rehearsed by symbolic experiences in the form of musical rites and performances involving his whole people. He would have found it easier to respect than comprehend, but it would have led to an idea of the overall texture of the cosmos. In his own person, he would have been part of a genuine theory of cosmology, one he had absorbed by heart, that was responsive to his emotions, and one that could act on his aspirations and dreams. This kind of participation in ultimate things, now extremely difficult for anyone who has not graduated in astrophysics, was then possible to some degree for everyone, and nowhere could it be vulgarized.

That is what is meant here by mythical knowledge. It was understood only by a very few, it appealed to many, and it is forever intractable for those who approach it through "mathematics for the million" or by speculations on the unconscious. In other words, this is a selective and difficult approach, employing the means at hand and much thought, limited surely, but resistant to falsification.

How, in former times, essential knowledge was transmitted on two or more intellectual levels can be learned from Germaine Dieterlen's introduction to Marcel Griaule's *Conversations with Ogotemmêli*, which deals with Dogon education and with the personal experience of the members of *La Mission Griaule*, who had to wait sixteen years before the sage old men of the tribe decided to "open the door."[3] The description is revealing enough to be quoted in full:

> In African societies which have preserved their traditional organization the number of persons who are trained in this knowledge is quite considerable. This they call "deep knowledge" in contrast with "simple knowledge" which is regarded as "only a beginning in the understanding of beliefs and customs" that people who are not fully instructed in the *cosmogony* possess. There are various reasons for the silence that is generally observed on this subject. To a natural

[3] M. Griaule, *Conversations with Ogotemmêli* (1965), pp. xiv–xvii.

reserve before strangers who, even when sympathetic, remain unconsciously imbued with a feeling of superiority, one must add the present situation of rapid change in African societies through contact with mechanization and the influence of school teaching. But among groups where tradition is still vigorous, this knowledge, which is expressly characterized as esoteric, is only secret in the following sense. It is in fact open to all who show a will to understand so long as, by their social position and moral conduct, they are judged worthy of it. Thus every family head, every priest, every grown-up person responsible for some small fraction of social life can, as part of the social group, acquire knowledge on condition that he has the patience and, as the African phrase has it, "he comes to sit by the side of the competent elders" over the period and in the state of mind necessary. Then he will receive answers to all his questions, *but it will take years*. Instruction begun in childhood during assemblies and rituals of the age-sets continues in fact throughout life.

These various aspects of African civilization gradually became clear in the course of intensive studies undertaken among several of the peoples of Mali and Upper Volta over more than a decade. In the case of the Dogon, concerning whom there have already been numerous publications, these studies have made possible the elaboration of a synthesis covering the greater part of their activities.

We should now record the important occurrence during the field expedition of 1947 which led to the writing of this particular study. From 1931 the Dogon had answered questions and commented on observations made during previous field trips on the basis of the interpretation of facts which they call "la parole de face"; this is the "simple knowledge" which they give in the first instance to all enquirers. Publications of information obtained before the studies in 1948 relate to this first level of interpretation.

But the Dogon came to recognize the great perseverance of Marcel Griaule and his team in their enquiries, and that it was becoming increasingly difficult to answer the multiplicity of questions without moving on to a different level. They appreciated our eagerness for an understanding which earlier explanations had certainly not satisfied, and which was clearly more important to us than anything else. Griaule had also shown a constant interest in the daily life of the Dogon, appreciating their efforts to exploit a difficult country where there was a serious lack of water in the dry season, and our relationships, which had thus extended beyond those of ethnographical enquiry, became more and more trusting and affectionate. In the light of all this, the Dogon took their own decision, of which we learned only later when they told us themselves. The elders of the lineages of the double village of Ogol and the most important totemic priests of the region of Sanga met together and decided that the more esoteric aspects of their religion should be fully revealed to Professor Griaule. To begin this they chose one of their best informed members, Ogotommêli, who, as will be seen in the introduction, arranged the first interview. This first exposition lasted exactly the number of days recorded in *Dieu d'Eau,* in which the meandering flow of information is faithfully reported. Although we knew nothing of it at the time, the progress of this instruction by Ogotemmêli was being reported on daily to the council of elders and priests.

The seriousness and importance of providing this exposé of Dogon belief was all the greater because the Dogon elders knew perfectly well that in doing so they were opening the door, not merely to these thirty days of information, but to later and more intensive work which was to extend over months and years. They never withdrew from this decision, and we should like to express here our grateful thanks to them. After Ogotemmêli's death, others carried on the work. And since Professor Griaule's death they have continued with the same patience and eagerness to complete the task they had undertaken. These later enquiries have made possible the publication of the many further studies cited in the bibliography, and the preparation of a detailed treatise entitled *Le Renard Pâle,* the first part of which is now in press. And in 1963, as this is written, the investigation still continues.

A Guide for the Perplexed

Tout-puissants étrangers, inévitables astres . . .

VALÉRY, *La Jeune Parque*

THIS BOOK is highly unconventional, and often the flow of the tales will be interrupted to put in words of guidance, in the fashion of the Middle Ages, to emphasize salient points.

To begin with, there is no system that can be presented in modern analytical terms. There is no key, and there are no principles from which a presentation can be deduced. The structure comes from a time when there was no such thing as a system in our sense, and it would be unfair to search for one. There could hardly have been one among people who committed all their ideas to memory.

It can be considered a pure structure of numbers. From the beginning we considered calling this essay "Art of the Fugue." And that excludes any "world-picture," a point that cannot be stressed strongly enough. Any effort to use a diagram is bound to lead into contradiction. It is a matter of times and rhythms.

The subject has the nature of a hologram, something that has to be present as a whole to the mind.[1] Archaic thought is cosmological first and last; it faces the gravest implications of a cosmos in ways which reverberate in later classic philosophy. The chief implication is a profound awareness that the fabric of the cosmos is not only determined, but overdetermined and in a way that does not permit the simple location of any of its agents, whether simple magic or

[1] In optics, "hologram" is the interference pattern of light with itself; i.e., every part of an image is displayed at every point, as if every point looked at every source of light.

astrology, forces, gods, numbers, planetary powers, Platonic Forms, Aristotelian Essences or Stoic Substances. Physical reality here cannot be analytical in the Cartesian sense; it cannot be reduced to concreteness even if misplaced. Being is change, motion and rhythm, the irresistible circle of time, the incidence of the "right moment," as determined by the skies.

There are many events, described with appropriate terrestrial imagery, that do not, however, happen on earth. In this book there is mention of floods. In tradition, not one but three floods are registered, one being the biblical flood, equivalents of which are mentioned in Sumerian and Babylonian annals. The efforts of pious archaeologists to connect the biblical narrative with geophysical events are highly conjectural. There have been floods in Mesopotamia causing grievous loss of life. There still are in the river plains of China and elsewhere, but none of the total nature that the Bible describes.

There are tales, too, of cataclysmic deluges throughout the great continental masses, in Asia and America, *told by peoples who have never seen the sea, or lakes, or great rivers.* The floods the Greeks described, like the flood of Deucalion, are as "mythical" as the narrative of Genesis. Greece is not submersible, unless by tsunamis. Deucalion and his wife landed on Mount Parnassus, high above Delphi, the "Navel of the Earth," and were the only survivors of this flood, the second, sent by Zeus in order to destroy the men of one world-age. Classical authors disagreed on the specification of which world-age. Ovid voted for the Iron Age. Plato's Solon keeps his conversation with the Egyptian priest on a mythical level, and his discussion of the two types of world destruction, by fire or water, is astronomical.

The "floods" refer to an old astronomical image, based on an abstract geometry. That this is not an "easy picture" is not to be wondered at, considering the objective difficulty of the science of astronomy. But although a modern reader does not expect a text on celestial mechanics to read like a lullaby, he insists on his capacity to understand mythical "images" instantly, because he can respect as "scientific" only page-long approximation formulas, and the like.

He does not think of the possibility that equally relevant knowledge might once have been expressed in everyday language. He never suspects such a possibility, although the visible accomplishments of ancient cultures—to mention only the pyramids, or metallurgy—should be a cogent reason for concluding that serious and intelligent men were at work behind the stage, men who were bound to have used a technical terminology.

Thus, archaic "imagery" is strictly verbal, representing a specific type of scientific language, which must not be taken at its face value nor accepted as expressing more or less childish "beliefs." Cosmic phenomena and rules were articulated in the language, or terminology, of myth, where each key word was at least as "dark" as the equations and convergent series by means of which our modern scientific grammar is built up. To state it briefly, as we are going to do, is not to explain it—far from it.

First, what was the "earth"? In the most general sense, the "earth" was the ideal plane laid through the ecliptic. The "dry earth," in a more specific sense, was the ideal plane going through the celestial equator. The equator thus divided two halves of the zodiac which ran on the ecliptic, $23\frac{1}{2}°$ inclined to the equator, one half being "dry land" (the northern band of the zodiac, reaching from the vernal to the autumnal equinox), the other representing the "waters below" the equinoctial plane (the southern arc of the zodiac, reaching from the autumnal equinox, via the winter solstice, to the vernal equinox). The terms "vernal equinox," "winter solstice," etc., are used intentionally because myth deals with time, periods of time which correspond to angular measures, and not with tracts in space.

This could be neglected were it not for the fact that the equinoctial "points"—and therefore, the solstitial ones, too—do not remain forever where they should in order to make celestial goings-on easier to understand, namely, at the same spot with respect to the sphere of the fixed stars. Instead, they stubbornly move along the ecliptic in the opposite direction to the yearly course of the sun, that is, against the "right" sequence of the zodiacal signs (Taurus→Aries→Pisces, instead of Pisces→Aries→Taurus).

This phenomenon is called the Precession of the Equinoxes, and it was conceived as causing the rise and the cataclysmic fall of ages of the world. Its cause is a bad habit of the axis of our globe, which turns around in the manner of a spinning top, its tip being in the center of our small earth-ball, whence our earth axis, prolonged to the celestial North Pole, describes a circle around the North Pole of the ecliptic, the true "center" of the planetary system, the radius of this circle being of the same magnitude as the obliquity of the ecliptic with respect to the equator: 23½°. The time which this prolonged axis needs to circumscribe the ecliptical North Pole is roughly 26,000 years, during which period it points to one star after another: around 3000 B.C. the Pole star was alpha Draconis; at the time of the Greeks it was beta Ursae Minoris; for the time being it is alpha Ursae Minoris; in A.D. 14,000 it will be Vega. The equinoxes, the points of intersection of ecliptic and equator, swinging from the spinning axis of the earth, move with the same speed of 26,000 years along the ecliptic.

The sun's position among the constellations at the vernal equinox was the pointer that indicated the "hours" of the precessional cycle —very long hours indeed, the equinoctial sun occupying each zodiacal constellation for about 2,200 years. The constellation that rose in the east just before the sun (that is, rose heliacally) marked the "place" where the sun rested. At this time it was known as the sun's "carrier," and as the main "pillar" of the sky, the vernal equinox being recognized as the fiducial point of the "system," determining the first degree of the sun's yearly circle, and the first day of the year. (When we say, it was "recognized," we mean that it was spelled "carrier" or "pillar," and the like: it must be kept in mind that we are dealing with a specific terminology, and not with vague and primitively rude "beliefs.") At Time Zero (say, 5000 B.C.—there are reasons for this approximate date), the sun was in Gemini; it moved ever so slowly from Gemini into Taurus, then Aries, then Pisces, which it still occupies and will for some centuries more. The advent of Christ the Fish marks our age. It was hailed by Virgil, shortly before Anno Domini: "a new great order of centuries is now being born . . ." which earned Virgil the strange title of prophet of Christianity. The preceding age, that of Aries, had been heralded by Moses coming down from Mount Sinai as "two-horned," that is, crowned with the Ram's horns, while his flock disobediently insisted upon dancing around the "Golden Calf" that was, rather, a "Golden Bull," Taurus.

Thus, the revolving heavens gave the key, the events of our globe receding into insignificance. Attention was focused on the supernal presences, away from the phenomenal chaos around us. What moved in heaven of its own motion, the planets in their weeks and years, took on ever more awesome dignity. They were the Persons of True Becoming. The zodiac was where things really happened, for the planets, the true inhabitants, knew what they were doing, and mankind was only passive to their behest. It is revealing to look at the figure drawn by a West Sudanese Dogon at the request of Professor Zahan, showing the world egg, with the "inhabited world" between the tropics, "le cylindre ou rectangle du monde."[2] The Dogon are fully aware of the fact that the region between the *terrestrial* tropics is not the best of inhabitable quarters, and so were their teachers of far-off times, the archaic scientists who coined the terminology of myth. What counted was the zodiacal band between the celestial tropics, delivering the houses, and the inns, the "masks" (prosôpa), and the disguises to the much traveling and "shape-shifting" planets.

How far this point of view was from modern indifference can hardly be appreciated except by those who can see the dimensions of the historical chasm that opened with the adoption of the Copernican doctrine. What had been for Sir Thomas Browne an *o altitudo* crowded with religious emotions, presences and presentiments has become a platitude that could at best inspire a Russian cosmonaut with the triumphant observation: "I have been up in the sky, and nowhere did I find God." Astronomy has come down into the realm of exterior ballistics, a subject for the adventures of the Space Patrol.

[2] D. Zahan and S. de Ganay, "Études sur la cosmologie des Dogon," *Africa 21* (1951), p. 14.

A diagram of the Precession of the Equinoxes. The symmetrical drawing shows that the phenomenon occurs at both poles.

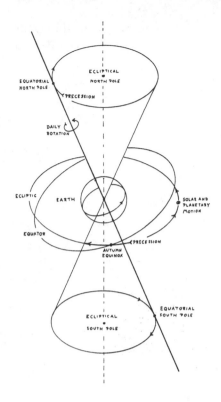

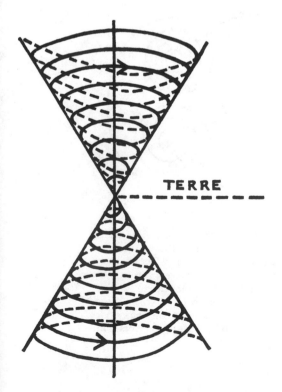

TERRE

"The internal motion of the cosmic tree," according to North-West Africans. "In the firmament that motion marks the rotation of the stars above the earth and below the earth, around the fixed poles indicated by the axis formed by the elements in the middle of the cosmic tree."

The ways of the Demiurge during creation,
according to the Bambara.

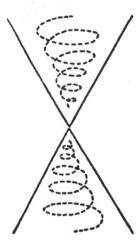

"In order to make heaven and earth, the demiurge
stretched himself into a conical helix; the turnings-back
of that spiral are marked graphically by the sides of two
angles which represent also the space on high and the
space below."

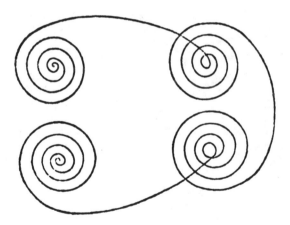

"In order to mix the four elements (air, fire, water,
earth) of which all things are formed, and to distribute
them down to the borders of space which he had determined by girdling it, the demiurge travels through the
universe turning on himself. These movements are figured by four spirals bound one to the other which represent at the same time the circular voyage, the four angles
of the world in which the mixing of the elements takes
place, and the motion of matter."

Mount Meru, the world mountain, rising from the sea, surmounted by holy radiation, with sun and moon circling around it, as depicted in an old Buddhist cave sanctuary in Chinese Turkestan.

The collapse of the hourglass-shaped Meru, caused by Buddha's death, with sun and moon rolling down; the moon shows the hare contained in it. Many collapsing world-pillars, unhinged mill-trees, and the like have been mentioned in this book, and this is one of the few pieces of pictorial evidence for a crumbling *skambha*.

One might say that it takes a wrenching effort of the imagination to restore in us the capacity for wonder of an Aristotle. But it would be misleading to talk of "us" generally, because the average Babylonian or Greek showed as little inclination to wonder at order and law in nature as our average contemporaries do. It has been and will be the mark of a true scientific mind to search for, and to wonder at, the invariable structure of number behind the manifold appearances. (It needs the adequate "expectation," the firm confidence in "sense"—and "sense" does mean number and order for us, since the birth of high civilization—to discover the periodical system of the elements or, further on, Balmer's series.) Whence it is much easier for a great scientist—for instance, Galileo, Kepler, or Newton—to appreciate master feats of early mathematicians than it is for the average humanist of all ages. No professional historian of culture is likely to understand better the intellectual frame of mind of the Maya than the astronomer Hans Ludendorff has done.

It is not so much the enormous number of new facts established by scientists in the many centuries between antiquity and the 20th century which separates us from the outlook of our great scientific ancestors but the "deteriorated" expectations ruling our time. Kepler's quest, were he living today, would be to discover a modified perspective from which to rediscover the *Harmonice Mundi* on another scale. But, after all, what else if not such a quest for the establishment of a new kind of cosmos is the work on the "general field formula"? This time, the cosmos, as covered by the coming-to-be formula, will be understandable and will make "sense" only for the best mathematicians, to the complete exclusion of the common people, and it will hardly be a "meaningful" universe such as the archaic one had been.

To come back to the key words of ancient cosmology: if the words "flat earth" do not correspond in any way to the fancies of the flat-earth fanatics who still infest the fringes of our society and who in the guise of a few preacher-friars made life miserable for Columbus, so the name of "true earth" (or of "the inhabited world") did not in any way denote our physical geoid for the archaics. It applies to the band of the zodiac, two dozen degrees right and left of the ecliptic, to the tracks of the "true inhabitants" of this world, namely, the planets. It comprises their various oscillations and curlicues from their courses, and also the "dragon," well known from very early times, which causes eclipses by swallowing the sun and moon.

On the zodiacal band, there are four essential points which dominate the four seasons of the year. They are, in fact, in church liturgy the *quatuor tempora* marked with special abstinences. They correspond to the two solstices and the two equinoxes. The solstice is the "turning back" of the sun at the lowest point of winter and at the highest point of summer. The two equinoxes, vernal and autumnal, are those that cut the year in half, with an equal balance of night and day, for they are the two intersections of the equator with the ecliptic. Those four points together made up the four pillars, or corners, of what was called the "quadrangular earth."

This is an essential feature that needs more attention. We have said above that "earth," in the most general sense, meant the ideal plane laid through the ecliptic; meanwhile we are prepared to improve the definition: "earth" is the ideal plane going through the four points of the year, the equinoxes and the solstices. Since the four constellations rising heliacally at the two equinoxes and the two solstices determine and define an "earth," it is *termed* quadrangular (and by no means "believed" to be quadrangular by "primitive" Chinese, and so on). And since constellations rule the four corners of the quadrangular earth only temporarily, such an "earth" can rightly be said to perish, and a new earth to rise from the waters, with four new constellations rising at the four points of the year. Virgil says: "*Iam redit et Virgo . . .*" (already the Virgin is returning). (It is important to remember the vernal equinox as the fiducial point; it is from this fact that a new earth is termed to rise from the waters. In reality, only the new vernal equinoctial constellation climbs from the sea onto the dry land above the equator[3]— the inverse happens diametrically opposite. A constellation that

[3] In a similar sense, Petronius' Trimalchio says about the month of May: "totus coelus taurulus fiat" ("the whole sky turns into a little bull").

ceases to mark the autumnal equinox, gliding below the equator, is drowned.) This "formula" will make it easier to understand the myth of Deucalion, in which the devastating waves of the flood were ordered back by Triton's blowing the conch: the conch had been invented by Aigokeros, i.e., Capricornus, who ruled the winter solstice in the world-age when Aries "carried" the sun.

At Time Zero, the two equinoctial "hinges" of the world had been Gemini and Sagittarius, spanning between them the arch of the Milky Way: both bicorporeal signs[4]—and so were Pisces, and Virgo with her ear of wheat, at the two other corners—to mark the idea that the way (the Milky Way itself) was open between earth and heaven, the way up and the way down where men and gods could meet in that Golden Age. As will be shown later, the exceptional virtue of the Golden Age was precisely that the crossroads of ecliptic and equator coincided with the crossroads of ecliptic and Galaxy, namely in Gemini and Sagittarius, both constellations "standing" firmly at two of the four corners of the quadrangular earth.

At the "top," in the center high above the "dry" plane of the equator, was the Pole star. At the opposite top, or rather in the depth of the waters below, unobserved from our latitudes, was the southern pole, thought to be Canopus, by far the brightest star of these regions, more remarkable than the Southern Cross.

This brief sketch of archaic theory indicates—to repeat—that geography in our sense was never meant, but a cosmography of the kind needed even now by navigators. Ptolemy, the great geographer of antiquity, had been thinking of nothing else. His *Geography* is a set of coordinates drawn from the skies, and transferred onto an uncouth outline map of our globe, with a catalogue of earthly distances added on by sailors and travelers to pinpoint, or confirm, the positions of countries around the Mediterranean world. It was an uncouth outline map, for it covered only a few countries known around the Mediterranean region. Nothing was shown beyond the latitude 16° south of the equator and 63° north, cor-

---

[4] These constellations were, originally, called "bicorporeal" for reasons very different from those given by Ptolemy's *Tetrabiblos 1.11.*

responding to Iceland. Nothing west beyond the Canary Islands or east of the easternmost city of China, an arc of longitude fixed for simplicity at 180°, twelve equinoctial hours from end to end, the breadth over the whole latitudes being nine equinoctial hours. A large part of the space is blank and the limits are assigned, as they should be, astronomically. This is what the ancients knew after a thousand years of exploration, and they handed it down to the Renaissance. They called it the *oikoumene*, the inhabited earth.

One may well understand how the archaics gave this name for purely astronomical reasons to the zodiacal band, about as wide in degrees but embracing the whole globe. The world, the cosmos, was above, revolving majestically in twenty-four hours, and it lent itself to the passionate exploration of cosmographers through the starlit night. Astrology was the inevitable outcome of astronomy through those ages. The early Greeks derived their mathematics from astronomy. In those centuries, their insatiable curiosity developed a knowledge of our earth and the events on it which drove them to create the beginnings of our science. But soon after Aristotle, the Stoics reverted to the oriental pattern and reinstalled astrology. Three centuries of pre-Socratic thought had equipped them with an interest in physics, but with it they had nowhere to go. They still had no experimental science as we mean it. What they needed was an interpretation of influences, to go with the all-in-all that the cosmos has to be. Stoic physics was a seductive presentment of a field theory, but it was a counterfeit. Nothing was to come of it because the true implications of the archaic cosmos, no less than those of the Platonic, were incompatible with anything that our physics can think of. In Stoic physics there is no simple location, no analytical space.

It should be understood once and for all that the gulf between the archaic world and ours was as wide as science itself. Prodigies of exactitude and computation could not bridge it. Only the astronomical map could. Whitehead has summed it up succinctly: "Our science has been founded on simple location and misplaced concreteness." Modern physics has turned the original words into queries. For Newton, it had the force of evidence: "No person endowed with a capacity of rational understanding will believe that a thing acts where it is not." Newton himself put the first query, by stating the theory of gravitation—mathematically irresistible, physically unexplainable. He could only accept it: "I do not understand it, and I am going to feign no suppositions." The answer was to come only with Einstein. It amounted to pure mathematical rationalization, which did away with simple location, and with concreteness altogether. The edifice of Descartes lay in ruins. Nonetheless, the mind of civilized man clung to both principles invincibly, as being equal to common sense. It was a model case of habit having become second nature. The birth of experimental physics was a decisive factor in the change.

No such common sense obtained once upon a time, when time was the only reality, and space had still to be discovered—or invented—by Parmenides after 500 B.C. (See G. de Santillana, "Prologue to Parmenides," in *Reflections on Men and Ideas* [1968], pp. 82–119.)

The task then was to recover from the remote past an utterly lost science, linked to an equally lost culture—one in which anthropologists have seen only illiterate "primitive man." It was as if the legendary "Cathedrale Engloutie" emerged from the depth of prehistory with its bells still ringing.

The problem was also clear: this lost science, immensely sophisticated, had no "system," no systematic key that could be a basis for teaching. It existed before systems could be thought of. It was, to repeat, a spontaneously generated "Art of the Fugue." That is why it took us so many years to work it out.

The archaics' vision of the universe appears to have left out all ideas of the earth suspended, or floating, in space. Whether or not this was really so cannot be decided yet: there are peculiar rumors to be heard about innumerable "Brahma-Eggs," that is, spheres like our own, in India. The Maori of New Zealand claimed, as the Pythagoreans had done, that every star had mountains and plains, and was inhabited like the earth. Varahamihira (5th century A.D.)

even stated that the earth was suspended between magnets.[5] For the time being, one must continue to assume that the earth was simply the center of the world, and a sphere, and that there was no trace of Galilean relativism which is so natural to us, posing so many problems of motion. The Greeks still had the old idea, but they asked themselves questions about it. What moved was the sky, but questions about the sky posed abstruse problems. The greatest one was, of course, the slow motion of the tilt of the sky, described above, which went through a Great Year of 26,000 years.

The Greek astronomers had enough instrumentation and data to detect the motion, which is immensely slow, and they saw that it applied to the whole of the sky. Hipparchus in 127 B.C. called it the Precession of the Equinoxes. There is good reason to assume that he actually rediscovered this, that it had been known some thousand years previously, and that on it the Archaic Age based its long-range computation of time. Modern archaeological scholars have been singularly obtuse about the idea because they have cultivated a pristine ignorance of astronomical thought, some of them actually ignorant of the Precession itself. The split between the two cultures begins right here. But apart from this, although the scholars unanimously cling to the accepted conventions about the tempo of historical evolution, they widely disagree when it comes to judging the evidence in detail. The verdicts concerning the familiarity of ancient Near Eastern astronomers with the Precession depend, indeed, on arbitrary factors: namely, on the different scholarly opinions about the difficulty of the task. Ernst Dittrich, for instance, remarked that one should not expect much astronomical knowledge from Mesopotamians around 2000 B.C. "Probably they knew only superficially the geometry of the motions of sun and moon. Thus, if we examine the simple, easily observable motions by means of which one could work out chronological determinants with very little mathematical knowledge, we find only the

[5] *Pancasiddhantika*, chapter XII (Thibaut trans., p. 69): "The round ball of the earth, composed of the five elements, abides in space in the midst of the starry sphere, like a piece of iron suspended between magnets."

Precession."[6] There was also a learned Italian Church dignitary, Domenico Testa, who snatched at this curious argument to prove that the world had been created ex nihilo, as described in the first book of Moses, an event that supposedly happened around 4000 B.C. If the Egyptians had had a background of many millennia to reckon with, who, he asked, could have been unaware of the Precession? "The very sweepers of their observatories would have known."[7] Hence time could not have begun before 4000, Q.E.D.

The comparison of the views just quoted with those upheld by the majority of modern scholars shows that one's own subjective opinions about what is easy and what is difficult might not be the most secure basis for a serious historiography of science. As Hans Ludendorff once pointed out, it is an unsound approach to Maya astronomy to start from preconceived convictions about what the Maya could have known and what they could not possibly have known: one should, instead, draw conclusions only from the data as given in the inscriptions and codices.[8] That this had to be stressed explicitly reveals the steady decline of scientific ethics.

We today are aware of the Precession as the gentle tilting of our globe, an irrelevant one at that. As the GI said, lost in the depth of jungle misery, when his friends took refuge in their daydreams: "When I close my eyes, I see only a mule's behind. Also when I don't." This is, as it were, today's vision of reality. Today, the Precession is a well-established fact. The space-time continuum does not affect it. It is by now only a boring complication. It has lost relevance for our affairs, whereas once it was the only majestic secular motion that our ancestors could keep in mind when they looked for a great cycle which could affect humanity as a whole. But then our ancestors were astronomers and astrologers. They believed that the sliding of the sun along the equinoctial point affected the frame of the cosmos and determined a succession of

[6] "Gibt es astronomische Fixpunkte in der ältesten babylonischen Chronologie?" OLZ *15* (1912), col. 104.

[7] *Il Zodiaco di Dendera Illustrato* (1822), p. 17.

[8] "Zur astronomischen Deutung der Maya-Inschriften," SPAW (1936), p. 85.

world-ages under different zodiacal signs. They had found a large peg on which to hang their thoughts about cosmic time, which brought all things in fateful order. Today, that order has lapsed, like the idea of the cosmos itself. There is only history, which has been felicitously defined as "one damn thing after another."

And yet, were history really understood in this admittedly flat sense of things happening one after another to the same stock of people, we should be better off than we are now, when we almost dare not admit the assumption from which this book starts, that our ancestors of the high and far-off times were endowed with minds wholly comparable to ours, and were capable of rational processes —always given the means at hand. It is enough to say that this flies in the face of a custom which has become already a second nature.

Our period may some day be called the Darwinian period, just as we talk of the Newtonian period of two centuries ago. The simple idea of evolution, which it is no longer thought necessary to examine, spreads like a tent over all those ages that lead from primitivism into civilization. Gradually, we are told, step by step, men produced the arts and crafts, this and that, until they emerged into the light of history.

Those soporific words "gradually" and "step by step," repeated incessantly, are aimed at covering an ignorance which is both vast and surprising. One should like to inquire: which steps? But then one is lulled, overwhelmed and stupefied by the gradualness of it all, which is at best a platitude, only good for pacifying the mind, since no one is willing to imagine that civilization appeared in a thunderclap.

One could find a key in a brilliant TV production on the Stonehenge problem given a few years ago. With the resources of the puissant techniques of ubiquity, various authorities were called to the screen to discuss the possible meaning of the astronomical alignments and polygons discovered in the ancient Megalith since 1906, when Sir Norman Lockyer, the famous astronomer, published the results of his first investigation. Specialists, from prehistorians to astronomers, expressed their doubts and wonderments down to the last one, a distinguished archaeologist who had been working on the monument itself for many years. He had more fundamental doubts. How could one not realize, he said, that the builders of Stonehenge were barbarians, "howling barbarians" who were, to say the least, utterly incapable of working out complex astronomical cycles and over many years at that? The uncertain coincidences must be due to chance. And then, with perverse irony, the midwinter sun of the solstice appeared on the screen rising exactly behind the Heel Stone, as predicted. The "mere" coincidences had been in fact ruled out, since Gerald Hawkins, a young astronomer unconcerned with historical problems, had run the positions through a computer and discovered more alignments than had been dreamed of. Here was the whole paradox. Howling barbarians who painted their faces blue must have known more astronomy than their customs and table manners could have warranted. The lazy word "evolution" had blinded us to the real complexities of the past.

That key term "gradualness" should be understood to apply to a vastly different time scale than that considered by the history of mankind. Human history taken as a whole in that frame, even raciation itself, is only an evolutionary episode. In that whole, Cro-Magnon man is the last link. All of protohistory is a last-minute flickering.

But while the biologists were wondering, something great had come upon the scene, arriving from unexpected quarters. Sir James George Frazer was a highly respected classical scholar who, while editing the *Description of Greece* by Pausanias, was impressed with the number of beliefs, practices, cults and superstitions spread over the classical landscape of Greece in classical times. This led him to search deeper into the half-forgotten strata of history, and out of it came his *Golden Bough*. The historian had turned ethnologist, and extended his investigations to the whole globe. Suddenly, an immense amount of material became available about fertility cults as the universal form of earliest religion, and about primitive magic connected with it. This appeared to be the humus from which civilization had grown—simple deities of the seasons, a dim multitude of peasants copulating in the furrows and building up rituals of fertility with human sacrifice. Added to this, in political circles, there came the vision of war as both inherent in human nature and ennobling—the law of natural selection applied to nations and races. Thus, many materials and much history went to build the temple of evolutionism. But as the theory moved on, its high-minded aspects began to wane; psychoanalysis moved in as a tidal wave. For if the struggle for life (and the religions of the life force) can explain so much, the unconscious can explain *anything*. As we know today only too well.

The universal and uniform concept of gradualness thus defeated itself. Those key words (gradualness and evolution) come from the earth sciences in the first place, where they had a precise meaning. Crystallization and upthrust, erosion and geosynclinals are the result of forces acting constantly in accordance with physical laws. They provided the backdrop for Darwin's great scenario. When it comes to the evolution of life, the terms become less precise in meaning, though still acceptable. Genetics and natural selection stand for natural law, and events are determined by the rolling of the dice over long ages. But we cannot say much about the why and the how of *this* instead of *that* specific form, about where species, types, cultures branched off. Animal evolution remains an overall historical hypothesis supported by sufficient data—and by the lack of any alternative. In detail, it raises an appalling number of questions to which we have no answer. Our ignorance remains vast, but it is not surprising.

And then we come to history, and the evolutionary idea reappears, coming in as something natural, with all scale lost. The accretion of plausible ideas goes on, its flow invisibly carried by "natural law" since the time of Spencer. It all remains within an unexamined kind of *Naturphilosophie*. For if we stopped to think, we would agree that as far as human "fate" is concerned organic evolution ceased before the time when history, or even prehistory, began. We are on another time scale. This is no longer nature acting on man, but man on nature. People like to think of a constancy of laws which apply to us. But man is a law unto himself.

When, riding on the surf of the general "evolutionism," Ernst Haeckel and his faithful followers proposed to solve the "world riddles" once and for all, Rudolf Virchow[9] warned time and again of an evil "monkey wind" blowing round; he reminded his colleagues of the index of excavated "prehistoric" skulls and pointed to the unchanged quantity of brain owned by the species Homo sapiens. But his contemporaries paid no heed to his admonitions; least of all the humanists who applied, without blinking, the strictly biological scheme of the evolution of organisms to the cultural history of the single species Homo sapiens.

In later centuries historians may declare all of us insane, because this incredible blunder was not detected at once and was not refuted with adequate determination. Mistaking cultural history for a process of gradual evolution, we have deprived ourselves of every reasonable insight into the nature of culture. It goes without saying that the still more modern habit of replacing "culture" by "society" has blocked the last narrow path to understanding history. Our ignorance not only remained vast, but became pretentious as well.

A glimpse at some *Pensées* might show the abyss that yawns between us and a serious thinker of those golden days before the outbreak of "evolution." This is what Pascal asked: "What are our natural principles but principles of custom? In children they are those which they have received from the habits of their fathers, as hunting in animals. A different custom will cause different natural principles." And: "Custom is a second nature which destroys the former. But what is nature? For is custom not natural? I am much afraid that nature is itself only a first custom, as custom is a second nature."[10]

This kind of question, aimed with precision at the true problematical spots, would have been enough to make hash of social anthropology two centuries ago, and also of anthropological sociology. Although fully aware of the knot of frightening problems

---

[9] In several of his addresses to the "Versammlungen deutscher Naturforscher und Arzte."

[10] *Pensées*, nos. 92, 93 (Trotter trans. [1941], p. 36).

arising from the results of the most modern neurophysiology—the building up of microneurons in the brain after birth, etc.—we are by no means entitled to feign any hypotheses beyond saying that the master brain who will, sooner or later, fashion a new philosophical anthropology deserving the title, one that will account for all the new implications, will find himself up against these same few questions of Pascal.

Some words have still to be said about the problem that is at the very root of the many misunderstandings, that of translation. Most of the texts were written—if they were ever originally written—in remote and half-obliterated languages from the far past. The task of translation has been taken over by a guild of dedicated, highly specialized philologists who have had to reconstruct the dictionaries and grammars of these languages. It would be bad grace to dismiss their efforts, but one must take into account several layers of error: (1) personal or systematic errors, arising from their preconceptions and from well-implanted prejudices (psychological and philosophical) of their age; (2) the very structure of our own language, of the architecture of our own verbal system, of which very few individuals are aware. There was once a splendid article by Erwin Schroedinger, with the title "Are there quantum jumps?" which laid bare many such misunderstandings inside the well-worked area of modern physics.[11] And all this ties up with another major source of error that comes from the underestimation of the thinkers of the far past. We instinctively dismiss the idea that five to ten thousand years ago there may very well have been thinkers of the order of Kepler, Gauss, or Einstein, working with the means at hand.

In other words, we must take language seriously. Imprecise language discloses the lack of precision of thought. We have learned to take the language of Archimedes or Eudoxos seriously, simply because it *can* translate directly into modern forms of thought. This should extend to forms of thought utterly different from ours in appearance. Take that great endeavor on the hieroglyphic language,

[11] *British Journal for the Philosophy of Science 3* (1952), pp. 112ff.

embodied in the imposing Egyptian dictionary of Erman-Grapow. For our simple word "heaven" it shows thirty-seven terms whose nuances are left to the translator and used according to his lights. So the elaborate instructions in the Book of the Dead, referring to the soul's celestial voyage, translate into "mystical" talk, and must be treated as holy mumbo jumbo. But then, modern translators believe so firmly in their own invention, according to which the underworld has to be looked for in the interior of our globe—instead of in the sky—that even 370 specific astronomical terms would not cause them to stumble.

One small example may indicate the way in which texts are "improved." In the inscriptions of Dendera, published by Dümichen, the goddess Hathor is called "lady of every joy." For once, Dümichen adds: "Literally . . . 'the lady of every heart circuit.' "[12] This is not to say that the Egyptians had discovered the circulation of the blood. But the determinative sign for "heart" often figures as the plumb bob at the end of a plumb line coming from a well-known astronomical or surveying device, the *merkhet*. Evidently, "heart" is something very specific, as it were the "center of gravity."[13] And this may lead in quite another direction. The Arabs preserved a name for Canopus—besides calling the star *Kalb at-taiman* ("heart of the south"):[14] *Suhail el-wezn*, "Canopus Ponderosus," the heavy-weighing Canopus, a name promptly declared meaningless by the experts, but which could well have belonged to an archaic system in which Canopus was the weight at the end of the plumb line, as befitted its important position as a heavy star at the South Pole of the "waters below." Here is a chain of inferences which might or might not be valid, but it is allowable to test it, and no inference at all would come from the "lady of every joy." The line seems to state that Hathor (= Hat Hor, "House of Horus") "rules" the revolution of a specific celestial body—whether or not Canopus is alluded to—or, if we can trust the trans-

---

[12] Hon-t, rer het-neb; see J. Duemichen, "Die Bauurkunde der Tempelanlagen von Edfu," Aeg.Z. *9* (1871), p. 28.

[13] See Aeg.Wb. *2*, pp. 55f. for the sign of the heart (*ib*) as expressing generally "the middle, the center."

[14] S. Mowinckel, *Die Sternnamen im Alten Testament* (1928), p. 12.

lation "every," the revolution of all celestial bodies. As concerns the identity of the ruling lady, the greater possibility speaks for Sirius, but Venus cannot be excluded; in Mexico, too, Venus is called "heart of the earth." The reader is invited to imagine for himself what many thousands of such pseudo-primitive or poetic interpretations must lead to: a disfigured interpretation of Egyptian intellectual life.

The problem of astrology—The greatest gap between archaic thinking and modern thinking is in the use of astrology. By this is not meant the common or judicial astrology which has become once again a fad and a fashion among the ignorant public, an escape from official science, and for the vulgar another kind of black art of vast prestige but with principles equally uncomprehended. It is necessary to go back to archaic times, to a universe totally unsuspecting of our science and of the experimental method on which it is founded, unaware of the awful art of separation which distinguishes the verifiable from the unverifiable. This was a time, rich in another knowledge which was later lost, that searched for other principles. It gave the *lingua franca* of the past. Its knowledge was of cosmic correspondences, which found their proof and seal of truth in a specific determinism, nay overdeterminism, subject to forces completely without locality. The fascination and rigor of Number made it mandatory that the correspondences be exact in many forms (Kepler in this sense is the last Archaic). The multiplicity of relations seen or intuited brought the idea to a focus in which the universe appeared determined not on one but on many levels at once. This was the signature of "panmathematizing ideation." This idea may well have led up to a pre-established harmony on an infinite number of levels. Leibniz has shown us how far it could go, given modern tools: the universe conceived all at once, complete with its individual destinies for all time, out of an "effulguration" of the divine mind. Some prehistoric or protohistoric Pythagorean Leibniz, whose existence is far from inconceivable, may well have cherished this impossible dream, going to the limit more innocently than our own historic sage. Starting from the power of

Number, a whole logic is thinkable in this view. *Fata regunt orbem, certa stant omnia lege.*

The only thinker of Antiquity who could be proof against this temptation was Aristotle, for he thought that forms were only potential in the beginning, and came into actualization only in the course of their lifetime, thus undergoing their fate as individuals. But that is because Aristotle refused mathematics from the start. He had the grounds for opposing universal synchronicity (the word and the idea were invented by C. G. Jung, replacing space with time, which goes to show that the archaic scheme has more lives than a cat).

Yet, here again, Dante comes to the fore as a witness; for, by art of Gramarye, as the simple used to say, he spans the whole itinerary, or shall we say the *cheminement de la pensée*, between two world epochs. An Aristotelian to the core, steeped in the discipline of Thomism, hence by inheritance anti-mathematical, his spirit in its sweep understands the stars, in the sense of their Pythagorean implications. In his ascent to the realm of heaven, he encounters his friend and onetime companion of his wanton and romantic youth, Charles Martel (*Paradiso*, VIII.34–37), who tells him what it means to be of the elect: "We circle in one orbit, at one pace, with one thirst, along with the heavenly Princes whom thou didst once address from the world"—"You Who by Understanding Move the Third Heaven." This is one of his early poems, a celebrated one at that, and it relates to the heavenly intelligences in a spirit of unrestrained Platonic worship. The progress of his song through the three realms will show him more and more wrapped in Platonic harmonies, much as he had dreamed of in his youth, and it will actually confirm his belief in astrology as a divine grant which keeps nature in order. Thus, the requirements of both doctrines have been saved: the arrangement of nature by genus and species (Aristotle) and the free development of one's own self (Aquinas) in a kind of Plotinian compromise overshadowed by the "Harmony of the Spheres." Such was Dante's own inimitable "art of Gramarye."

CHAPTER V

The Unfolding in India

They reckon ill who leave me
out.
When Me they fly, I am the
wings;
I am the doubter and the doubt
I am the hymn the Brahmin sings.

EMERSON, *Brahma*

THE PARALLEL between the Tale of Kai Khusrau and the final plot of the vast Hindu epic, the *Mahabharata*, has received attention for over a century. It was noticed by the great Orientalist James Darmesteter. The translators of Firdausi are not unaware of it, and they analyze the last phase of events as follows:

The legend of Kai Khusrau's melancholy, his expedition into the mountains, and his attainment to heaven without having tasted death has its parallel in the *Mahabharata*, where Yudhishthira, the eldest of the five Pandavas, becoming weary of the world, resolves to retire from the sovereignty and acquire merit by pilgrimage. On hearing of his intentions his four brothers—Bhima, Arjuna, and the twins Nakula and Sahadeva—resolve to follow his example and accompany him. Yudhishthira appoints successors to his various kingdoms. The citizens and the inhabitants of the provinces, hearing the king's words, became filled with anxiety and disapproved of them. "This should never be done"—said they unto the king. The monarch, well versed with the changes brought about by time, did not listen to their counsels. Possessed of righteous soul, he persuaded the people to sanction his views . . . Then Dharma's son, Yudhishthira, the King of Pandavas, casting off his ornaments, wore barks of trees . . . The five brothers, with Draupadi forming the sixth [she was the joint wife of the brothers], and a dog forming the seventh, set out on their journey. The citizens and the ladies of the royal household followed them for some distance . . . The denizens of the city then returned [exactly as Kai Khusrau's subjects had done]. The seven pilgrims meanwhile had set out upon their journey. They first wandered eastward, then southward, and then westward. Lastly they faced northward and crossed the Himalaya. Then they beheld before them a vast desert of sand and beyond it Mount Meru. One by one the pilgrims sank exhausted and expired, first Draupadi, then the twins, then Arjuna, then Bhima; but Yudhishthira, who never even looked back at his fallen comrades, still pressed on and, followed by the faithful *dog who turns out to be Dharma* (the Law), in disguise, entered Heaven in his mortal body, not having tasted death.

Among minor common traits, Warner stresses particularly these:

Both journey into the mountains with a devoted band, the number of them is the same in both cases, and both are accompanied by a divine being, for the part of the dog in the Indian legend is indicated in the Iranian as being taken by Surush, the angel of Urmuzd. In both, the leaders pass deathless into Heaven, and in both their mortal comrades perish. One legend therefore must be derived from the other, or else, and this seems to be the better opinion, they must be referred to a common origin of great antiquity.[1]

Of great antiquity these legends must be, indeed; otherwise there would not be a very similar end ascribed to Enoch and to Quetzalcouatl. In fact, just as Kai Khusrau's paladins did not listen to the Shah's advice not to remain with him until his ascension—the crowd had been left behind, anyhow—so Enoch

urged his retinue to turn back: "Go ye home, lest death overtake you, if you follow me farther." Most of them—800,000 there were—heeded his words and went back, but a number remained with him for six days. . . On the sixth day of the journey, he said to those still accompanying him, "Go ye home, for on the morrow I shall ascend to heaven, and whoever will then be near me, he will die." Nevertheless, some of his companions remained with him, saying: "Whithersoever thou goest, we will go. By the living God, death alone shall part us." On the seventh day Enoch was carried into the heavens in a fiery chariot drawn by fiery chargers. The day thereafter, the kings who had turned back in good time sent messengers to inquire into the fate of the men who had refused to separate themselves from Enoch, for they had noted the number of them. They found snow and great hailstones upon the spot whence Enoch had risen, and, when they searched beneath, they discovered the bodies of all who had remained

---

[1] Firdausi, *Shahnama* (Warner trans.), vol. *4*, pp. 136ff.

behind with Enoch. He alone was not among them; he was on high in heaven.[2]

Quetzalcouatl's paladins, "the slaves, the dwarves, the hunch backed . . . they died there from the cold . . . , upon all of them fell the snow," in the mountain pass between Popocatepetl and Iztacte-petl.[3] Quetzalcouatl, lamenting, and utterly lonely, had some more stations to pass, before he took off on his serpent raft, announcing he would come back, someday, "to judge the living and the dead" (appendix #3).

Were it only the dry fact of Yudhishthira's ascension, and the end of his companions high up in the mountains, we might have avoided the maze of the *Mahabharata* altogether. But, labyrinthine as this epic of twelve volumes truly is—and the same goes for the *Puranas*—Indian myth offers keys to secret chambers to be had nowhere else. The *Mahabharata* tells of the war of the Pandavas and the Kauravas, that is the Pandu brothers and the Kuru brothers, who correspond to the Iranians and Turanians, to the sons of Ka-leva and the people of Untamo, etc. Thus far the general situation is not foreign to us. But the epic states unmistakably that this tre-mendous war was fought during the *interval between the Dvapara and the Kali Yuga*.[4]

This "dawn" between two world-ages can be specified further. The real soul and force on the side of the Pandavas is Krishna—in the words of Arjuna: "He, who was our strength, our might, our heroism, our prowess, our prosperity, our brightness, has left us, and departed."[5] Now Krishna ("the Black") is the most outstanding avatar of Vishnu. And it is only when Krishna has been shot in the heel (or the sole of his foot), the only vulnerable spot of his body, by the hunter Jara (= old age) that the Pandavas, too, resolve to depart—just as Kai Khusrau did after the death of Kai Ka'us. There was Kai Khusrau's statement: "And now I deem it better to de-

[2] L. Ginzberg, *Legends of the Jews* (1954), vol. *1*, pp. 129ff.

[3] E. Seler, *Einige Kapitel aus dem Geschichtswerk des Fray B. de Sahagún* (1927), p. 290.

[4] Mbh. *1.2* (Roy trans., vol. *1*, p. 18). See H. Jacobi's *Mahabharata* (1903), p. 2.

[5] *Vishnu Purana* 5.38 (trans. H. H. Wilson [1840; 3d ed. 1961], p. 484).

part . . . Because this Kaian crown and throne will pass." And this happens at the following crucial point:

> When that portion of Vishnu (that had been born by Vasudeva and Devaki) returned to heaven, then the Kali age commenced. As long as the earth was touched by his sacred feet, the Kali age could not affect it. As soon as the incarnation of the eternal Vishnu had departed, the son of Dharma, Yudhishthira, with his brethren, abdicated the sovereignty . . . The day that Krishna shall have departed from the earth will be the first of the Kali age . . . it will continue for 360,000 years of mortals.[6]

And as Krishna is reunited with Vishnu, as Arjuna returns into Indra,[7] and Balarama into the Shesha-Serpent, so it will happen to the other heroes. Thus, when Yudhishthira is finally rejoined with his whole Pandu-Family in heaven, the poet Sauti explains,

> "That the various heroes, after exhausting their Karma, become reunited with that deity of which they were avatars."[8]

Yudhishthira is reunited with Dharma, disguised as a faithful dog.[9] Seen from this vantage point, the Finnish epic appears as a last dim and apparently meaningless reflection. Kullervo goes with the black dog Musti, the only living soul left from his home, into the forest where he throws himself upon his sword.

Now what about Krishna, most beloved deity of the Hinduistic Pantheon? Some of his innumerable deeds and victorious adventures before his "departure" will look familiar.

[6] *Vishnu Purana* 4.24 (Wilson trans., p. 390). Cf. 5.38, pp. 481f.: "and on the same day that Krishna departed from the earth the powerful dark-bodied Kali age descended. The ocean rose, and submerged the whole of Dvaraka," i.e., the town which Krishna himself had built, as told in *Vishnu Purana* 5.23, p. 449.

[7] See *Vishnu Purana* 5.12 (Wilson trans., p. 422), where Indra tells Krishna, "A portion of me has been born as Arjuna."

[8] Mbh. *18.5* (Swargarohanika Parva) (Roy trans., vol. *12*, pp. 287–90). See also Jacobi, p. 191.

[9] Arrived at the last stage of deterioration, we find Dharma, the Dog, in a fairy tale from Albania: The youngest daughter of a king—her two sisters resemble Regan and Goneril—offers to go to war in her father's place, asking for three suits only, and for the paternal blessing. "Then the king procured three male suits, and gave her his blessing, and this blessing changed into a little dog and went with the princess." (J. G. von Hahn: *Griechische und Albanische Märchen* [1918], vol. 2, p. 146.).

Young Krishna is the persecuted nephew of a cruel uncle, Kansa (or Kamsa), both being, as Keith[10] styles it, "protagonists in a ritual contest." This is not modestly understating it, but grossly misleading. Kansa is an Asura (appendix #4), and Krishna is a Deva, and that means, again, that the affair concerns the great divine "Parties" (Iranians-Turanians, and the like). The uncle, warned beforehand through prophecies about the danger coming from the eighth son of Devaki and Vasudeva, kills six children of this couple, but the seventh (Balarama) and eighth (Krishna) are saved and live with herdsmen. There young Krishna performs some of the deeds of the "Strong Boy."

If Kullervo, three days old, destroyed his cradle, we might expect something spectacular from Krishna, and we are not disappointed:

> On one occasion, whilst Madhusudana was asleep underneath the wagon, he cried for the breast, and kicking up his feet he overturned the vehicle, and all the pots and pans were upset and broken. The cowherds and their wives, hearing the noise, came exclaiming: "Ah! ah!" and they found the child sleeping on his back. "Who could have upset the wagon?" said the cowherds. "This child," replied some boys, who witnessed the circumstance; "we saw him," said they, "crying, and kicking the wagon with his feet, and so it was overturned: no one else had any thing to do with it." The cowherds were exceedingly astonished at this account.[11]

One day the child repeatedly disobeyed his mother and she became angry.

> Fastening a cord round his waist, she tied him to the wooden mortar Ulukhala, and being in a great passion, she said to him, "Now, you naughty boy, get away from hence if you can." She then went to her domestic affairs. As soon as she had departed, the lotus-eyed Krishna, endeavouring to extricate himself, pulled the mortar after him to the space between the two ariuna trees that grew near together. Having dragged the mortar between these trees, it became wedged awry there, and as Krishna pulled it through, it pulled down the trunks of the trees. Hearing the crackling noise, the people of Vraja came to see what was the matter, and there they beheld the two large trees,

[10] A. B. Keith, *Indian Mythology* (1917), p. 126. For the deeds of Krishna, see pp. 174ff.
[11] *Vishnu Purana* 5.6 (Wilson trans., p. 406f.).

with shattered stems and broken branches, prostrate on the ground, with the child fixed between them, with a rope round his belly, laughing, and showing his little white teeth, just budded . . . The elders of the cowherds . . . looked upon these circumstances with alarm, considering them of evil omen. "We cannot remain in this place," said they, "let us go to some other part of the forest."

Thus, they go to Vrindavana, exactly where the child had wished. The Harivamsha explains the move to Vrindavana in this way:

Krishna converts the hairs of his body into hundreds of wolves, who so harass and alarm the inhabitants of Vraja—the said cowherds—, that they determine to abandon their homes.[12]

In the Indian myth, for once, the episode of Krishna's hairs turning into hundreds of wolves seems a mere trifle, compared with Kullervo's wolves which "he sang to cattle, and he changed the bears to oxen," the more so, as Krishna's only "harass and alarm" the cowherds. These wild beasts, however, indispensable to the "Urkind," whether Kullervo or Dionysos—see above, p. 30—are present in Krishna's story, and this is remarkable enough.

Kansa,[13] hearing of the deeds of Krishna and Rama, determines to have the boys brought to his capital Mathura and there to procure their death, if he cannot slay them before. Needless to say, all is in vain: Krishna kills Kansa and all his soldiers, and places Kansa's father on the throne.

Krishna does not pretend to be a fool, the smiling one. He merely insists again and again on being a simple mortal when everybody wishes to adore him as the highest god, which he is. Nor is he known particularly as an "avenger." He was delegated from higher quarters to free the earth—"overburdened" as it was with Asura—as he had done time and again in his former avatars. Krishna belongs here, however, because Indian tradition has preserved the consciousness of the cosmic frame, and it is this alone that gives mean-

---

12 *Vishnu Purana* 5.6 (Wilson trans., pp. 406f.).

13 That "uncle"—really "the great Asura Kalanemi who was killed by the powerful Vishnu . . . revived in Kansa, the son of Ugrasena" (*Vishnu Purana* 5.1 [Wilson trans., p. 396]).

ing to the incidence of war and the notion of crime and punishment as they appear in myth.

It is useful to keep philosophy and mythology carefully separated, and yet the many gods and heroes who avenge their fathers —beginning with "Horus-the-avenger-of-his-father" and "Ninurta who has avenged his father"—have their function destined to them, as has the long line of wicked uncles. These figures pay reparation and atonement to each other for their mutual injustice *in the order of time*, as Anaximander said. Anaximander was a philosopher. Despite its fantastic language the Indian epic has an affinity with his thought. Vishnu returns regularly in his capacity of "avenger," collecting the "reparations" of the bad uncle "according to the order of time." In the *Mahabharata* he does so under the name of Krishna, but he will come again in the shape of another avatar to clean the earth of the Asura who overburden it. The Asura, too, grow into "overbearing characters" strictly according to the order of time. Under the name of Kalki the Vishnu figure is expected to introduce a new Krita Yuga (Golden Age), when our present Kali Yuga has come to its miserable end.

It is this regular returning of avatars of Vishnu which helps clarify matters. Because it is Vishnu's function to return as avenger at fixed intervals of time, there is no need in the epic to emphasize the revenge taken by Krishna on Uncle Kansa. But in the West, where the continuity of cosmic processes as told by myth has been forgotten—along with the knowledge that gods are stars—the very same revenge is given great importance because it is an unrepeated event accomplished by one figure, whether hero or god, and this hero or god is, moreover, understood to be the creation of some imaginative poet. The introduction of Indian tradition makes it possible to rediscover the context in which such characters as Saxo's Amlethus, such typically unlucky fellows as Kullervo, have significance. Once it is fully realized that "the day Krishna shall have departed from the earth will be the first of the Kali Yuga," the proper perspective is established. Our hero stands precisely on the threshold between a closed age and a new Time Zero. In fact, he closes the old one.

The most inconspicuous details become significant when observed from this point of view. For instance Saxo, without giving it much thought, divided the biography of Amlethus in two parts (incidentally involving the hero in bigamy), in the same way as Firdausi told us nine-tenths of Kai Khusrau's adventures in the book on Kai Ka'us. This is actually the more puzzling of the two as Firdausi states: "For from today new feasts and customs date / Because tonight is born Shah Kai Khusrau." Firdausi, who was well versed in astrology, insisted on the Shah's birthday because, in the astrological sense, birth is the decisive moment. But here, and in related cases where chronology is at issue, it is the moment of death, of leaving the stage, that counts. Krishna's departure gives the scheme away. Al-Biruni, in his chapter on "The Festivals of the Months of the Persians," describing the festival *Naurôz* ("New Day") in the first month of spring, writes:

> On the 6th day of Farwardîn, the day Khurdâdh, is the Great Naurôz, for the Persians a feast of great importance. On this day—they say—God finished the creation, for it is the last of the six days . . . On this day God created Saturn, therefore its most lucky hours are those of Saturn. On the same day—they say—the Sors Zarathustrae came to hold communion with God, and *Kaikhusrau ascended into the air*. On the same day the happy lots are distributed among the people of the earth. Therefore the Persians call it "the day of hope."[14]

The so-called Kaianian Dynasty—the "Heroes" according to Al-Biruni's *Chronology*[15]—succeeding the first Pishdadian Dynasty ("the Just"), is supposed to have started with Kai Kubad, his son Kai Ka'us, and the latter's grandson Kai Khusrau, and to have ended with Sikander, Alexander the Great, with whose death a new era actually began. But it is obvious that something new begins with Kai Khusrau's assumption into heaven. Thus, the Warners state that with our Shah "the old epic cycle of the poem comes to an end, and up to this point the Kaianian may be regarded as the complement of the Pishdadian dynasty."[16]

[14] Al-Biruni, *The Chronology of Ancient Nations* (trans. C. E. Sachau [1879], p. 201).

[15] P. 112.

[16] Firdausi, *Shahnama* (Warner trans.), vol. 2, pp. 8f.

In his introduction to the Firdausi translation, however, the Warners claim that the poem is divided into two periods, one mythic, the other historic:[17]

> This distinction is based not so much on the nature of the subject-matter as on the names of the chief characters. At a certain point in the poem the names cease to be mythic and become historic. The mythic period extends from the beginning of the narrative down to the reigns of the last two Shahs of the Kaianian dynasty . . . The Shahs in question are Dara, son of Darab, better known as Darius Codomanus, and Sikander (Alexander)."[18]

Firdausi makes it clear that the mythic period ends only with the death of Alexander, the two last Shahs being Darius Codomanus and Alexander who overcame him. After him begins the "historic" period of the poem. In other words, "history" begins only when the Iranian empire vanishes from the scene, to be replaced by the successors of Alexander. To remove from history the great and solidly historical feats of Darius I, Xerxes, Cambyses, etc., is paradoxical for a poem which is meant to celebrate the Iranian empire. Presumably Firdausi meant that so long as the Zoroastrian religion reigned, time was holy and thus belonged to myth rather than ordinary history. This is confirmed by a strange statement of the Warners: "Rightly or wrongly, Zoroastrian tradition couples Alexander with Zahhak and Afrasiyab as one of the three arch-enemies of the faith."[19]

The great myths of the Avestan religion have overcome chronology and reshaped it to their purpose. The true kings of Persia have disappeared notwithstanding their glory, and are replaced by mythical rulers and mythical struggles. Kai Khusrau rehearses a "Jamshyd" role in his beginnings, and with his ascent to heaven—the date of which marks New Year from now on—the Holy Empire

[17] The time structure is a very complicated one, and we cannot manage with a subdivision of two "periods" at all, the less so, as the reigns of the Shahs overlap with the rather miraculous lifetimes of the "heroes" or Paladins (Rustam, Zal, etc.). The same goes for the "primordial" emperors of China and their "vassals." But God protect us from meddling with lists of alleged "kings" from whichever area, but particularly from the Iranian tables!

[18] Firdausi, vol. *1*, pp. 49f.

[19] Firdausi, vol. *1*, p. 59.

really comes to a close. The struggle has been between gods and demons throughout.

We have been following the story of powers coming to an end, embodied first in the Iranian then in the Indian "kings," a story which is differently emphasized by two different legends. Each legend has a disturbing similarity to the other, and each removes the narration from any known classic pattern, forcing the events to a catastrophic conclusion which is clearly commanded by Time itself, and by a very different chain of causes than that indicated by the actual sequence of events in the texts.

To avoid misunderstanding it should be emphasized that it is not possible yet to know precisely who is who, or to make positive identifications such as saying that Brjam is Yudhishthira or Krishna. But the hints provided by Iranians and Indians may lead to a better understanding of Kullervo ("Kaleva is reborn in him"), and may indicate that the feat of the doggish fool Brutus in driving out the Kings was significant on a higher level than the political. This is not to deny that the Kings were expelled, but rather to point to a special set of firmly coined "figures of speech" derived from "large" changes or shifts (such as the onset of Kali Yuga) that could be, and were, applied to minor historical events.

CHAPTER VI

Amlodhi's Quern

The stone which the builders refused
is become the head stone of the corner.

Psalm CXVIII.22; Luke XX.17

Whosoever shall fall upon that stone shall be broken,
but on whomsoever it shall fall, it will grind him to powder.

Luke XX.18

WITH SUGGESTIVE INSIGHTS from other continents, it is time to take a fresh look at Shakespeare's Gentle Prince, a cultivated, searching intellectual, the glass of fashion and the mold of form in the Danish Court, who was known once upon a time as a personage of no ordinary power, of universal position, and, in the North, as the owner of a formidable mill.

Well trained by the Church, Saxo could write excellent and ornate Latin, a rare achievement in his time. Though inspired by his patriotism to write the great chronicles of his own country, he was in Denmark an isolated if respectable fish in a small provincial pond. He remained oriented to the cultural pole of his times, which was Iceland. From there he had to draw most of his materials even if he helped to Danicize them, as we see in the story of Hamlet where all the features point toward a local dynastic story. But what he drew from Iceland were pieces of already "historical" lore. He could not draw, as did Snorri Sturluson, on the resources of a high position at the very center of Iceland's rich bilingual culture, and on the experience of a wide-ranging and adventurous life. He could never have formed, like Snorri, the great project of reorganizing the corpus of pagan and skaldic tradition inside an already Christian frame. Saxo seems to have known Icelandic fairly well, but not enough to understand the precious and convoluted language of ancient poetry. He was unsure of his bearings and simply arranged his story as best he could even though the name of Hamlet's father, Orvendel (see appendix #2), should have been sufficient to warn him of its derivation from high myth. It is Snorri who provides a piece of decisive information: and it appears, as earlier noted, in chapter 16 of his *Skaldskaparmal* ("Poetical Diction"), a collection of *kenningar*, or turns of speech from ancient bards. It is couched in a language that even modern scholars can translate only tentatively. Appendix #5 contains a discussion of the many versions. The one quoted again here is that of Gollancz (p. xi), which appears to be the most carefully translated:

> T'is said, sang Snaebjörn, that far out, off yonder ness, the Nine Maids of the Island Mill stir amain the host-cruel skerry-quern—they who in ages past ground Hamlet's meal. The good chieftain furrows the hull's lair with his ship's beaked prow. Here the sea is called Amlodhi's Mill.

The Mill is thus not only very great and ancient, but it must also be central to the original Hamlet story. It reappears in the *Skaldskaparmal*, where Snorri explains why a kenning for gold is "Frodhi's meal."[1] Frodhi appears in the chronicles, but his name is really an alias of Freyr, one of the great *Vanir* or Titans of Norse myth. But Snorri, who likes to give things a historical ring as befits his Christian upbringing, fixed his Frodhi to "the same time when Emperor Augustus established peace in the whole world, and when Christ was born." Under King Frodhi the general state of things was similar to that of the Golden Age, and it was called "Frodhi's peace." Saxo follows suit and attributes unsuspectingly a duration of thirty years to this peace.[2]

---

[1] *Skaldskap. 42*, according to Brodeur (1929), pp. 163–69, and Neckel and Niedner (*Thule 20* [1942]), pp. 195f. The other translators of Snorri's *Edda* cannot agree on the manner of dividing the work into chapters, if they do not desist from doing so at all, as R. B. Anderson (1880), pp. 206–13, parts of whose translation we quote here. (Simrock [n.d.], pp. 89–93, makes it chapter 63.)

[2] P. Herrmann, *Die Heldensagen des Saxo Grammaticus* (1922), pp. 376ff.

Now Frodhi happened to be the owner of a huge mill, or quern, that no human strength could budge. Its name was Grotte, "the crusher." We are not told how he got it, it just happened, as in a fairy tale. He traveled around looking for someone who could work it, and in Sweden he recruited two giant maidens, Fenja and Menja, who were able to work the Grotte. It was a magic mill, and Frodhi told them to grind out gold, peace and happiness. So they did. But Frodhi in his greed drove them night and day. He allowed them rest only for so long as it took to recite a certain verse. One night, when everybody else was sleeping, the giantess Menja in her anger stopped work, and sang a dire song.

This obscure prophetic imprecation, as Muellenhoff has shown, is the oldest extant document of skaldic literature, antedating Snorri's tale by far. It contains the biography of the grim sisters:

> *Frode! you were not / Wary enough,—*
> *You friend of men,— / When maids you bought!*
> *At their strength you looked, / And at their fair faces,*
> *But you asked no questions / About their descent.*

> *Hard was Hrungner / And his father;*
> *Yet was Thjasse / Stronger than they,*
> *And Ide and Orner / Our friends, and*
> *The mountain-giants' brothers, / Who fostered us two.*

> *Not would Grotte have come / From the mountains gray,*
> *Nor this hard stone / Out from the earth;*
> *The maids of the mountain-giants / Would not thus be grinding*
> *If we two knew / Nothing of the mill.*

> *Such were our deeds / In former days,*
> *That we heroes brave / Were thought to be.*
> *With spears sharp / Heroes we pierced,*
> *So the gore did run / And our swords grew red.*

> *Now we are come / To the house of the king,*
> *No one us pities. / Bond-women are we.*
> *Dirt eats our feet / Our limbs are cold,*
> *The peace-giver we turn. / Hard it is at Frode's.*

Now hold shall the hands / The lances hard,
The weapons bloody,— / Wake now, Frode! Wake now, Frode!
If you would listen / To our songs,—
To sayings old.

Fire I see burn / East of the burg,—
The war news are awake. / That is called warning.
A host hither / Hastily approaches
To burn the king's / Lofty dwelling.

No longer you will sit / On the throne of Hleidra
And rule o'er red / Rings and the mill.
Now must we grind / With all our might,
No warmth will we get / From the blood of the slain.

Now my father's daughter / Bravely turns the mill.
The death of many / Men she sees.
Now broke the large / Braces 'neath the mill,—
The iron-bound braces. / Let us yet grind!

Let us yet grind! / Yrsa's son
Shall on Frode revenge / Halfdan's death.
He shall Yrsa's / Offspring be named,
And yet Yrsa's brother. / Both of us know it.

However obscure the prophecy, it brought its own fulfillment.
The maidens ground out for Frodhi's "a sudden host," and that
very day Mysingr, the Sea-King, landed and killed Frodhi. Mysingr
("son of the Mouse"—see appendix #6) loaded Grotte on his ship,
and with him he also took the giantesses. He ordered them to grind
again. But this time they ground out salt.

"And at midnight they asked whether Mysingr were not weary
of salt. He bade them grind longer. They had ground but a little
while, when down sank the ship,"

> *"the huge props flew off the bin,*
> *the iron rivets burst,*
> *the shaft tree shivered,*

the bin shot down,
the massy mill-stone rent in twain."[3]

"And from that time there has been a whirlpool in the sea where the water falls through the hole in the mill-stone. It was then that the sea became salt."

Here ends Snorri's tale (appendix #7). Three fundamental and far-reaching themes have been set: the broken mill, the whirlpool, the salt. As for the curse of the miller women, it stands out alone like a megalith abandoned in the landscape. But surprisingly it can also be found, already looking strange, in the world of Homer, two thousand years before.[4]

It is the last night, in the *Odyssey* (*20*.103–19, Rouse trans.), which precedes the decisive confrontation. Odysseus has landed in Ithaca and is hiding under Athena's magic spell which protects him from recognition. Just as in Snorri, everybody sleeps. Odysseus prays to Zeus to send him an encouraging sign before the great ordeal.

> Straightaway he thundered from shining Olympus, from on high from the place of the clouds; and goodly Odysseus was glad. Moreover, a woman, a grinder at the mill, uttered a voice of omen from within the house hard by, where stood the mills of the shepherd of the people. At these handmills twelve women in all plied their task, making meal of barley and of wheat, the marrow of men. Now all the others were asleep, for they had ground out their task of grain, but one alone rested not yet, being the weakest of all. She now stayed her quern and spake a word, a sign to her Lord [*epos phato sema anakti*]. "Father Zeus, who rulest over gods and man, loudly hast thou thundered from the starry sky, yet nowhere is there a cloud to be seen: this is surely a portent thou art showing to some mortal. Fulfil now, I pray thee, even to miserable me, the word that I shall speak. May the wooers, on this day, for the last and latest time make their sweet feasting in the halls of Odysseus! They that have loosened my knees with cruel toil to grind their barley meal, may they now sup their last!"

"The weakest of all," yet a giant figure in her own right. In the tight and shapely structure of the narrative, the episode is fitted

---

[3] These five verses are taken from Gollancz (p. xiii), the three previous and the two last lines from Brodeur (pp. 162f.); otherwise, we followed the Anderson translation.

[4] It was J. G. von Hahn (*Sagwissenschaftliche Studien* [1876], pp. 401f.) who first pointed to the similarity of the episodes in Snorri's *Edda* and in the *Odyssey*.

The *Carta Marina* of Olaus Magnus (16th century) shows the "horrenda caribdis," i.e., the Maelstrom, on the lower right, with ships, destructive sea-animals, and icebergs on the left.

The whirlpool, here called "Norvegianus Vortex," but usually spoken of as "gurges mirabilis" by Athanasius Kircher, as depicted in his *Mundus Subterraneus*.

Kircher's rather curious conception of the subterranean flow of rivers may have been evoked by Socrates' last tale, but transposed to a strictly geological level. This drawing illustrates the subterranean connection between the whirlpool west of Norway and the Baltic Sea.

with art, yet it stands out like a cyclopean stone embedded in a house. There are many such things in Homer.

Going back to Grotte, the name has an interesting story. It is still used today in Norwegian for the "axle-block," the round block of wood which fills the hole in the millstone, and in which the end of the mill axle is fixed. In the Färöer as well as in the Shetland dialect, it stands for "the nave in the millstone." The original Sanskrit *nabhi* covers both "nave" and "navel," and this point should be kept in mind. In the story, it is obviously the nave that counts, for it created a hole when the mill tree sprang out of it, and the whirlpool formed in the hole. But "navel of the sea" was an ancient name for great whirlpools. Gollancz, with sound instinct, saw the connection right away:

> Indeed, one cannot help thinking of a possible reference to the marvellous Maelstroem, the greatest of all whirlpools, one of the wonders of the world; *Umbilicus maris* according to the old geographers, "*gurges mirabilis omnium totius orbis terrarum celeberrimus et maximus*" as Fr. Athanasius Kircher describes it in his fascinating folio "Mundus Subterraneus." According to Kircher, it was supposed that every whirlpool formed around a central rock: a great cavern opened beneath; down this cavern the water rushed; the whirling was produced as in a basin emptying through a central hole. Kircher gives a curious picture of this theory, with special reference to the Maelstroem.[5]

Clearly, the Mill is not a "chose transitory," as lawyers say in their jargon. It must belong to the permanent equipment of the ancient universe. It recurs all the time, even if its connotations are rarely pleasant. From another corner of memory, there come the lines of Burns' "John Barleycorn":

> *They wasted o'er a scorching flame*
> *The marrow of his bones*
> *But a miller used him worst of all*
> *For he crushed him between two stones.*

The mock tragedy of the yearly rural feast is part of the immense lore on fertility rites that Frazer has unfolded, with the ritual lamentations over the death of Tammuz, Adonis, the "Grain-Osiris" of Egypt; and no one would deny that the Tammuz festival

[5] I. Gollancz, *Hamlet in Iceland* (1898), p. xiv.

was a seasonal ritual celebrating the death and rebirth of vegetation. It has entered the commonplaces of our knowledge. But was this the original meaning? An irresistible preconception leads to the thought that when peasant rites are found tied to vegetation, there is the most elementary and primitive level of myth from which all others derive. It carries, too, its own moral tidings: "if the grain die not . . . ," which led on to higher religious thought.

In truly archaic cults, however, such as that of the Ssabians of Harran, reflected also in Ibn Wa'shijja's "Book of Nabataean Agriculture," the death and grinding up of Tammuz is celebrated and lamented by the images of all the planetary gods gathered in the temple of the Sun suspended "between earth and heaven," in the same way as they once cried and lamented over the passing of Jamshyd (or Jambushad as they then called him). This is a strange and unusual note, very un-agrarian, which deserves more careful study.

But this leads back to the Norse myth of the Mill, and in fact to Snorri himself, who in his "Fooling of Gylfi" commented on a verse from the *Vafthrudnismal* which has been much discussed since. In this ancient poem, the end of Ymer is recounted. Ymer is the "initial" world giant from whose scattered body the world was made. Snorri states that Ymer's blood caused a flood which drowned all giants except Bergelmer, who, with his wife, "betook himself upon his *ludr* and remained there, and from there the race of giants are descended." The word *ludr*, as Snaebjörn said, stands for Mill. But in *Vafthrudnismal* (ch. 35), Odin asks the wise giant Vafthrudner of the oldest event he can think of, and gets this answer: "Countless ages ere the earth was shapen, Bergelmer was born. The first thing I remember—is when he *á var ludr um lagidr*" (appendix #8). Rydberg renders the words as "laid on a mill," and understands them as "laid under a millstone." Accordingly, he explains Snaebjörn's *lidmeldr*, which the great mill grinds, as "limb grist."[6] As will appear later, there is a different interpretation to propose.

---

[6] V. Rydberg, *Teutonic Mythology* (1907), p. 575.

The problem, however, keeps turning up. In the *Lokasenna* (43ff.), Freyr, the original master of Grotte, is brought directly into action. The occasion is a banquet to which Aegir invited the gods. Loke uninvited made his appearance there to mix harm into the ale of the gods and to embitter their pleasure. But when Loke taunts Freyr, Byggver, the faithful retainer, becomes angry on his master's behalf:

> *"Had I the ancestry*
> *of Ingunar Freyr*
> *and so honoured a seat*
> *know I would grind you*
> *finer than marrow, you evil crow,*
> *and crush you limb by limb."*

To which Loke replies:

> *"What little boy is that*
> *whom I see wag his tail*
> *and eat like a parasite?*
> *Near Freyr's ears*
> *always you are*
> *and clatter 'neath the mill-stone."*

There are several more clues which hint that this mill upon which Bergelmer was "heaved" was a very distinct if unattractive mythological feature, and they cannot be dealt with here. But if it should be remarked that Bergelmer was not in a state to produce offspring for the giants, if he really was laid under the millstone, there is also an example from Mexico, the "jewel-bone" or "sacrificial bone" which Xolotl or Quetzalcouatl procures from the "underworld," bringing it to Tamoanchan (the so-called "House of descending"). There, the goddess Ciua couatl or Quilaztli grinds the precious bone on the grindstone, and the ground substance is put into the jewel bowl (*chalchiuhapaztli*). Several gods maltreat themselves, making blood flow from their penises on the "meal." Out of this mixture mankind is fashioned.

These stories may not be in exquisite taste, but at least they are grotesque and contorted enough to rid us of reliance on the natural or intuitive understanding of artless tales sung by rustics dancing on the green. Real cosmological similes are anything but intuitive.

One question remains from this discussion. Who was Snaebjörn, that dim figure, a few of whose lines have revealed so much? The scholars have gone searching, and have unearthed a veritable treasure in the ancient "Book of Iceland Settlements." It links the poet with the first discovery of America. In that book, writes Gollancz:

> There is a vivid picture of a tenth-century Arctic adventurer, Snaebjörn by name, who went on a perilous expedition to find the unknown land, "Gunnbjörn's Reef," after having wrought vengeance, as became a chivalrous gentleman of the period, on the murderer of a fair kinswoman. It is generally accepted, and there can be little doubt, that this Snaebjörn is identical with the poet Snaebjörn.
>
> His family history is not without interest. His great-grandfather, Eywind the Easterling, so called because he had come to the Hebrides from Sweden, married the daughter of Cearbhall, Lord of Ossory, who ruled as King of Dublin from 882 to 888, "one of the principal sovereigns of Europe at the time when Iceland was peopled by the noblemen and others who fled from the tyranny of Harold Harfagr." Cearbhall was descended from Connla, the grandson of Crimhthann Cosgach, the victorious King of Ireland, who is said to have flourished about a century before the Christian era. Lann or Flann, the half-sister of Cearbhall, was married to Malachy I., King of Ireland, whose daughter Cearbhall had married. Flann was the mother of King Sionna and of the Lady Gormflaith, whom a cruel fate pursued; a king's daughter, the wife of three kings, [she was] forced at last to beg for bread from door to door. About the date of Snaebjörn's Arctic expedition (circa 980), his cousin, Ari Marson, is said to have landed on "White Man's Land," or "Great Iceland,"—that part of the coast of North America which extends from Chesapeake Bay, including North and South Carolina, Georgia, and Florida,—and became famous as one of the earliest discoverers of the New World.[7]

Thus Snaebjörn, as a member of an Irish royal family, typifies the mutual influence of Celtic and Scandinavian culture, between A.D. 800 and 1000, that influence which has been traced into the Eddic songs by Vigfusson in his *Corpus Poeticum Boreale*. The

---

[7] Gollancz, pp. xviif.

Hamlet story itself typifies that exchange. For an earlier and simpler form of it may have been brought to Iceland from Ireland, whither the Vikings had originally taken the story of the great Orvendel's son.

This places Hamlet within the circle not only of Norse tradition, but of that prodigious treasury of archaic myth which is Celtic Ireland, from which many lines have been traced to the Near East. The universality of the Hamlet figure becomes more understandable.

CHAPTER VII

The Many-Colored Cover

> There is a mill which grinds by itself, swings of itself, and scatters the dust a hundred versts away. And there is a golden pole with a golden cage on top which is also the Nail of the North. And there is a very wise tomcat which climbs up and down this pole. When he climbs down, he sings songs; and when he climbs up, he tells tales.
>
> *Tale of the Ostyaks of the Irtysh*

THE KALEVALA is vaguely known by the general public as the national epic of Finland. It is a tale of wild fancy, enticing absurdity and wonderfully primitive traits, actually magical and cosmological throughout. It is all the more important in that the Ugro-Finnic tradition has different roots from Indo-European ones.

Until the 19th century the epic existed only in fragments entrusted to oral transmission among peasants. From 1820 to 1849, Dr. Elias Lönnrot undertook to collect them in writing, wandering from place to place in the most remote districts, living with the peasantry, and putting together what he heard into some kind of tentative sequence. Some of the most valuable songs were discovered in the regions of Archangel and Olonetz in the Far North, which now belong again to Russia. The 1849 final edition of Lönnrot comprises 22,793 verses in fifty runes or songs. A large amount of new material has been discovered since.

The poem has taken its name from Kaleva, a mysterious ancestral personage who appears nowhere in the tale. The heroes are his three sons: Vainamoinen,[1] "old and truthful," the master of magic song; Ilmarinen, the primeval smith, the inventor of iron, who can forge more things than are found on land or sea; and the "beloved," or "lively," Lemminkainen, a sort of Arctic Don Juan. Kullervo, the Hamlet-like one whose story was told earlier, fair-haired Kullervo "with the bluest of blue stockings," is another "son of Kaleva," but his adventures seem to unfold separately. They tie up only at one point with Ilmarinen, and seem to belong to a different frame of time, to another world-age.

It is time now to deal with the main line of events. The epic opens with a very poetical theory of the origin of the World. The virgin daughter of the air, Ilmatar, descends to the surface of the waters, where she remains floating for seven hundred years until Ukko, the Finnish Zeus, sends his bird to her. The bird makes its nest on the knees of Ilmatar and lays in it seven eggs, out of which the visible world comes. But this world remains empty and sterile until Vainamoinen is born of the virgin and the waters. Old since birth, he plays the role, as it were, of "midwife" to nature by causing her to create animals and trees by his magic song. An inferior magician from Lapland, Youkahainen, challenges him in song and is sung step by step into the ground, until he rescues himself by promising Vainamoinen his sister, the lovely Aino. But the girl will not have Vainamoinen, he looks too old. She wanders off in despair and finally comes to a lake. She swims to a rock, seeking death; "when she stood upon the summit, on the stone of many colors, in the waves it sank beneath her." Vainamoinen tries to fish for her, she swims into his net as a salmon, mocks him for not recognizing her, and then escapes forever. Vainamoinen decides to look for another bride, and embarks upon his quest. His goal is the country of Pohjola, the "North country," a misty land "cruel to heroes," strong in magic, vaguely identified with Lapland. Events unfold as in a dream, with surrealistic irrelevance. The artlessness, the wayward charm and the bright nonsense suggest Jack and the Beanstalk, but behind them appear the fossil elements of a tale as old as the world—at least the world of man's consciousness—

---

[1] The name is Väinämöinen, due to vowel harmonization, but we had pity on the typesetter.

whose meaning and thread were lost long ago. The pristine archaic themes remain standing like monumental ruins.

The main sequence is built around the forging and the conquest of a great mill, called the Sampo (rune *10* deals with the forging, runes *39–42* with the stealing of the Sampo).

Comparetti's studies have shown that the Sampo adventure is a distinct unit (like Odysseus' voyage to the underworld), "a mythic formation which has remained without any action that can be narrated" and which was then fitted more or less coherently into the rest of the tradition.[2] Folk legend has lost its meaning, and treats the Sampo as some vague magic dispenser of bounty, a kind of Cornucopia, but the original story is quite definite.

Vainamoinen, "sage and truthful," conjurer of highest standing, is cast upon the shore of Pohjola much as Odysseus lands on Skyra after his shipwreck. He is received hospitably by Louhi, the Mistress (also called the Whore) of Pohjola, who asks him to build for her the Sampo, without explanation. He tells her that only Ilmarinen, the primeval smith, can do it, so she sends Vainamoinen home on a ship to fetch him. Ilmarinen, who addresses his "brother" and boon companion rather flippantly as a liar and a vain chatterer, is not interested in the prospect, so Vainamoinen, ancient of days and wise among the wise, has recourse to an unworthy trick. He lures the smith with a story of a tall pine, which, he says, is growing

> *Near where Osmo's field is bordered.*
> *On the crown the moon is shining,*
> *In the boughs the Bear is resting.*

Ilmarinen does not believe him; they both go there, to the edge of Osmo's field,

> *Then the smith his steps arrested,*
> *In amazement at the pine-tree,*
> *With the Great Bear in the branches,*
> *And the moon upon its summit.*

Ilmarinen promptly climbs up the tree to grasp the stars.

[2] D. Comparetti, *The Traditional Poetry of the Finns* (1898).

> *Then the aged Vainamoinen,*
> *Lifted up his voice in singing:*
> *"Awake, oh Wind, oh Whirlwind*
> *Rage with great rage, oh heavens,*
> *Within thy boat, wind, place him*
> *Within thy ship, oh east wind*
> *With all thy swiftness sweep him*
> *To Pohjola the gloomy."[3]*

> *Then the smith, e'en Ilmarinen*
> *Journeyed forth, and hurried onwards,*
> *On the tempest forth he floated,*
> *On the pathway of the breezes,*
> *Over moon, and under sunray,*
> *On the shoulders of the Great Bear*
> *Till he reached the halls of Pohja,*
> *Baths of Sariola the gloomy.*

In this utterly unintended manner, Ilmarinen lands in Pohjola, and not even the dogs are barking, which astonishes Louhi most of all. She showed herself hospitable,

> *Gave the hero drink in plenty,*
> *And she feasted him profusely,*

then spoke to him thus:

> *"O thou smith, o Ilmarinen,*
> *Thou the great primeval craftsman,*
> *If you can but forge a Sampo,*
> *With its many-coloured cover,*
> *From the tips of swan's white wing-plumes,*
> *From the milk of barren heifer,*
> *From a little grain of barley*
> *From the wool of sheep in summer,[4]*
> *Will you then accept this maiden*
> *As reward, my charming daughter?"*

[3] The magic spell, published in the *Variants* and translated by Comparetti, was sung by Ontrei in 1855.

[4] See the epigraph to the *Introduction*, p. 1.

Ilmarinen agrees to the proposal, and looks around three days for a proper spot on which to erect his smithy, "in the outer fields of Pohja." The next three days his servants keep working the bellows.

> On the first day of their labour
> He himself, smith Ilmarinen,
> Stooped him down, intently gazing,
> To the bottom of the furnace,
> If perchance amid the fire
> Something brilliant had developed.
> From the flames there rose a crossbow,
> Golden bow from out the furnace;
> 'Twas a gold bow tipped with silver,
> And the shaft shone bright with copper.
> And the bow was fair to gaze on,
> But of evil disposition
> And a head each day demanded,
> And on feast-days two demanded,
> He himself, smith Ilmarinen,
> Was not much delighted with it,
> So he broke the bow to pieces,
> Cast it back into the furnace.

The next day, Ilmarinen looks in anew,

> And a boat rose from the furnace,
> From the heat rose up a red boat,
> And the prow was golden-coloured,
> And the rowlocks were of copper.
>
> And the boat was fair to gaze on,
> But of evil disposition;
> It would go to needless combat,
> And would fight when cause was lacking.

Ilmarinen casts the boat back into the fire, and on the following day he gazes anew at the bottom of the furnace,

And a heifer then rose upward,
With her horns all golden-shining,
With the Bear-stars on her forehead;
On her head appeared the Sun-disc.

And the cow was fair to gaze on,
But of evil disposition;
Always sleeping in the forest,
On the ground her milk she wasted.
Therefore did smith Ilmarinen
Take no slightest pleasure in her,
And he cut the cow to fragments,
Cast her back into the furnace.

The fourth day:

And a plough rose from the furnace,
With the ploughshare golden-shining,
Golden share, and frame of copper,
And the handles tipped with silver.
And the plough was fair to gaze on,
But of evil disposition,
Ploughing up the village cornfields,
Ploughing up the open meadows,
Therefore did smith Ilmarinen
Take no slightest pleasure in it.
And he broke the plough to pieces,
Cast it back into the furnace,
Called the winds to work the bellows
To the utmost of their power.

Then the winds arose in fury,
Blew the east wind, blew the west wind,
And the south wind yet more strongly,
And the north wind howled and blustered.

Thus they blew one day, a second,
And upon the third day likewise.
Fire was flashing from the windows,

From the doors the sparks were flying
And the dust arose to heaven,
With the clouds the smoke was mingled.
Then again smith Ilmarinen,
On the evening of the third day,
Stooped him down, and gazed intently
To the bottom of the furnace,
And he saw the Sampo forming,
With its many-coloured cover.

Thereupon smith Ilmarinen,
He the great primeval craftsman,
Welded it and hammered at it,
Heaped his rapid blows upon it,
Formed with cunning art the Sampo.

And on one side was a corn-mill,
On another side a salt-mill,
And upon the third a coin-mill.

Now was grinding the new Sampo,
And revolved the pictured cover,
Chestfuls did it grind till evening,
First for food it ground a chestful,
And another ground for barter,
And a third it ground for storage.

Now rejoiced the Crone of Pohja,
And conveyed the bulky Sampo,
To the rocky hills of Pohja,
And within the Mount of Copper,
And behind nine locks secured it.

There it struck its roots around it,
Fathoms nine in depth that measured,
One in Mother Earth deep-rooted,
In the strand the next was planted,
In the nearest mount the third one.

Ilmarinen does not gain his reward, not yet. He returns without a bride. For a long while we hear nothing at all about the Sampo. Other things happen: adventures, death, and resuscitation of Lemminkainen, then Vainamoinen's adventures in the belly of the ogre. This last story deserves telling. Vainamoinen set about building a boat, but when it came to putting in the prow and the stern, he found he needed three words in his rune that he did not know, however much he sought for them. In vain he looked on the heads of the swallows, on the necks of the swans, on the backs of the geese, under the tongues of the reindeer.[5] He found a number of words, but not those he needed. Then he thought of seeking them in the realm of Death, Tuonela, but in vain. He escaped back to the world of the living only thanks to potent magic. He was still missing his three runes. He was then told by a shepherd to search in the mouth of Antero Vipunen, the giant ogre. The road, he was told, went over swords and sharpened axes.

Ilmarinen made shoes, shirt and gloves of iron for him, but warned him that he would find the great Vipunen dead. Nevertheless, the hero went. The giant lay underground, and trees grew over his head. Vainamoinen found his way to the giant's mouth, and planted his iron staff in it. The giant awoke and suddenly opened his huge mouth. Vainamoinen slipped into it and was swallowed. As soon as he reached the enormous stomach, he thought of getting out. He built himself a raft and floated on it up and down inside the giant. The giant felt tickled and told him in many and no uncertain words where he might go, but he did not yield any runes. Then Vainamoinen built a smithy and began to hammer his iron on an anvil, torturing the entrails of Vipunen, who howled out magic songs to curse him away. But Vainamoinen said, thank you, he was very comfortable and would not go unless he got the secret words. Then Vipunen at last unlocked the treasure of his powerful runes.

---

[5] In the Eddic lay of Sigrdrifa, the valkyria enumerates the places where can be found *hugruna*, i.e., the runes that give wisdom and knowledge, among which are the following: the shield of the sun, the ear and hoof of his horses, the wheel of Rognir's chariot, Sleipnir's teeth and Bragi's tongue, the beak of the eagle, the clutch of the bear, the paw of the wolf, the nail of the Norns, the head of the bridge, etc. (Sigrdr. vs. 13–17).

Many days and nights he sang, and the sun and the moon and the waves of the sea and the waterfalls stood still to hear him. Vainamoinen treasured them all and finally agreed to come out. Vipunen opened his great jaws, and the hero issued forth to go and build his boat at last.

The story then switches abruptly to introduce Kullervo, his adventures, incest and suicide. When Kullervo incidentally kills the wife that Ilmarinen had bought so dearly in Pohjola, the tale returns again to Ilmarinen's plight. He forges for himself "Pandora," a woman of gold. Finding no pleasure with her, he returns to Pohjola and asks for the second daughter of Louhi. He is refused. Ilmarinen then captures the girl, but she is so spiteful and unfaithful that he changes her into a gull. Then he visits Vainamoinen, who asks for news from Pohjola. Everything is fine there, says Ilmarinen, thanks to the Sampo. They decide, therefore, to get hold of the Sampo, even against Louhi's will. The two of them go by boat, although Ilmarinen is much more in favor of the land route, and Lemminkainen joins them. The boat gets stuck on the shoulder of a huge pike. Vainamoinen kills the fish and constructs out of his jawbones (appendix #10) the Kantele, a harp which nobody can play properly except Vainamoinen himself. There follows a completely Orphic chapter about Vainamoinen's Kantele music, the whole world falling under its spell. Finally, they arrive at Pohjola, and Louhi, as was to be expected, will not part with the Sampo, nor will she share it with the heroes. Vainamoinen then plays the Kantele until all the people of Pohjola are plunged in sleep. Then the brothers go about stealing the Sampo, which turns out to be a difficult task.

> *Then the aged Vainamoinen*
> *Gently set himself to singing*
> *At the copper mountain's entrance,*
> *There beside the stony fortress,*
> *And the castle doors were shaken,*
> *And the iron hinges trembled.*

Thereupon smith Ilmarinen,
Aided by the other heroes,
Overspread the locks with butter,
And with bacon rubbed the hinges,
That the doors should make no jarring,
And the hinges make no creaking.
Then the locks he turned with fingers,
And the bars and bolts he lifted,
And he broke the locks to pieces,
And the mighty doors were opened.

Then the mighty Vainamoinen
Spoke aloud the words that follow:
"O thou lively son of Lempi,
Of my friends the most illustrious,
Come thou here to take the Sampo,
And to seize the pictured cover."

Then the lively Lemminkainen,
He the handsome Kaukomieli,
Always eager, though unbidden,
Ready, though men did not praise him,
Came to carry off the Sampo,
And to seize the pictured cover . . .

Lemminkainen pushed against it,
Turned himself, and pushed against it,
On the ground his knees down-pressing,
But he could not move the Sampo,
Could not stir the pictured cover,
For the roots were rooted firmly,
In the depths nine fathoms under.

There was then a bull in Pohja,
Which had grown to size enormous,
And his sides were sleek and fattened,
And his sinews from the strongest;

Horns he had in length a fathom,
One half more his muzzle's thickness,
So they led him from the meadow,
On the border of the ploughed field,
Up they ploughed the roots of Sampo
Those which fixed the pictured cover,
Then began to move the Sampo,
And to sway the pictured cover.

Then the aged Vainamoinen,
Secondly, smith Ilmarinen,
Third, the lively Lemminkainen,
Carried forth the mighty Sampo,
Forth from Pohjola's stone mountain,
From within the hill of copper,
To the boat away they bore it,
And within the ship they stowed it.

In the boat they stowed the Sampo,
In the hold the pictured cover,
Pushed the boat into the water,
In the waves its sides descended.

Asked the smith, said Ilmarinen,
And he spoke the words which follow:
"Whither shall we bear the Sampo,
Whither now we shall convey it,
Take it from this evil country,
From the wretched land of Pohja?"

Vainamoinen, old and steadfast,
Answered in the words which follow:
"Thither will we bear the Sampo,
And will take the pictured cover,
To the misty island's headland,
At the end of shady island.
There in safety can we keep it,

> *There it can remain for ever,*
> *There's a little spot remaining,*
> *Yet a little plot left over,*
> *Where they eat not and they fight not,*
> *Whither swordsmen never wander.*

The Sampo, then, is brought on board the ship—just as Mysingr the pirate brought Grotte on board his boat—and the heroes row away as fast as possible. Lemminkainen wants music—you can row far better with it, he claims. Vainamoinen demurs, so Lempi's son sings quite by himself, with a voice loud but hardly musical, indeed, for

> *On a stump a crane was sitting,*
> *On a mound from swamp arising,*
> *And his toe-bones he was counting,*
> *And his feet he was uplifting,*
> *And was terrified extremely*
> *At the song of Lemminkainen.*

> *Left the crane his strange employment,*
> *With his harsh voice screamed in terror,*
> *Over Pohjola in terror,*
> *And upon his coming thither,*
> *When he reached the swamp of Pohja,*
> *Screaming still, and screaming harshly,*
> *Screaming at his very loudest,*
> *Waked in Pohjola the people,*
> *And aroused the evil nation.*

Thus, pursuit begins; impediment after magic impediment is thrown across their path by Louhi, wretched hostess of Pohjola, but Vainamoinen overcomes them. He causes her warship to be wrecked upon a cliff which he has conjured forth, but on that occasion his beloved Kantele, the harp, sinks to the bottom of the sea. Finally, Louhi changes herself into a huge eagle which fills all the space between waves and clouds, and she snatches the Sampo away.

From the boat she dragged the Sampo,
Down she pulled the pictured cover,
From the red boat's hold she pulled it,
'Mid the blue lake's waters cast it,
And the Sampo broke to pieces,
And was smashed the picture cover.

Fragments of the colored cover are floating on the surface of the sea. Vainamoinen collects many of them, but Louhi gets only one small piece; hence Lapland is poor, Suomi (Finland) well off and fertile. Vainamoinen sows the fragments of Sampo, and trees came out of it:

From these seeds the plant is sprouting
Lasting welfare is commencing,
Here is ploughing, here is sowing,
Here is every kind of increase.
Thence there comes the lovely sunlight,
O'er the mighty plains of Suomi,
And the lovely land of Suomi.

Vainamoinen constructs a new Kantele, of birchwood this time, and with the hairs of a young maiden as strings—but the strings come last. Before that he asks,

"Now the frame I have constructed,
From the trunk for lasting pleasure,
Whence shall now the screws be fashioned,
Whence shall come the pegs to suit me?

'Twas an oak with equal branches,
And on every branch an acorn,
In the acorns golden kernels,
On each kernel sat a cuckoo.
When the cuckoos all were calling,
In the call five tones were sounding
Gold from out their mouths was flowing,
Silver too they scattered round them,

On a hill the gold was flowing,
On the ground there flowed the silver,
And from this he made the harp-screws,
And the pegs from that provided."

Once more, Vainamoinen begins to play on his irresistible instrument, but this time Louhi manages to capture sun and moon. She was able to do so because

. . . the moon came from his dwelling,
Standing on a crooked birch-tree,
And the sun came from his castle,
Sitting on a fir-tree's summit,
To the kantele to listen,
Filled with wonder and rejoicing.

The grasping Louhi hides sun and moon in an iron mountain. Ilmarinen forges a substitute sun and moon, but they will not shine properly. Eventually, Louhi sets free the luminaries, since she has become afraid of the heroes; repeatedly she complains that her strength has left her with the Sampo.

But time is running out, too, on the ancient Vainamoinen. All that is left for him to do is kindle a new fire, and he does. Beginning far back, he had sung all there was to sing.

Day by day he sang unwearied,
Night by night discoursed unceasing,
Sang the songs of by-gone ages,
Hidden words of ancient wisdom,
Songs which all the children sing not,
All beyond men's comprehension,
In these ages of misfortune,
When the race is near its ending.

Now a Miraculous Child was born, heralding a new era. Vainamoinen knew that there was not room for both of them in the world. If the child lived, he must go. He said good-bye to his country,

And began his songs of magic,
For the last time sang them loudly,
Sang himself a boat of copper,
With a copper deck provided.
In the stern himself he seated,
Sailing o'er the sparkling billows,
Still he sang as he was sailing:

"May the time pass quickly o'er us,
One day passes, comes another,
And again shall I be needed,
Men will look for me and miss me,
To construct another Sampo,
And another harp to make me,
Make another moon for gleaming,
And another sun for shining.
When the sun and moon are absent,
In the air no joy remaineth."

Then the aged Vainamoinen
Went upon his journey singing,
Sailing in his boat of copper,
In his vessel made of copper,
Sailed away to loftier regions,
To the land beneath the heavens.

Actually, there are more runes which tell of Vainamoinen's departure, as we learn from Haavio. He plunges

to the depths of the sea;
to the lowest sea
to the lowest bowels of the earth
to the lowest regions of the heavens
to the doors of the great mouth of death.

Or, he sailed

into the throat of the maelstroem
into the mouth of the maelstroem,
into the gullet of the maelstroem,
into the maw of the monster of the sea.

This is the Vortex that swallows all waters, the one that comes of the destruction of Grotte, which must be dealt with later. Its Norse name is Hvergelmer; its most ancient name is Eridu. But that name belongs to another story and world.

It is difficult for moderns to grasp the quality of that ancient recitation, the *laulo*, of only a few notes going on interminably with freely improvised verbal "cadenzas," yet with a core of formulas rigidly preserved in the canonic form. It is not actually folk poetry in the accepted sense even though its "copyists," its "printers" and its "publishers" are only peasants with an iron memory.[6] An old *laulaja* who recited the origin of the world told Lönnrot: "You and I know that this is the real Truth about how the world began." He said this after centuries of Christendom, never doubting, for the essence of the rune was an incantation, sung or murmured (cf. the German *raunen*), which brings things back to their actual beginning, to the "deep origins." To heal a wound from a sword, the *laulaja* had to sing the rune of the "origin of iron," and one wrong word would have ruined its power. In this way fragments of ageless antiquity remained embedded in living folk poetry. Those whom the Greeks called the "nameless ones," *typhlòs anèr*, who had preserved the epic rhapsodies, reach out to meet us almost in our days, in those humble villages of the Far North, their names of our own time: Arhippa Perttunen, Simana of Mekrijärvi, Okoi of Audista, Ontrei, the Pack Peddler.

Out of the whole bewildering story, one thing is established beyond controversy, that the Sampo is nothing but heaven itself. The fixed adjective *kirjokansi*, "many-coloured," did apply to the cover of the heavenly vault in Finnish folk poetry, as Comparetti and others showed long ago. As for the name Sampo, it resisted the efforts of linguists, until it was found that the word was derived from the Sanskrit *skambha*, pillar, pole.[7] Because it "grinds," Sampo is obviously a mill. But the mill tree is also the world axis, so the inquiry returns to the Norse mill, and to the complex of meanings involved in the difficult word *ludr* (with radical *r*) which

---

[6] M. Haavio, *Vainamoinen, Eternal Sage* (1952), p. 40 (quoting Setälä).
[7] See chapter VIII.

stands for the timbers of the mill and reappears as "loor," a wind instrument. This involves time both ways: the setting and the scansion of time. This does not present embarrassing ambiguity, but a richer meaning, which must have appeared heaven-sent to early thinkers.

The Sampo is—or was—the dispenser of all good things and this is delightfully underscored by the many variants which insist that because most of it fell into the sea, the sea is richer than the land. Men were bound to compare the teeming life of Arctic waters with the barren land in the Far North. But the Sampo did undergo a catastrophe as it was being moved, and that clinches the parallel with Grotte. The astronomical idea underlying these strange representations has been described in the *Intermezzo*, and will be taken up again in chapter IX.

CHAPTER VIII

Shamans and Smiths

Of this base metal may be filed a
key
That will unlock the door they
howl without.

OMAR KHAYYAM

IN ADDITION to the Sampo, there are many myths embedded in the *Kalevala*'s narrative sequence whose analysis would yield surprises. There was the contest of Vainamoinen with Youkahainen (see p. 97), a malevolent Lapp magician who seems to be his constant opponent. Youkahainen tries to overcome the ancient sage by asking cosmogonic riddles, but Vainamoinen "sings" the Lapp step by step into the bog up to his throat, and sings his magic formulae "backwards" to free him only when the Lapp has promised him Aino, his only sister. There was also the tale of Vainamoinen searching in the dead giant's belly for three lost runes. These, unless they are treated as "just so stories," look very much like "erratic boulders" deposited in Finland by the glacial movement of time. For once, it is possible to trace the archaic formation back to Egypt.[1] A young Egyptian called Setna (or Seton Chamwese) wanted to steal the magic book of Thot from the corpse of Nefer-ka Ptah, one of the great Egyptian gods, who was often portrayed as a mummy. Ptah, however, was awake and asked him: "Are you able to take this book away with the help of a knowing scribe, or do you want to overcome me at checkerboards? Will you play 'Fifty-Two'?" Setna agreed, and the board with its "dogs" (pieces) being

---

[1] G. Roeder, *Altaegyptische Erzählungen und Märchen* (1927), p. 149; A. Wiedemann, *Herodots Zweites Buch* (1890), p. 455.

brought up, Nefer-ka Ptah won a game, spoke a formula, laid the checkerboard upon Setna's head and made him sink into the ground up to his hips. On the third time, he made him sink up to his ears; then Setna cried aloud for his brother, who saved him.

There is also a Finnish folktale which repeats the well-known Babylonian story of Etana and the Eagle.[2] Here, instead of the King, it is the "Son of the Widow" (no reason is given for this epithet, which appears to belong to Perceval in the first line, but we find it again in later Masonic tradition)[3] who is taken up into the air by a griffin and sees the earth growing smaller and smaller under him. When the earth appears "no bigger than a pea" (analogous similes are to be found also in Etana), the griffin plunges straightaway to the bottom of the sea, where the hero finds a certain object for which he had looked everywhere, and finally he is restored to land. This looks like the full story of what in the Babylonian cuneiform is interrupted halfway through because the tablet is broken off: it might be the first version of the legend of Alexander exploring the Three Realms.

The anomalous position of Kullervo in the *Kalevala* remains a puzzle. Where Lönnrot put him in the sequence, he remains a displaced person, seeming to come, as was noted, from another age. There are many such incongruities. On the basis of the variants discovered, it has been boldly suggested that he was supposed to appear only after the departure of Vainamoinen—in fact, that he himself is the nameless Miraculous Child who compelled Vainamoinen to quit the scene, and that would be why the two never met. The people now understand the Child to be Christ himself, but that is the normal transforming influence of the Church. The Child in Virgil's Fourth Eclogue was also later thought to be Christ, and Virgil earned a reputation as a magician on the strength

[2] See M. Haavio, *Der Etanamythos in Finnland* (1955), pp. 8-12; also S. Langdon, *The Legend of Etana and the Eagle* (1932), pp. 46-50.

[3] Such words have long lives. At the height of Pickett's Charge at Gettysburg, the first man over the wall was Gen. Armistead, who fell into the breach mortally wounded. To those who picked him up, the general kept repeating: "I am a Son of the Widow"—obviously the password of a secret military brotherhood that his captors did not understand, nor the historian either.

of his supposed prophecy. Actually, the mother says of Vainamoinen's and her illegitimate little son (*50*.199f.): "He shall be a
mighty conqueror, strong as even Vainamoinen." This struck the
English editor of the *Kalevala*, the more so as the baby is also called
"the two-week-old Kaleva." For Kullervo is both in Finnish and in
Esthonian tradition *the* son of Kaleva—"Kalevanpoika" or "Kalevipoeg"—much more explicitly than the other heroes, who are
"sons" only generically. It would fit into the mythical picture, for
reasons which will soon be evident, to have a time-bound tragic
avatar of Vainamoinen, following upon the timeless sage.

But then, who is Kaleva? He is a mysterious entity that shines
by his absence, and yet is the eponymous presence through the
whole poem. The connotation of "giant" is attached to him: in some
of the Finnish versions of the Old Testament, the gigantic Rephaim
and Enakim are called "children of Kaleva." But there are many
reasons for understanding the word as smith.[4] Kaleva might be a
smith even more primeval than Ilmarinen. There is a strange line
in the spell describing the origin of iron: "Poor Iron, man Kaleva,
at that time thou wast neither great nor small." In any case, the
current notion that Kaleva is a "personification" of Finland, a sort
of Britannia with her trident, can be dismissed as unserious. Those
were no times for rhetorical figures. Kaleva remains for the present
a significant void. But Setälä notes that the Russian *bylini*, the
close neighbors of the Esthonian runes, sing the feats of Kolyvanovic, the son of Kolyvan, and say next to nothing of Kolyvan
himself.

The Russian texts give the full name as Samson Kolyvanovic,
just as in Finland it is Kullervo Kalevanpoika. Here perhaps by
chance a name turns up which runs like a barely visible thread
through the whole tradition. We have Samson in the *Kalevala* right
in the first rune; his name is Sampsa Pellervoinen, who "sows the
trees" and also helps Vainamoinen to cut them down.[5] His name,

[4] E. N. Setälä, "Kullervo-Hamlet," FUF 7 (1907), p. 249. See also K. Krohn,
Kalevalastudien 1. Einleitung (1924), pp. 93–101.

[5] Krohn suggests deriving Sampsa from Sampo. Comparetti would like it the
other way around. Neither is convinced or convincing, but they both show that
the name of Samson is a rarity which has to be accounted for.

"the man of the field" or the "earth-begotten," shows him to be a rural deity, which might translate into the Greek Triptolemus, or the Etruscan Aruns Velthymnus. One can no longer tell what his role was in the original order of the poem. It is enough that he is there. The lore of the Mill begins to extend beyond reach. It will be no surprise, then, to find Lykophron, the master mythologist, speaking of Zeus the Miller (435). With it, paradoxically, goes again the name of Mylinos, "Miller," given to the leader of the Battle of the Giants against the Gods. The struggle was seen obviously as one for the control of the Mill of Heaven.

It is, then, maybe not by chance that the name of Samson appears in the Far North. For Samson himself, Samson Agonistes, should have a place of honor among the giant Heroes of the Mill. He is in fact the first one in our literature. We are told (Judges XVI.21) how he ground away, "eyeless in Gaza, at the mill, with slaves," until his cruel captors unbound him to "make sport" for them in their temple, and with his last strength he took hold of the middle pillars and brought the temple crushing down on the heads of the Philistines. Like Menja, he had taken his revenge.

But Samson leads beyond the confines of this topic into a world-wide context. He brings more abstruse concepts into play. He had better be reserved for the next chapter.

Now, at the end of the strange story of the Sampo, one is entitled to ask: does all this make much sense by itself? Is it relevant at all beyond literary history? Comparetti, the great old scholar who in the last century tackled the difficult study of Finnish poetry, set himself a neat and classic philological question. Would it help us to understand the birth of the Homeric poems? Yes, he says. Yet he admits that the Homeric question remains open. In other words, the famous "commission of Orphic and Pythagorean experts set up by Pisistratus to collect the scattered rhapsodies" can hardly have produced by itself any more than Lönnrot could, such works as the *Iliad* and the *Odyssey*. Hence in conclusion there must still have been a Homer. Which goes to show that the conventional idea of epic genius ends up in mystery even for the comparative philologist. But Comparetti is prompt to point out that those experts were no scholars in today's sense, belonging as they did to a period when myth, poetry and intellectual creation were all one.

It might have been better perhaps to take the question from the other end. Supposing that Lönnrot had been himself some kind of "Orphic and Pythagorean" in the old sense, might he not have produced a better reconstruction than the—surely intelligent—stitching together to which he had to limit himself? Was he not hampered by his ignorance of the archaic background? Firdausi *did* actually know the astrological doctrines through which his scattered sources made coherent sense, and this is undoubtedly what allowed him to weld his *Shahnama* into a real whole. Lönnrot was not, but the "short songs" of Finnish peasant tradition were too far removed from the original thought for anyone to recapture it. His successors who unearthed a bewildering number of variants to every single rune have left the confusion intact. Instead of forcing the bulky piece into an arbitrary whole, the Finnish Folklore Fellows (F. F. for short) have taken up comparative mythology, the only means by which order can be established eventually.

As concerns Homer and the presupposed Homeric rhapsodies, this is dangerous territory. Not so much because Homer belongs in fee simple to the redoubtable guild of Homeric scholars—Comparetti as a respectable member of the guild could afford deviations—but essentially because it is not fitting to try to reduce to a "scheme" what remains a prodigious and subtle work of art, the limpidity and immediacy of which should not be spoiled. It is unfortunately a common prejudice that to work out high theoretical allusions contained in the text reduces the text to an irrelevant conundrum, whereas, for instance, the Catalogue of Ships studied literally reveals hidden beauties to the reader. It is enough to suggest here that Homer found pre-existent materials at hand, squared blocks and well-cut ashlars, which he transformed into poetry. One of those prefabricated pieces, the Curse of the Miller Woman, is located in chapter VI, and there is more such evidence to come. Homer's craft lay really in reshaping and humanizing these materials so well that they became inconspicuous. In the case of the Greek tragedies more is known, thanks to Apollodorus. His "Library" of myths, supplemented by Frazer's wonderful notes, shows that the "Library" provided the "book" for every tragedy, those that we have and those that are lost, those written and those never written. Yet it took an Aeschylus or a Sophocles to transform the meaning, to make out of it a work of art.

Much closer to hand, and better known, are the sources of the *Divine Comedy*—history, philosophy and myth, measures and intervals—which provide a virtually complete structure without gaps. Yet because of this, Dante is all the more a true creator. Clearly, it is the very idea of "poet," *poiētēs*, which has to be redefined in moving closer to traditional sources. *Veteres docti poetae* as Ovid said, himself not the least of them. "Learned" is the key word, not in theoretical tropes and allegories, but in the living substances of mythical doctrine.

But here again common usage is misleading. Today, a learned man is usually one who understands what it is all about. Dante was certainly one. But was it so in remote ages? There is reason to doubt it. An esoteric doctrine, as defined by Aristotle, is one which is learned long before being understood. Much of the education of Chinese scholars was until very recently along those lines. Understanding remained something apart. It might never come at all, and at best would come when the learning was complete. There were other ways.

One can give an extreme case from Rome. Athenaeus[6] says that there was a much-applauded mime, Memphis by name, who in a brief dance was said to convey faultlessly the whole essence of the Pythagorean doctrine. It is not said that he understood it: he may have had an inkling, and the rest was his extraordinarily sharpened sense of expression. He had, so to speak, a morphological understanding that he could only express in action. His public understood surely no more than he: but they would be strict and unforgiving judges. *Dictum sapienti sat*, the wise would say. But here even the one word was missing. His spectators would shout deliriously nonetheless, in their own demotic language: "I dig you, Jack." And for

---

[6] *Deipnosophistai 1.20d.* See also Lucian's *De Saltatione 70.*

the slightest lapse from the exact form, they were ready with eggs and overripe tomatoes. Here is a case of true communication which does not need understanding. It takes place only through the form, *morphē*. In mystery rites there were things which "could not be said" (*arrhēta*) but could only be acted out.

Such happenings must be kept in mind when trying to determine how well the poet understood the material handed on to him. Creative misunderstanding may have been of the essence of his "freedom": but strict respect was there nonetheless. The rune of the "origin of iron" (the ninth of the *Kalevala*) was incomprehensible to the *laulaja*, yet he knew he had to recite this "deep origin" to control the lethal powers of cold iron. Magic and mantic implications were present always in the grim business of the smith, as they were in the high business of the poet. Understanding lay beyond them.

Every era, of course, has freely invented its own ballads, romances, songs and fables to entertain it. That is another matter. This concerns the poet, *poiētēs*, as he was understood in early times. There was an original complex of meaning which comprised the words poet, *vates*, prophet, seer. Every knowledge and law, Vico wrote with a flash of genius two centuries ago, must once upon a time have been "serious poetry," *poesia seriosa*. It is in this sense that Aristotle in a sophisticated age still refers respectfully to "the grave testimony of [early] poets."

Now that documents of the earliest ages of writing are available, one is struck with a wholly unexpected feature. Those first predecessors of ours, instead of indulging their whims with childlike freedom, behave like worried and doubting commentators: they always try an exegesis of a dimly understood tradition. They move among technical terms whose meaning is half lost to them, they deal with words which appear on this earliest horizon already "tottering with age" as J. H. Breasted says, words soon to vanish from our ken. Long before poetry can begin, there were generations of strange scholiasts.

The experts have noted the uncertainty prevailing in the successors of old texts, the attempts in them to establish correct names and their significance from obsolete formulas and ideograms. S. Schott, dealing with early star lists of Egypt,[7] points to the perplexity of later generations concerning the names of constellations, even those of the "greatest gods of the Decans, Orion and Sothis, who in Ancient Egyptian are called by the names of old hieroglyphs, without anybody knowing, in historical times, what these hieroglyphs had meant, once upon a time. During the whole long history of these names we meet attempts at interpretation." This last sentence goes for every ancient text, not only for the names contained therein: there is no end of commentaries on the Pyramid Texts, the Coffin Texts and the Book of the Dead,[8] on the *Rigveda*, the *I-Ging*, just as on the Old Testament.[9] W. von Soden regrets that we depend on the documents of the "Renaissance of Sumerian culture" (around 2100 B.C.) instead of having the real, old material at our disposal.[10] The mere fact that Sumerian was the language of the educated Babylonian and Assyrian, the existence of the many Sumerian-Akkadian "dictionaries" and the numerous translations of the Gilgamesh epic betray the activity of several academies responsible for the officially recognized text editions. One can almost see the scholars puzzling and frowning over the texts. And in Mexico it was the same. In Chimalpahin's *Memorial Breve* we find notes such as "In the year '5-house' certain old men explained some pictographs to the effect that king Hueymac of Tollan [the mythical Golden Age city] had died."[11] This took place before the coming of the Spaniards. The Greek "Renaissance," no less than those of the previous millennia in the Near East, was the result of such an antiquarian effort. Hesiod still bears the mark of it.

These few notions should be present in any ideas about "transmission." The word need in no way imply "understanding" on

[7] W. Gundel, *Dekane und Dekansternbilder* (1936), p. 5.

[8] See, for example, G. Roeder, *Urkunden zur Religion des Alten Aegypten* (1915), pp. 185f., 199f., 224.

[9] J. Dowson (*A Classical Dictionary of Hindu Mythology*, p. 60) bluntly calls the Brahmanas "a Hindu Talmud."

[10] "Licht und Finsternis in der sumerischen und babylonisch-assyrischen Religion," *Studium Generale 13* (1960), p. 647.

[11] Chimalpahin, *Memorial Breve*, trans. W. Lehmann and G. Kutscher (1958), p. 10.

the part of those who transmit, and this is true from early ages down to contemporary minstrels. As has been pointed out, it is easy to slip into ordinary literary history if the origins are not seriously investigated. Does the tale of the Sampo have a wider interest than this? A few handsome cosmic motifs scattered through the tale of magic might still have reached Finland through the "corridors of Time" from other cultures without any meaning attached. In short, it might all be "folk poetry" in the usual sense.

The editors of the *Kalevala* themselves insistently described the background as "shamanistic," by which they simply understood some kind of primitive "religion." It corresponded in their minds to primeval, instinctive magic, to be found in all five continents, associated with the tribal "medicine man." Then came Frazer to introduce the cleavage between "magic" and "religion" as distinct forms, to complicate matters further. Shamanism remained until recently a catchword of an uncertain sort—a portmanteau term for specialists, a vague notion for the public, of the kind that gives one the pleasant impression of understanding what it is all about—like that other too-famous term, *mana.* One of the present authors is willing to admit ruefully that he once stressed the link of Pythagoras and Epimenides with Thracian shamans, with no more thought than to show that there was much in them of the ageless medicine man.[12] This was several years ago, and it seemed to correspond to the state of the art. It is no longer so. To have uncovered the inadmissibility of the general usage of the term is the merit of Laszlo Vajda's short but dense and logical study on the subject.[13] Vajda has shown that no historical verdict based on such generalities is valid. It is inadmissible to reduce shamanism to memories of Eskimo *angekoks* or to a "technique of induced ecstasy," or to derive such phenomena from the Asiatic North where, undeniably, this particular kind of queerness is fostered.

"Shaman" is a Tungusian word. Shamanism has its epicenter in Ural-Altaic Asia, but it is a very complex phenomenon of *culture*

---

[12] G. de Santillana, *The Origins of Scientific Thought* (1961), p. 54.

[13] "Zur Phaseologischen Stellung des Schamanismus," in *Ural-Altaische Jahrbücher 31* (1959), pp. 456–85.

which can be explained neither by psychologists nor by sociologists, but only by way of historical ethnology. To put it in a few words, a shaman is elected by spirits, meaning that he cannot choose his profession. Epileptics and mentally unhinged persons are obvious privileged candidates. Once elected, the future shaman goes to "school." Older shamans teach him his trade, and only after the concluding ceremony of his education is he accepted. This is, so to speak, the visible part of his education. The real shamanistic initiation of the soul happens in the world of spirits—while his body lies unconscious in his tent for days—who dismember the candidate in the most thorough and drastic manner and sew him together afterwards with iron wire, or reforge him, so that he becomes a new being capable of feats which go beyond the human. The duties of a shaman are to heal diseases which are caused by hostile spirits who have entered the body of the patient, or which occur because the soul has left the body and cannot find the way back. Often the shaman is responsible for guiding the souls of the deceased to the abode of the dead, as he also escorts the souls of sacrificed animals to the sky. His help is needed, too, when the hunting season is bad; he must find out where the game is. In order to find out all the things which he is expected to know, the shaman has to ascend to the highest sky to get the information from his god—or go into the underworld. On his way he has to fight hostile spirits, and/or rival shamans, and tremendous duels are fought. Both combatants have with them their helping spirits in animal form, and much shape-shifting takes place. In fact, these fantastic duels form the bulk of shamanistic stories. The last echoes are the so-called "magic-flights" in fairy tales. The shaman's soul ascends to the sky when he is in a state of ecstasy; in order to get into this state, he needs his drum which serves him as a "horse," the drumstick as a "whip."[14] Now, the "frame" within which the shaman proper acts, that is, the

[14] The shamans also use as a "main artery" a stream flowing through all levels of the sky, and they identify it with the Yenissei—a conception which will become clearer at a later point of this inquiry. (U. Holmberg, *Finno-Ugric and Siberian Mythology* [1964], pp. 307f.).

world conception of Ural-Altaic shamanism, has been successfully traced back to India (under its Hinduistic and Buddhistic aspects, including Tibetan Lamaism and Bön-po) as well as to Iran. When reading Radloff's many volumes, one runs into insufficiently disguised Bodhisatvas at every corner (Manjirae = Manjusri; Maiterae, or Maidere = Maitreya, etc.), but the best organized material has been provided by Uno Holmberg (Uno Harva),[15] who has been quoted here and will be quoted frequently.

This world conception, however, with its three "domains," with seven or nine skies, one above the other, and with corresponding "underworlds," with the "world-pillar" running through the center of the whole system, crowned by the "north Nail," or "World Nail" (Polaris), goes farther back than Indian and Iranian culture, namely to the most ancient Near East, whence India and Iran derived their idea of a "cosmos"—a cosmos being in itself by no means an obvious assumption. The shaman climbing the "stairs" or notches of his post or tree, pretending that his soul ascends at the same time to the highest sky, does the very same thing as the Mesopotamian priest did when mounting to the top of his seven-storied pyramid, the ziqqurat, representing the planetary spheres.[16]

[15] See the bibliography.

[16] Nine skies, instead of seven, within the sphere of fixed stars, result from the habit of including among the planets the (invisible) "head" and "tail" of the "Dragon," which is to say the lunar nodes, conjunctions or oppositions in the vicinity of which cause the eclipses of Sun and Moon; the revolution of these "draconitic points" is c. 18½ years. This notion, upheld in medieval Islamic astrology, is Indian, but apparently not of Indian origin, as will come out eventually. Reuter, *Germanische Himmelskunde* (1934), pp. 291ff., thinks that the Teutonic idea of nine planets including the draconitic points goes back to the common "Urzeit" of Indo-Europeans, and refers to Luise Troje, *Die 13 und 12 im Traktat Pelliot* (1925), pp. 7f., 25, 149f. Even if the "Dragon" should go back to this time, we do not take the Indo-Europeans, whether united or not, for the inventors of this idea. As concerns Islamic and Indian tradition, see the most thorough and thoughtful inquiries by Willy Hartner, "The Pseudoplanetary Nodes of the Moon's Orbit in Hindu and Islamic Iconographies," in *Ars Islamica 5* (1938), Pt. 1; *Le Problème de la planète Kaïd* (1955); "Zur Astrologischen Symbolik des 'Wade Cup'," in *Festschrift Küchnel* (1959), pp. 234-43. Whether we shall find the time to deal in the appropriate form with the tripartite Universe in this essay remains doubtful. This much can be safely stated: it goes back to "The Ways of Anu, Enlil, and Ea" in Babylonian astronomy.

From the majestic temple at Borobudur in Java to the graceful *stupas* which dot the Indian landscape, stretches a schematized reminder of the seven heavens, the seven notches, the seven levels. Says Uno Holmberg: "This pattern of seven levels can hardly be imagined as the invention of Turko-Tatar populations. To the investigator, the origin of the Gods ruling those various levels is no mystery, for they point clearly to the planetary gods of Babylon, which already in their far-away point of origin, ruled over seven superposed starry circles."[17] This was also the considered conclusion, years ago, of Paul Mus. To have taken the conception of several skies and underworlds as natural, ergo primitive, was a grievous blunder which distorted the historical outlook of the last two centuries. It stems from the fact that philologists and Orientalists have lost all contact with astronomical imagination, or even the fundamentals of astronomy. When they find something which savors undeniably of astronomical lore, they find a way to label it under "prelogical thought" or the like.

But even apart from the celestial "ladder," and the sky-travel of the shaman's soul, a close look at shamanistic items always discloses very ancient patterns. For instance, the *drum*, the most powerful device of the shaman, representing the Universe in a specific way, is the unmistakable grandchild of the bronze lilissu drum of the Mesopotamian Kalu-priest (responsible for music, and serving the god Enki/Ea).[18] The *cover* of the lilissu drum must come from a black bull, "which represents Taurus in heaven," says Thureau-Dangin.[19] Going further, W. F. Albright and P. E. Dumont[20]

---

[17] *Der Baum des Lebens* (1922), p. 123.

[18] See B. Meissner, *Babylonien und Assyrien* (1925), vol. 2, p. 66.

[19] *Rituels accadiens* (1921), p. 2. See also E. Ebeling, *Tod und Leben nach den Vorstellungen der Babylonier* (1931), for a cuneiform text in which the hide is explicitly said to be Anu (p. 29), and C. Bezold, *Babylonisch-Assyrisches Glossar* (1926), p. 210 s.v. "sugugalu, 'the hide of the great bull,' an emblem of Anu." We might point, once more, to the figure of speech used by Petronius' Trimalchio, who, talking of the month of May, states: "Totus coelus taurulus fiat" ("the whole heaven turns into a little bull").

[20] "A Parallel between Indian and Babylonian Sacrificial Ritual," JAOS *54* (1934), pp. 107–28.

compared the sacrifice of the Mesopotamian bull, the hide of which was to cover the lilissu drum, with the Indian Ashvamedha, a huge horse sacrifice which only the most successful king (always a Kshatrya) could afford. They found that the Indian horse must have the Krittika, the Pleiades, on his forehead, and this too, according to Albright, is what the Akkadian text prescribes concerning the bull. This should be enough to indicate the *level* of phenomena brought into play.

The striking of the drum covered with that specific bull hide was meant as a contact with heaven at its most significant point, and in the Age of Taurus (c. 4000–2000 B.C.) this was also explicitly said to represent Anu, now casually identified as "God of Heaven." But Anu was a far more exact entity. In cuneiform script, Anu is written with one wedge, which stands for the number 1 and also for 60 in the sexagesimal system (the Pythagoreans would have said, he stands for the One and the Decad). All this does not mean some symbolic or mystical, least of all magical quality or quantity, but the fundamental time measure of celestial events (that is, motions).[21] Striking the drum was to involve (this time, yes, magically) the essential Time and Place in heaven.

It is not clear whether or not the Siberian shamans were still aware of this past. The amount of highly relevant star lore collected by Holmberg, and the innumerable figures of definitely astronomical character found on shamanistic drums could very well allow for much more insight than the ethnologists assume, but this is irrelevant at this point. What is plain and relevant is that the Siberian shamans did not *invent* the zodiac, and all that goes with it.

There is no need for a detailed inspection of Chinese mythical drums, merely a few lines from an "Ocean of Stories":

> In the Eastern Sea, there is to be found an animal which looks like an ox. Its appearance is green, and it has no horns. It has one foot only. When it moves into the water or out of it, it causes wind or rain. Its

21 Compare the sexagesimal round of days in customary notation of the oracle bones of Shang China, 15th century B.C., about which Needham states that it is "probably an example of Babylonian influence on China" (*Science and Civilisation in China* [1962], vol. 4, Pt. 1, p. 181).

shining is similar to that of the sun and the moon. The noise it makes is like the thunder. Its name is K'uei. The great Huang-ti, having captured it, made a drum out of its skin.[22]

This looks prima facie like the description of an ancient case of delirium tremens, but the context makes it sober enough. This is a kind of Unnatural Natural History which has small regard for living species, but deals with events from another realm. The One-Legged Being, in particular, can be followed through many appearances beginning with the Hunrakán of the Mayas, whose very name means "one-leg." From it comes our "hurricane," so there is no wonder that he disposes of wind, rain, thunder and lightning in lavish amounts. But he is not for all that a mere weather god, since he is one aspect of Tezcatlipoca himself, and the true original One-Leg that looks down from the starry sky—but his name is not appropriate yet.

And so back by unexpected ways to mythical drums and their conceivable use. A lot more might be found by exploring that incredible storehouse of archaic thought miraculously preserved among the Mande peoples of West Sudan.[23] In the large and complicated creation myth of the Mande, there are two drums. The first was brought down from heaven by the bardic ancestor, shortly after the Ark (with the eight twin-ancestors) had landed on the primeval field. This drum was made from Faro's skull and was used for producing rain. (The experts style Faro usually "le Moni-

---

[22] M. Granet, *Danses et légendes de la Chine ancienne* (1959), p. 509. Such imagery is by no means unique. E.g., the *Taittiriya Sanhita* says: "The pressing stone [of the Soma-press] is the penis of the sacrificial horse, Soma is his seed; when he opens his mouth, he causes lightning, when he shivers, it thunders, when he urinates, it rains" (7.5.25.2 = *Shatapatha Brahmana 10.6.4.1* = *Brihad Aranyaka Upanishad 1.1*; see R. Pischel and K. F. Geldner, *Vedische Studien*, vol. 1, p. 86).

It will come out later why it is important to supplement these strange utterances with the statement of the *Shatapatha Brahmana*: "In the water having its origin is the horse," which sounds ever so inconspicuous until E. Sieg (*Die Sagenstoffe des Rigveda*, p. 98) obliges the non-Sanskritist by giving the Sanskrit words in transcription, i.e., "*apsuyanir vā asvah*"; *apsu* is something more specific than just water; it is, in fact, the very same topos as the Babylonian *apsu* (Sumerian: *abzu*).

[23] In East Africa, the drum occupied the place that the Tabernacle had in the Old Testament, as Harald von Sicard has shown in *Ngoma Lungundu: Eine afrikanische Bundeslade* (1952).

teur," thus avoiding mislabeling him as culture hero, savior or god.) The first sanctuary was built, and the "First Word" revealed (30 words there were) to mankind through the mouth of one of the twin-ancestors, who "talked the whole night, ceasing only when he saw the sun and Sirius rising at the same time." When the "Second Word" was to be revealed (consisting of 50 words this time), and again connected with the heliacal rising of Sirius, the ancestor "decided to sacrifice in the sanctuary on the hill the first twins of mixed sex. He asked the bard to make an arm-drum with the skin of the twins.[24] The tree, from which he carved the drum, grew on the hill and symbolized Faro's only leg."[25]

Here again are important one-legged characters, of whom there are a bewildering number with various functions all over the world. It is not necessary to enter that jungle, except to note that the temporary mock-king of Siam, who was set up for yearly expiatory ceremonies, also had to stand on one leg upon a golden dais during all the coronation ceremonies, and he had the fine-sounding title of "Lord of the Celestial Armies."[26] The Chinese K'uei is then no isolated character. The Chinese myth is more explicit than the others and becomes more understandable because the Chinese were extremely sky-conscious. Their sinful monsters are thrown into pits or banished to strange mountain regions for the sin of having upset the calendar.

As for K'uei himself, engagingly introduced as a green oxlike creature of the Eastern Sea, he will grow more bewildering as his nature unfolds. Marcel Granet writes that the Emperor Shun made K'uei "master of music"—actually ordered no less a power than the Sun (Chong-li) to fetch him from the bush and bring him to court, because K'uei alone had the talent to bring into harmony

---

[24] It is an hourglass-shaped drum, with two skins, said "to recall the two geographic areas, Kaba and Akka, and the narrow central part of the drum is the river itself [Niger] and hence Faro's journey."

[25] Germaine Dieterlen, "The Mande Creation Story," *Africa 27* (1957), pp. 124–38; cf. JSA 25 (1955), pp. 39–76. See also Marcel Griaule, "Symbolisme des tambours soudanais," *Mélanges historiques offerts à M. Masson 1* (1955), pp. 79–86; Griaule and Dieterlen, *Signes Graphiques Soudanais* (1951), p. 19.

[26] W. Deonna, *Un divertissement de table "à cloche-pied"* (1959), p. 33. See J. Frazer, *The Dying God* (Pt. III of *The Golden Bough*), pp. 149f.

the six pipes and the seven modes, and Shun, who wanted to bring peace to the empire, stood by the opinion that "music is the essence of heaven and earth."[27] K'uei also could cause the "hundred animals" to dance by touching the musical stone, and he helped Yü the Great, that indefatigable earth-mover among the Five First Emperors, to accomplish his labor of regulating the "rivers." And it turns out that he was not only Master of the Dance, but Master of the Forge as well. He must have been a remarkable companion for Yü the Great, whose dancing pattern (the Step of Yü) "performed" the Big Dipper.[28]

Enough of drums, and of their shamanic use. They have at least ceased to seem like tribal tom-toms. They are connected with time, rhythm and motion in heaven.

Moving now to another great theme, in fact a very great one, it is possible to trace back the significance of the *blacksmith* in Asiatic shamanism, particularly the celestial blacksmith who is the legitimate heir to the divine "archi-tekton" of the cosmos. Several representatives of this type, whom we call Deus Faber, still have both functions, being architects and smiths at the same time, e.g., the Greek Hephaistos, who builds the starry houses for the gods and forges masterworks, and the Koshar-wa-Hasis of Ras Shamra, who builds Baal's palace and forges masterworks also.

The Yakuts claim: "Smith and Shaman come from the same nest," and they add: "the Smith is the older brother of the Shaman,"[29] which might be valid also for Vainamoinen, coupled with Ilmarinen, who is said to have "hammered together the roof of the sky." It is the primeval Smith who made the Sampo, as we know, and forged sky and luminaries in Esthonia. It is no idle fancy that the representative of the celestial smith, the King, is himself frequently titled "Smith." Jenghiz Khan had the title "Smith"[30] and

---

[27] Granet, *Danses et légendes*, pp. 311, 505–508.

[28] We are indebted for this last piece of information to Professor N. Sivin.

[29] P. W. Schmidt, *Die asiatischen Hirtenvölker* (1954), pp. 346f. Concerning the terrestrial blacksmith: the many iron pieces which belong to the costume of a shaman can be forged only by a blacksmith of the 9th generation, i.e., eight of his direct ancestors must have been in the profession. A smith who dared forge a shamanistic outfit without having those ancestors would be torn by bird-spirits.

[30] A. Alföldi, "Smith As a Title of Dignity" (in Hungarian), in *Magyar Nyelv* 28 (1932), pp. 205–20.

the standard of the Persian Empire was the stylized leather apron of the Smith Kavag (appendix #11). The Chinese mythical emperors Huang-ti and Yü are such unmistakable smiths that Marcel Granet drew historic-sociological conclusions all the way, forgetting the while that Huang-ti, the Yellow Emperor, is acknowledged to be Saturn. And just as the Persian Shahs held their royal jubilee festival after having reigned thirty years, which is the Saturnian revolution, so the Egyptian Pharaoh also celebrated his jubilee after thirty years, true to the "inventor" of this festival, Ptah, who *is* the Egyptian Saturn, and also Deus Faber. It was necessary to enter this subject in depth abruptly and lay stress on these few selected data, because otherwise the charming and harmless-looking Finnish runes would not be seen for what they are, the badly damaged fragments of a once whole and "multicolored cover." It does no harm to stamp Vainamoinen a "shaman" as long as one remains aware of the background of shamanism. In fact, there is again a vision in depth from seeing that Vainamoinen has discarded the drum which remains the one instrument of his Lapp cousins; he has created the harp, and this means that he must be seen as the Orpheus of the North.

   Last survivals are not easily recognized. It needs experience, and it cannot be expected that an unsuspecting reader of "folk poetry" would spot well-known divine characters when they come his way clad in Longfellow's meter. For instance, in reading *Kalevala* 9.107ff., it is not easy to discover the mighty Iranian God of Time, Zurvan akarana, who is portrayed as standing upon the world egg, carrying in his hands the tools of the architect:

> *Then was born smith Ilmarinen*
> *Thus was born and thus was nurtured*
> *Born upon a hill of charcoal,*
> *Reared upon a plain of charcoal,*
> *In his hands a copper hammer,*
> *And his little pincers likewise.*
> *Ilmari was born at night time,*
> *And at day he built his smithy.*

Since Christendom was very successful in destroying old traditions, Altaic and Siberian survivals are often found in far better shape than Finnish runes, but even the Lapps still speak of *"Waralden ol-may*, 'World Man' . . . and this is the same as Saturnus."[31] Nor are Jupiter and Mars absent, the former being called Hora Galles (Thorkarl), the latter Bieka Galles, the "Wind Man."[32] Voguls, Yakuts and Mongols tell of God's seven sons, or seven gods (or nine), among whom are a "Scribe Man,"[33] and a "Man observing the World." The latter has been compared straightaway with Kullervo by Karl Kerényi,[34] who claims his name to be the literal translation of *Avalokiteshvara*, the very great Bodhisatva, known in China as Kuan-yin, literally "deserving (musical) modes." One wonders whether this "World-observer" does not go back much farther: to Gilgamesh. We have to keep in mind that the Babylonians called their texts after their opening words; e.g., the Creation Epic they called *Enuma elish*, i.e., "When above"; accordingly, what we call the Epic of Gilgamesh was with them *Sha naqba imurū*, "Who saw everything." Such are the bewildering changes rung by time on great and familiar themes. And there is more. Actually, when still young, this Vogulian "World-observing Man"—Avalokiteshvara himself, this great and worshiped deity of Buddhist countries— was, like Kullervo, a much-plagued orphan, first in the house of his uncle, then in the house of "the Russian" and in that "of the Samoyed." After years of misery—quite specific "measured" mis-ery[35]—he kills all his tormentors. The revenge he takes is for

---

[31] This information comes from Johan Radulf (1723), quoted by K. Krohn, "Priapkultus," FUF 6 (1906), p. 168, who identifies Waralden olmay with Freyr. G. Dumézil, *La Saga de Hadingus* (1950), identifies him with Njordr.

[32] K. Krohn, "Windgott und Windzauber," FUF 7 (1907), pp. 173f., where the god is once called Ilmaris.

[33] The Ostyaks talk even of a golden Book of Destiny, and Holmberg points out that the Ostyaks who have no writing are not likely to have hit upon such notions by themselves. (Holmberg, *Der Baum des Lebens*, p. 97). Cf. the entire chapter, "The Seven Gods of Fate" (pp. 113–33 of the same work) and Holmberg's *Finno-Ugric and Siberian Mythology*, p. 415.

[34] "Zum Urkind-Mythologem," in *Paideuma* 2 (1940), pp. 245ff. See now C. G. Jung and K. Kerényi, *Essays on a Science of Mythology* (1949), pp. 30–39.

[35] E.g., in the house of "the Russian," he is kept in the door hinge (in the English translation this and other details are blurred to insignificance), and dishwater is emptied upon him. To be damned to play door hinge is one of the hellish punishments in Egypt, because the hinge is supposed to turn in the victim's eye. As concerns the heart-warming custom of abusing one's celestial fellow travelers as sink

himself. His father, who had lowered down the beloved son from the sky in a cradle, remained aloft.

These hints will suffice for the time being. It does not matter whether or not pieces, or even the whole, of cosmological tradition came late to the Ural-Altaic populations, that is, whether Manicheism had a part in their propagation. The Manicheans took over the whole parcel of old traditions, changing only the signs, as happens with every Gnostic system. Gnostics never have been of the inventive sort. Their very title, derived from their key word, gives the scheme away, *gnōsis tēs hodou* = knowledge of the way. The "Way" which had to be learned by heart is that which leads outwards-upwards through the planetary spheres, past the threatening "watchtowers" of the zodiac to the desired timeless Light beyond the sphere of fixed stars, above the Pole star: beyond and above everything, where the unknown god (*agnostos theos*) resides eternally.[36]

This "Way" is not exactly the same for everybody, and the largest highroads are not everlasting, but the principle remains un-

---

or toilet: we find this in the Eddic *Lokasenna* (34), where Loke says of Njordr that he was used as chamber pot by Hymir's daughters; with the Polynesian case of Tawhaki, whose father Hema is abused in the very same manner, we deal in the chapter on Samson (see p. 175); this model of a "Horus-avenger-of-his-father" fulfills not only his filial duty, he does it by means of Amlethus' own "Net-trick." The Samoyed binds the pitiable "World-observing Man" to his sledge with an iron wire of thirty fathoms length. We do not know yet what this means precisely. We know that victorious characters use the vanquished as this or that vehicle, saddle horse, etc.—Marduk uses Tiamat as "ship," as does Osiris with Seth; Ninurta's "Elamitic chariot, carrying the corpse of Enmesharra" is drawn by "horses who are the death-demon of Zu" (Ebeling, *Tod und Leben*, p. 33); Tachma Rupa rides on Ahriman for thirty years around the two ends of the earth (Yasht *19.29*; Yasht *19*, the Zamyad Yasht, is the one dedicated to Hvarna)—but these code formulae have not yet been broken.

[36] Absurd as it sounds, the many Gnostic sects who hated nothing more than philosophers and mathematicians have never denied or doubted the validity of their "evil" teachings. Sick with disgust, they learned the routes of ascension through (or across) those abominable spheres ruled by number, created by the evil powers. Surely, their "Father of Greatness" would not have created such a thing as a cosmos. Tradition does use the most peculiar vehicles for its motion through historical time. Or should one say, tradition did use? Face to face with the outbreaking revolution of "simple souls" against whichever rational thought, there is small reason for hope that our contemporary gnostics will hand down any tradition at all.

changed. The shaman travels through the skies in the very same manner as the Pharaoh did, well equipped as he was with his Pyramid Text or his Coffin Text, which represented his indispensable timetable and contained the ordained addresses of every celestial individual whom he was expected to meet.[37] The Pharaoh relied upon his particular text as the less distinguished dead relied upon his copy of chapters from the Book of the Dead, and he was prepared (as was the shaman) to change shape into the Sata serpent, a centipede, or the semblance of whatever celestial "station" must be passed, and to recite the fitting formulae to overcome hostile beings.[38]

To sum it up—whether Shamanism is an old or a relatively young offshoot of ancient civilization is irrelevant. It is not primitive at all, but it belongs, as all our civilizations do, to the vast company of ungrateful heirs of some almost unbelievable Near Eastern ancestor who first dared to understand the world as created according to number, measure and weight.

If the Finnish runes and Altaic legends sound harmless enough, so do the popular traditions of most of the European countries, including Greece: the kind of mythology known through Bulfinch. But here at least there are additional less popular traditions which have preserved more of the severe spirit and style of old. So the (13th) Orphic Hymn to Kronos addresses the god as "Father of

[37] The ephemerides on the inner side of the coffin lids of the Middle Kingdom, and the astronomical ceilings in tombs of the New Kingdom, as well as the "Ramesside Star Clocks," made navigation still easier for the royal soul.

[38] Many of the heavenly creatures do all the damage they possibly can; they try, for instance, to rob the dead of his text without which he would be helpless, and generally their conduct, as described in the literature of the Hereafter, is weird. Thus, in chapter 32 of the Book of the Dead, the crocodile of the West is accused of eating certain stars; the properly equipped soul, however, knows how to play up to the celestial monsters, and the traveler addresses the Northern crocodile with the words: "Get thee back, for the goddess Serqet is in my interior and I have not yet brought her forth." The goddess Serqet is the constellation Scorpius. As concerns the Sata serpent, "whose years are infinite . . . who dwells at the farthest ends of the earth . . . who renews his youth everyday" (Book of the Dead, ch. 87), he makes himself suspect of representing the sphere of Saturn, whereas the centipede is not likely to fit any "body" besides Moon or Mercury; that it is no constellation is certain.

the blessed gods as well as of man, you of changeful counsel, . . . strong Titan who devours all and begets it anew [lit. "you who consume all and increase it contrariwise yourself"], you who hold the indestructible bond according to the *apeirona* (unlimited) order of Aiōn, Kronos father of all, wily-minded Kronos, offspring of Gaia and starry Ouranos . . . venerable Prometheus." Such sayings suddenly thrust information out of the usual patterns and show the true professional minds of ancient mythology working out their theorems. The only conventional attribute is "Son of Ouranos and Gaia." Kronos is termed a Titan, because the word "God" belongs properly to the Olympian generation, whereas Saturn's empire is not of "this world," any more than that of the Indian Asura and the king of the golden Krita Yuga, Varuna; and the formula is found still in the medieval "Kaiser-Sage." At the end of Thidrek's (Theodoric's) reign when there are only corpses left, a dwarf appears and asks the king to follow him; "your empire is no more in this world."[39] More puzzling, Kronos "is" that other Titan, Prometheus, that other adversary of the "gods," the Lighter of Fire. He "is" many more characters, too, but it will take some time to clear this up. We are at the heart of an "implex."

"Who holdest the unbreakable bond . . ." Assyrian Ninurta, too, holds "the bond of heaven and earth." We shall also hear of a magical invocation (see p. 147) that addresses Kronos as "founder of the world we live in." These words are, however, insufficient and ambiguous. Not only are translations imprecise generally, but in our times of accelerated decay of language even the best-intentioned reader is likely to overlook such words as "bond" or "to found." If instead he were to read "inch scale" and "to survey"—a divine foundation is every time a "temenos"—he would promptly react in a different manner. Kronos-Saturn has been and remains the one who owns the "inch scale," who gives the measures, continuously, because he is "the originator of times," as Macrobius says, although

[39] W. Grimm, *Die Deutsche Heldensage* (1957), p. 338: "Du solt mit mir gan.dyn reich ist nit me in dieser welt." The corresponding most popular folktale shows Theodoric of Verona ravished by a demon horse and cast headlong into the crater of Etna.

the poor man mistakes him for the sun for this very reason.[40] But "Helios the Titan" is not Apollo, quite explicitly.

Apart from this, apart also from Plutarch's report, according to which Kronos, sleeping in that golden cave in Ogygia, dreams what Zeus is planning,[41] there is an Orphic fragment of greater weight, preserved in Proclus' commentary on Plato's *Cratylus*.[42] The Orphic text being one of the delicate sort, we quote some sentences only:

> The greatest Kronos is giving from above the principles of intelligibility to the Demiurge [Zeus], and he presides over the whole "creation" [*demiourgia*]. That is why Zeus calls him "Demon" according to Orpheus, saying: "Set in motion our genus, excellent Demon!" And Kronos seems to have with him the highest causes of junctions and separations . . . he has become the cause of the continuation of begetting and propagation and the head of the whole genus of Titans from which originates the division of beings [*diairēsis tōn ontōn*].

The passage ends thus: "Also Nyx prophesies to him [i.e., occasionally] but the father does so continuously [*prosechōs*], and he gives him all the measures of the whole creation."[43]

In Proclus' style, the same phenomena which look simply flat and childish, mere "etymologizing," when handled by others, sound extremely difficult—which they actually are. So let us shortly compare how Macrobius deals with the responsibility of Kronos for the "division of beings" (*Sat. 1.8.6–7*). After having mentioned the current identification of Kronos (Saturn) and Chronos (Time), so often contested by philologists, Macrobius states:

> They say, that Saturn cut off the private parts of his father Caelus [Ouranos], threw them into the sea, and out of them Venus was born who, after the foam [*aphros*] from which she was formed, accepted the name of Aphrodite. From this they conclude that, when there was chaos, no time existed, insofar as time is a fixed measure derived from the revolution of the sky. Time begins there; and of this is be-

---

[40] *Sat.1.22.8: Saturnus ipse, qui auctor est temporum.*
[41] *De facie in orbe lunae* 941: "Hosa gar ho Zeus prodianoeitai, taut' oneiropolein ton Kronon."
[42] Fr. 155, Kern, p. 194.
[43] Kai panta ta metra tēs holēs demiourgias endidōsin. We might even say: Kronos "grants" him all the measures.

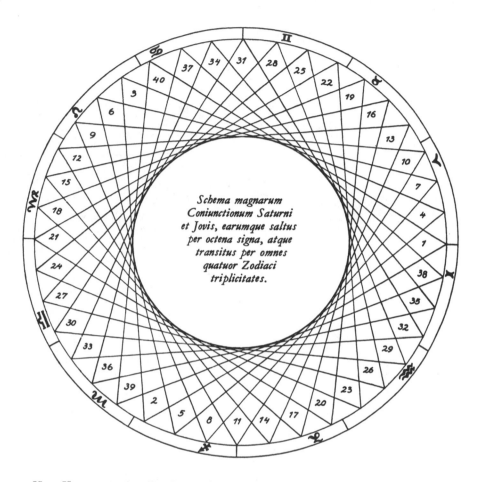

Schema magnarum
Coniunctionum Saturni
et Jovis, earumque saltus
per octena signa, atque
transitus per omnes
quatuor Zodiaci
triplicitates.

How Kronos continually gives to Zeus "all the measures of the whole creation":
Kepler's presentation of the Trigon built up by the Great Conjunctions of Saturn
and Jupiter every twenty years. The motion of this Trigon along the zodiacal
signs subdivided the cycle of the Precession, acting as a kind of vernier for this
great cycle. To go around the whole zodiac, it takes one angle of the Trigon
roughly 2400 years; to move from one sign of an elementary triplicity to the next
sign of the same element takes about 800 years.

lieved to have been born Kronos who is Chronos, as was said before [see appendix #12].[44]

The fact is that the "separation of the parents of the world," accomplished by means of the emasculation of Ouranos, stands for the establishing of the obliquity of the ecliptic: the beginning of measurable time. (The very same "event" was understood by Milton as the expulsion from Paradise [appendix #13].) And Saturn has been "appointed" to be the one who established it because he is the outermost planet, nearest to the sphere of fixed stars.[45] "This planet was taken for the one who communicated motion to the Universe and who was, so to speak, its king"; this is what Schlegel reports of China (*L'Uranographie Chinoise*, pp. 628ff.).

Saturn does give the measures: this is the essential point. How are we to reconcile it with Saturn the First King, the ruler of the Golden Age who is now asleep at the outer confines of the world? The conflict is only apparent, as will be seen. For now it is essential to recognize that, whether one has to do with the Mesopotamian Saturn, Enki/Ea, or with Ptah of Egypt, he is the "Lord of Measures"—spell it *mē* in Sumerian, *parshu* in Akkadian, *maat* in Egyptian. And the same goes for His Majesty, the Yellow Emperor of China—yellow, because the element earth belongs to Saturn— "Huang-ti established everywhere the order for the sun, the moon and the stars."[46] The melody remains the same. It might help to

[44] Ex quo intellegi volunt, cum chaos esset, tempora non fuisse, siquidem tempus est certa dimensio quae ex caeli conversione colligitur. Tempus coepit inde; ab ipso natus putatur Kronos qui, ut diximus, Chronos est.

[45] It is not hidden from us that the indestructible laws of philology do not allow for the identification of Kronos and Chronos, although in Greece to do so "was customary at all times" (M. Pohlenz, in RE *11*, col. 1986). We have, indeed, no acute reason to insist upon this generalizing identification—the "name" of a planet is a function of time and con-stellation—yet it seems advisable to emphasize, on the one hand, that technical terminology has its own laws and is not subject to the jurisdiction of linguists, and to point, on the other hand, to one of the Sanskrit names of Saturn, i.e., "Kāla," meaning "time" and "death," and "blue-black" (A. Scherer, *Gestirnnamen bei den indogermanischen Völkern* [1953], pp. 84f.)—a color which suits the planet perfectly, all over the world—and to point, moreover, to a passage from the Persian *Minokheird* (West trans. in R. Eisler, *Weltenmantel und Himmelszelt* [1910], p. 410): "The creator, Auharmazd (Jupiter) produced his creation . . . with the blessing of Unlimited Time (Zurvan akarana)."

[46] M. Granet, *Chinese Civilization* (1961), p. 12.

understand the general idea, but particularly the lucubration of Proclus, to have a look at the figure drawn by Kepler, which represents the moving triangle fabricated by "Great Conjunctions," that is, those of Saturn and Jupiter. One of these points needs roughly 2,400 years to move through the whole zodiac. The next chapter will show why this is of high importance: here it suffices to point to one possible manner in which measures are given "continuously."

Saturn, giver of the measures of the cosmos, remains the "Star of Law and Justice" in Babylon,[47] also the "Star of Nemesis" in Egypt,[48] the Ruler of Necessity and Retribution, in brief, the Emperor.[49] In China, Saturn has the title "Génie du pivot," as the god who presides over the Center, the same title which is given to the Pole star.[50] This is puzzling at first, and so is the laconic statement coming from Mexico: "In the year 2-Reed Tezcatlipoca changed into Mixcouatl, because Mixcouatl has his seat at the North pole and, being now Mixcouatl, he drilled fire with the fire sticks for the first time." It is not in the line of modern astronomy to establish any link connecting the planets with Polaris, or with any star, indeed, out of reach of the members of the zodiacal system. Yet such figures of speech were an essential part of the technical idiom of archaic astrology, and those experts in ancient cultures who could not understand such idioms have remained completely helpless in the face of the theory. What has Saturn, the far-out planet, to do with the pole? Yet, if he cannot be recognized as the "genie of the pivot," how is it possible to support Amlodhi's claim to be the legitimate owner of the Mill?

[47] P. Jensen, *Die Kosmologie der Babylonier* (1890), p. 115; Meissner, *Babylonien und Assyrien* (1925), vol. 2, pp. 145, 410; P. F. Gössmann, *Planetarium Babylonicum* (1950), 230.

[48] Achilles Tatius, see A. Bouché-Leclerq, *L'Astrologie Grecque* (1899), p. 94; W. Gundel, *Neue Astrologische Texte des Hermes Trismegistos* (1936), pp. 260, 316.

[49] "The title *basileus* is stereotyped with Kronos" (M. Mayer, in Roscher s.v. Kronos, col. 1458; see also Cornford in J. E. Harrison's *Themis*, p. 254). For China, see G. Schlegel, *L'Uranographie Chinoise* (1875), pp. 361, 630ff. Even the Tahitian text "Birth of the Heavenly Bodies" knows it: "Saturn was king" (T. Henry, *Ancient Tahiti* [1928], pp. 359ff.).

[50] Schlegel, *L'Uranographie Chinoise*, pp. 525, 628ff.

CHAPTER IX

Amlodhi the Titan and His Spinning Top

Tops of different sorts, and jointed dolls,
and fair golden
apples from the clear-voiced Hesperides . . .

Orpheus the Thracian

Though I am not by nature rash or splenetic
Yet there is in me something dangerous
Which let thy wisdom fear . . .

Hamlet, Act V

A REASONABLE CASE has been made for the extreme antiquity and continuity of certain traditions concerning the heavens. Even if Amlodhi's Quern, the Grotte and the Sampo as individual myths cannot be traced back beyond the Middle Ages, they are derived in different ways from that great and durable patrimony of astronomical tradition, the Middle East.

Now it is time to locate the origin of the image of the Mill, and further, what its alleged breakup and the coming into being of the Whirlpool can possibly mean.

The starting place is Greece. Cleomedes (c. A.D. 150), speaking of the northern latitudes, states (*1.7*): "The heavens there turn around in the way a millstone does." Al-Farghani in the East takes up the same idea, and his colleagues will supply the details. They call the star Kochab, beta Ursae Minoris, "mill peg," and the stars of the Little Bear, surrounding the North Pole, and Fas al-rahha (the hole of the mill peg) "because they represent, as it were, a hole (the axle ring) in which the mill axle turns, since the axle of the equator (the polar axis) is to be found in this region, fairly close

to the star Al-jadi (he-goat, Polaris: alpha Ursae Minoris)." These
are the words of the Arab cosmographer al-Kazvini. Ideler comments:[1]

> *Kotb,* the common name of the Pole, means really the axle of the
> movable upper millstone which goes through the lower fixed one,
> what is called the "mill-iron." On this ambiguity is founded the anal-
> ogy mentioned by Kazvini. The sphere of heaven was imagined as a
> turning millstone, and the North Pole as the axle bearing in which the
> mill-iron turns ... *Fas* is explained by Giggeo ... as *rima, scissura* etc.
> ... The *Fas al-rahha* of our text, which stands also in the Dresden
> globe beside the North Pole of the Equator, should therefore repre-
> sent the axle bearing.

Farther to the east, in India, the *Bhagavata Purana* tells us how
the virtuous prince Dhruva was appointed as Pole star.[2] The particular "virtue" of the prince, which alarmed even the gods, is
worth mentioning: he stood on one leg for more than a month,
motionless. This is what was announced to him: "The stars, and
their figures, and also the planets shall turn around you." Accordingly, Dhruva ascends to the highest pole, "to the exalted seat of
Vishnu, round which the starry spheres forever wander, like the
upright axle of the corn mill circled without end by the labouring
oxen."

The simile of the oxen driven around is not alien to the West. It
has remained in our languages thanks to the Latin *Septemtriones,*
the seven threshing oxen of Ursa Major: "that we are used to calling the Seven Oxen," according to Cicero's translation of Aratus.

On a more familiar level there is a remark by Trimalchio in Petronius (*Satyricon* 39): "Thus the orb of heaven turns around like
a millstone, and ever does something bad." It was not a foreign idea

[1] Ludwig Ideler: *Untersuchung über den Ursprung und die Bedeutung der Sternnamen* (1809), pp. 4, 17.

[2] F. Normann, *Mythen der Sterne* (1925), p. 208. See now *The Srimad-Bhagavatam of Krishna-Dwaipayana Vyasa* 5.3 (trans. J. N. Sanyal, vol. 2, pp. 248f.):
"Just as oxen, fastened to a post fixed in the center of a threshing floor, leaving
their own station, go round at shorter, middle or longer distances, similarly fixed
on the inside and outside of the circle of time, stars and planets exist, supporting
themselves on Dhruva; and propelled by the wind, they range in every direction
till the end of a kalpa."

to the ancients that the mills of the gods grind slowly, and that the result is usually pain.

Thus the image travels far and wide by many channels, reaches the North by way of Celtic-Scandinavian transmission and appears in Snaebjörn's account of his voyage of discovery in the Arctic. There should be added to those enigmatic lines of his what is known now of the background in Scandinavian lore. The nine grim goddesses who "once ground Amlodhi's meal," working now that "host-cruel skerry quern" beyond the edge of the world, are in their turn only the agents of a shadowy controlling power called Mundilfoeri, literally "the mover of the handle" (appendix #15). The word *mundil*, says Rydberg, "is never used in the old Norse literature about any other object than the sweep or handle with which the movable millstone is turned,"[3] and he is backed by Vigfusson's dictionary which says that "mundil" in "Mundilfoeri" clearly refers to "the veering round or revolution of the heavens."

The case is then established. But there is an ambiguity here which discloses further depths in the idea. " 'Moendull' comes from Sanskrit 'Manthati,' " says Rydberg, "it means to swing, twist, bore (from the root *manth-*, whence later Latin *mentula*), which occurs in several passages in the *Rigveda*. Its direct application always refers to the production of fire by friction."[4]

So it is, indeed. But Rydberg, after establishing the etymology, has not followed up the meaning. The locomotive engineers and airplane pilots of today who coined the term "joy stick" might have guessed. For the Sanskrit Pra-mantha is the male fire stick, or churn stick, which serves to make fire. And Pramantha has turned into the Greeks' Prometheus, a personage to whom it will be necessary

---

[3] V. Rydberg, *Teutonic Mythology* (1907), pp. 581ff. *Webster's New International Dictionary*, 2d ed., lists "mundle": A stick for stirring. Obsolete except for dialectical use. (We are indebted for this reference to Mrs. Jean Whitnack.)

[4] To term it "friction" is a nice way to shut out dangerous terms: actually, the Sanskrit radical *math*, *manth* means drilling in the strict sense, i.e., it involves alternate motion (see H. Grassmann, *Wörterbuch zum Rig-Veda* [1955], pp. 976f.) as we have it in the famous Amritamanthana, the Churning of the Milky Ocean, and this very quality of India's churn and fire drill has had far-reaching influence on cosmological conceptions.

to come back frequently. What seems to be deep confusion is in reality only two differing aspects of the same complex idea. The lighting of fire at the pole is part of that idea. But the reader is not the first to be perplexed by an imagery which allows for the presence of planets at the pole, even if it were only for the purpose of kindling the "fire" which was to last for a new age of the world, that world-age which the particular "Pramantha" was destined to rule. The handle, "moendull," and the fire drill are complementary: both have had great developments which superimpose on each other and on a multitude of myths. The obstacles which imagination has to overcome are the associations which are connected spontaneously with "fire," that is, the real burning fire in chimney or hearth, and the kind of "fire" associated with the mentioned "joy stick." Both are irrelevant as far as cosmological terminology is concerned, but they lent the linguistic vehicle which was used to carry the ideas of astronomy and alchemy.

It should be stated right now that *"fire" is actually a great circle reaching from the North Pole of the celestial sphere to its South Pole*, whence such strange utterances as *Rigveda* 5.13.6: "Agni! How the felly[5] the spokes, thus you surround the gods." (Agni is the so-called "fire-god," or the personified fire.) The *Atharva Veda* says, moreover, that the fire sticks belong to the *skambha*,[6] the world's axis, the very *skambha* from which the Sampo has been derived (see above, p. 111).

The identity of the Mill, in its many versions, with heaven is thus universally understood and accepted. But hitherto nobody seems to have wondered about the second part of the story, which also occurs in the many versions. How and why does it always happen that this Mill, the peg of which is Polaris, had to be wrecked or unhinged? Once the archaic mind had grasped the forever-enduring rotation, what caused it to think that the axle jumps out of the hole? What memory of catastrophic events has created this

---

[5] The rim of the wheel in which the spokes fit.

[6] *10.8.20.* Cf. RV *10.24.4* and *10.184.3* with Geldner's remark that in this stanza of the *Atharva Veda* the fire sticks are treated as a great secret and attributed to *skambha*.

story of destruction? Why should Vainamoinen (and he is not the only one) state explicitly that another Mill has to be constructed (see p. 110)? Why had Dhruva to be appointed to play Pole star—and for a given cycle?[7] For the story refers in no way to the creation of the world. One might even ask, as the alternative solution to Rydberg's challenging "limb-grist," whether Bergelmer was not heaved in the same manner "upon the millstone," that is, appointed to play Pole star (see above, p. 92).

The simple answer lies in the facts of the case. The Pole star *does* get out of place, and every few thousand years another star has to be chosen which best approximates that position. It is well known that the Great Pyramid, so carefully sighted, is not oriented at our Pole Star but at alpha Draconis, which occupied the position at the pole 5,000 years ago. But, as has been mentioned above (*Intermezzo*, p. 66), it is the more difficult for moderns to imagine that in those far-off ages men could keep track of such imperceptible shifting, as many of them are not aware of the mere facts. As Dr. Alexander Pogo, the Palomar astronomer, has written in frustration: "I give up quoting further examples of the obstinate belief of our Egyptologists in the immobility of the heavenly pole."[8]

Yet there is quite a collection of myths to show that once upon a time it was realized that the sphere of fixed stars is not meant to circle around the same peg forever and ever. Several myths tell how Polaris is shot down, or removed in some other way. That is reserved for an appendix (#15).

Most of these myths, however, come under a misleading name. They have been understood to deal with the end of the world. But there are extremely few "eschatological" myths entitled to this label. For example, the Twilight of the Gods is understood as the world's end, yet there is unambiguous testimony to the contrary from the Völuspa and other chapters of the *Edda*. What actually comes to an and is *a* world, in the sense of a world-age. The catas-

---

[7] The *Vishnu Purana 1.12* (cf. *2.8*, p. 187 of the Wilson translation) betrays the Indian predilection for huge and unrealistic numbers and periods: Dhruva is meant to last one *kalpa*—4,320,000 years.

[8] "Zum Problem der Identifikation der nördlichen Sternbilder der alten Aegypter," *ISIS 16* (1931), p. 103.

trophe cleans out the past, which is replaced by "a new heaven and a new earth," and ruled by a "new" Pole star. The biblical flood was also the end of a world, and Noah's adventure is rehearsed in many traditions and many forms all over the planet. The Greeks knew of three successive destructions.

Coherence will be re-established in this welter of traditions if it is realized that what is referred to is that grandest of heavenly phenomena, the Precession of the Equinoxes. The phenomenon has been dealt with in the *Intermezzo* already, but it is essential enough to be taken up more than once. Being so slow, and in a man's age so imperceptible, it has been taken for granted[9] that no one could have detected the Precession prior to Hipparchus' alleged discovery of the phenomenon, in 127 B.C. Hipparchus discovered and proved that the Precession turns around the pole of the ecliptic.[10] It is said that it must have taken an almost modern instrumentation to detect the motion over the brief space of a century, and this is certainly correct. Nobody claims, however, that the discovery was deduced from observations during one century. And the shift of 1 degree in 72 years, piling up over centuries, will produce appreciable shifts in certain crucial positions, if the observers have enough intentness of mind and know how to keep records. The technique of observation was relatively simple. It was based on the heliacal rising of stars, which remained a fundamental feature in Babylonian astronomy. The telescope of early times, as Sir Norman Lockyer has said, was the line of the horizon. If you came to realize that a certain star, which was wont to rise just before the equinoctial sun, was no longer visible on that day, it was clear that the gears of heaven had shifted. If that star was the last one of a given zodiac figure, it meant that the equinox was moving into a new figure. Nor is there any doubt—as was already said—that far antiquity was already aware of the shifting of the Pole star. But was it capable of con-

---

[9] I.e., during the last hundred years, at least. In former times, when the Humanities had not yet been "infected" by the biological scheme of evolution, the scholars showed better confidence in the capacities of the creators of high civilization.

[10] See Ptolemy, *Syntaxis* 7.3 (Manitius trans., vol. 2, pp. 16f.). The magnitude calculated by Hipparchus and accepted by Ptolemy was 1 degree in 100 years.

The Precession of the Poles

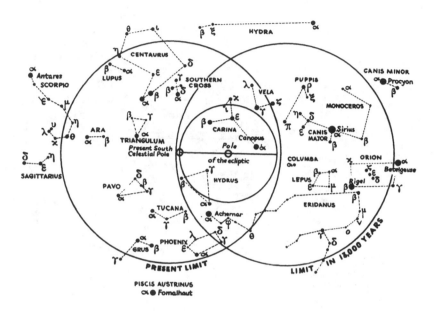

Limit of the northern circumpolar stars from Paris, today and in about 13,000 years' time (or again, 13,000 years ago). The center of the limit circle rotates around the pole of the ecliptic, and we have indicated its trajectory, which makes it possible to draw the limit circle for intermediate dates.

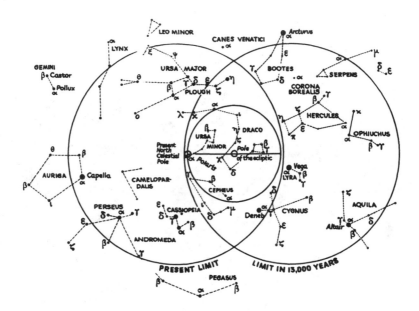

Limit of the southern circumpolar stars from Paris (that is to say, of those stars which never rise above the horizon at Paris in the course of diurnal motion), now and in 13,000 years' time (or 13,000 years ago).

necting both motions? This is where modern specialists, operating each from his own special angle of vision, have long hesitated.

What is the Precession? Very few have troubled to learn about it, yet to any man of our time, who knows the earth to be spinning around on her axis, the example of a spinning top with its inclined axis slowly shifting around in a circle makes the knowledge intuitive. Anyone who has played with a gyroscope will know all about the Precession. As soon as its axis is deflected from the vertical, the gyroscope will start that slow and obstinate movement around the compass which changes its direction while keeping its inclination constant. The earth, a spinning top with an axis inclined with respect to the sun's pull, behaves like a giant gyroscope, which performs a full revolution in 25.920 years.

Antiquity was not likely to grasp this, since dynamics came into this world only with Galileo. Hipparchus and Ptolemy could not understand the mechanism. They could only describe the motion. We must try to see through their eyes, and think only in terms of kinematics. Over a period of a thousand years ancient observers could discern in the secular shifting of the Great Gyroscope (it is here in fact that the word "secular" now used in mechanics originates) a motion through about ten degrees. Once attuned to the secular motion, they were able to detect, in the daily whirring of heaven around the pole, in its yearly turning in the round of the seasons, in the excruciatingly slow motion of the pole over the years, a point which seemed intrinsically more stable than the pole itself. It was the pole of the ecliptic,[11] often referred to as the Open Hole in Heaven because in that region there is no star to mark it. The symmetries of the machine took shape in their minds. And truly it was the time machine, as Plato understands it, the "moving image of eternity." The "mighty marching and the golden burning," cycle upon cycle, even down to shifts barely perceptible over the centuries, were the Generations of Time itself, the cyclical sym-

[11] See A. Bouché-Leclerq, *L'Astrologie Grecque* (1899), p. 122: "On sait que le pôle par excellence était pour les Chaldéens le pôle de l'écliptique, lequel est dans la constellation du Dragon." Cf. also A. Kircher, *Oedipus Aegyptiacus* (1653), vol. 2, pt. 2, p. 205: "Ponebant Aegyptii non Aequatorem, sed Zodiacum basis loco; ita ut centrum hemispherii utriusque non polum Mundi, sed polum Zodiaci referret."

bol of everlastingness: for, as Aristotle says, what is eternal is circular, and what is circular is eternal.

Yet this uniformly working time machine could be marked with important stations. The gyroscopic tilt causes a continual shifting of our celestial equator, which cuts the inclined circle of the ecliptic along a regular succession of points, moving uniformly from east to west. Now the points where the two circles cross are the equinoctial points. Hence the sun, moving on the ecliptic through the year, meets the equator on a point which shifts steadily with the years along the ring of zodiacal signs. This is what is meant by the Precession of the Equinoxes. They "precede" because they go against the order of the signs as the sun establishes this in its yearly march. The vernal equinox—we called it the "fiducial point" previously—which was traditionally the opening of spring and the beginning of the year, will take place in one sign after another. This gives great meaning to the change of signs in which the equinoctial sun happens to rise.

Some additional words of guidance may be called for here, where "signs" are mentioned—those "in" which the sun rises. For roughly two thousand years official terminology has used only zodiacal "signs," each of which occupies 30 degrees of the 360 degrees of the whole circle. These signs have the names of the zodiacal constellations, but constellations and signs are not congruent, the equinoctial sign ($= 1° - 30°$) being called Aries regardless of the constellation that actually rises before the equinoctial sun. In our time, the constellation rising heliacally on March 21 is Pisces, but the "sign" preserves the name Aries, and will continue to do so when in the future Aquarius rules the vernal equinox. So much for sign versus constellation.[12] As concerns the second ambiguous expression, namely, the

---

[12] Here, we leave out of consideration the much discussed question of exactly when signs of equal length were first introduced; allegedly it was very late (see below, p. 431, n. 1). The actual constellations differ widely in length—the huge Scorpion, e.g., covers many more degrees than 30, whereas the Ram is of modest dimensions. One would think that this lack of uniformity would have so hampered the ancient astronomers in making their calculations that they would have worked out a more convenient frame of coordinates in sheer self-defense.

sun's rising "in" a constellation (or a sign)—this means that the sun rises together with this constellation, making it invisible. There are several reasons for assuming that a constellation (and a planet which happened to be there), "in" which the equinoctial sun rose, was termed to be "sacrificed," "bound to the sacrificial post," and the like; and this might explain eventually why Christ, who opened the world-age in which Pisces rose heliacally in the spring, was understood as the sacrificed lamb. When Pisces is the last constellation visible in the east before sunrise, the sun rises together, i.e., "in," the constellation following next, the Ram.

Since the beginning of history, the vernal equinox has moved through Taurus, Aries, and Pisces. This is all that historic experience has shown mankind: a section of about one-quarter of the whole main circle of the machine. That it would come back full circle was at best an inference. It might also, for all men knew, have been part of an oscillation, back and forth, and in fact there were two schools of thought about it, and the oscillation theory seems to have exercised a greater attraction upon the mythographers of old.

For us, the Copernican system has stripped the Precession of its awesomeness, making it a purely earthly affair, the wobbles of an average planet's individual course. But *if*, as it appeared once, it was the mysteriously ordained behavior of the heavenly sphere, or the cosmos as a whole, then who could escape astrological emotion? For the Precession took on an overpowering significance. It became the vast impenetrable pattern of fate itself, with one world-age succeeding another, as the invisible pointer of the equinox slid along the signs, each age bringing with it the rise and downfall of astral configurations and rulerships, with their earthly consequences. Tales had to be told for the people about how successions of rulerships arose from an origin, and about the actual creation of the world, but for those in the know the origin was only a point in the precessional circle, like the $0 = 24$ of our dials. Our clocks today show two pointers only; but the tale-tellers of those bygone days, facing the immense and slow-moving machine of eternity, had to keep track of seven planetary pointers besides the daily revolution of the fixed sphere and of its secular motion in the opposite direction. All these motions meant parts of time and fate.

That things are not as they used to be, that the world is obviously going from bad to worse, seems to have been an established idea through the ages. The unhinging of the Mill is caused by the shifting of the world axis. Motion is the medium by which the wrecking is brought about. The Mill is "transported," be it Grotte or Sampo. The Grotte Song says explicitly that the giantesses first ground forth enemy action whereby the Mill was carried away and then, shortly afterwards, ground salt and wrecked the machine. It was the end of "Frodhi's peace"—the Golden Age. Even in Snaebjörn's famous lines, the grim goddesses "out at the edge of the world" are those "who ground Amlodhi's meal in ages past." They can hardly be doing it now, because the wrecked millstone is at the bottom of the sea, with its hole become the funnel of the whirlpool. So *that* Mill has been transferred to the waters, and it is now the sea itself which has become "Amlodhi's churn." The heavenly Mill has been readjusted, it goes on working in a new age. It churned once gold, then salt, and today sand and stones. But one cannot expect the rough Norse mythography to follow it in these legends, which are centered upon storm and wreck, the end of that first age.

Even Hesiod is far from clear about the early struggles and cataclysms; it is enough that in his *Works and Days* he marks a succession of five ages. A more coherent picture can only be built out of the convergence of several traditions, and this shall be the task of further chapters. But right now, there is at least one age designated as the first, when the Mill ground out peace and plenty. It is the Golden Age, in Latin tradition, *Saturnia regna*, the reign of Saturn; in Greek, Kronos. In this dim perplexing figure there is an extraordinary concordance throughout world myths. In India it was Yama; in the Old Persian *Avesta* it was Yima xsaēta,[13] a name which became in New Persian Jamshyd; in Latin Saeturnus, then Saturnus. Saturn

[13] See H. Collitz, "König Yima und Saturn," *Festschrift Pavry* (1933), pp. 86–108. See also A. Scherer, *Gestirnnamen bei den indogermanischen Völkern* (1953), p. 87.

or Kronos in many names had been known as the Ruler of the Golden Age, of that time when men knew not war and bloody sacrifices, not the inequality of classes—Lord of Justice and Measures, as Enki since Sumerian days, the Yellow Emperor and legislator in China.

If one wants to find the traces of his sunken Mill in classical mythology, they are not lacking.[14] The oldest is to be found where one would not expect it, in the Great Magical Papyrus of Paris, which is dated about the first half of the fourth century A.D.[15] In its recipes is the "much demanded Oracle of Kronos, the so-called Little Mill":

> Take two measures of sea salt and grind it with a handmill, repeating all the while the prayer that I give you, until the God appears. If you hear while praying the heavy tread of a man and the clanking of irons, this is the god that comes with his chains, carrying a sickle. Do not be afraid, for you are covered by the protection that I give you. Be wrapped in white linen such as the priests of Isis wear [here follow a number of magic rites]. The prayer to be said while grinding is as follows: I call upon thee, great and holy One, founder of the whole world we live in, who sufferest wrong at the hands of thy own son, thee whom Helios bound with iron chains, so that All should not come to confusion. Man-Woman, father of thunder and lightning, thou who rulest also those below the earth. [There follow more rites of protection, then the formula of dismissal]: Go, Lord of the World, First Father, return to your own place, so that the All remain well guarded. Be merciful, O Lord.[16]

[14] Although the *Telchines* are entitled to be investigated thoroughly, we can only mention them here: this strange family of "submarine magic spirits" and "demons of the depth of the sea"—they are followers of Poseidon in Rhodes—have invented the mill; i.e., their leader did so—*Mylas*, "the miller." Knowing beforehand, it was said, of the predestined flood which was to destroy Rhodes, these former inhabitants left for Lycia, Cyprus and Crete, the more so, as they also knew that Helios was going to take over the island after the flood. On the other hand, these envious creatures—they have the "evil eye," too—are accused of having ruined the whole vegetation of Rhodes by sprinkling it with Styx-water. As will come out later (see "Of Time and the Rivers," p. 200), the waters of Styx are not so easily to be had; that the *Telchines*, the "mill gods" (*theoi mylantioi*) had access to Styx proves beyond doubt that these earliest defoliators had turned, indeed, into citizens of the deep sea. See *Griechische Mythologie*, Preller-Robert (1964), vol. *1*, pp. 650ff.; M. Mayer, *Giganten und Titanen in der antiken Sage und Kunst* (1887), pp. 45, 98, 101; H. Usener, *Götternamen* (1948), pp. 198f.

[15] K. Preisendanz, *Papyri Graecae Magicae* (1928), vol. *1*, p. 64.

[16] 4.308ff., Preisendanz, vol. *1*, p. 173.

Sorcerers and conjurors are the most conservative people on earth. Theirs is not to reason why; they call upon the Power in terms they no longer understand, but they have to give an exact list of the archaic attributes of the fallen god, and even grind out sea salt from the Little Mill, the model of the whirlpool that marked his downfall. What had once been science has become with them pure technology, bent on preservation. A. Barb once coined a simile—he had revealed religion in mind, however, not science; dealing with the relation between magic practices and religion, he pointed to Matt. xxiv.28, Luke xvii.37: "Wheresoever the carcase is, there will the eagles be gathered together," and "Too many critical scholars have been ready to assume that the carcase is therefore a creation of the eagles. But eagles do not create; they disfigure, destroy and dispense what life has left, and we must not mistake the colourful display of decay for the blossoms and fruit of life."[17] Poignant as this image is, namely, in establishing the proper consecutio temporum, it leaves out of consideration the preserving function of magic and superstition: where would the historian of culture be left without those "eagles"?

For all the titles and attributes here listed, there is justification in archaic myth. Right here, only one point is of importance. The Lord of the Mill is declared to be Saturn/Kronos, he whom his son Zeus dethroned by throwing him off his chariot, and banished in "chains" to a blissful island, where he dwells in sleep, for being immortal he cannot die, but is thought to live a life-in-death, wrapped in funerary linen, until his time, say some, shall come to awaken again, and he will be reborn to us as a child.

[17] "St. Zacharias," *Journal of the Warburg and Courtauld Institutes 11* (1948), p. 95. It has not escaped his attention, by the way, that it should be vultures.

CHAPTER X

The Twilight of the Gods

T HERE WAS ONCE, then, a Golden Age. Why, how, did it come to an end? This has been a deep concern of mankind over time, refracted in a hundred myths, explained in so many ways which always expressed sorrow, nostalgia, despondency. Why did man lose the Garden of Eden? The answer has always been, because of some original sin. But the idea that man alone was able to commit sin, that Adam and Eve are the guilty ones, is not very old. The authors of the Old Testament had developed a certain conceit. Christianity then had to come to rescue and restore cosmic proportions, by insisting that God alone could offer himself in atonement.

In archaic times, this had seemed to be self-evident. The gods alone could run or wreck the universe. It is there that we should search for the origin of evil. For evil remains a mystery. It is not in nature. The faultless and all-powerful machine of the heavens should have yielded only harmony and perfection, the reign of justice and innocence, rivers flowing with milk and honey. It did, but that time did not last. Why did history begin to happen? History is always terrible. Philosophers from Plato to Hegel have offered their own lofty answer: pure Being was confronted of a necessity with Non-Being, and the result was Becoming, which is an uninsurable business. This was substantially the original answer of archaic times, but because of the lack of abstractions, it had to be derived in terms of heavenly motions.

Aristotle, the Master of Those Who Know, has cleared up this matter in a most important, yet little noted passage of Book Lambda of *Metaphysics* (1074b) where he talks about Kronos, Zeus, Aphrodite, etc.:

> Our forefathers in the most remote ages [*archaioi kai panpalaioi*] have handed down to their posterity a tradition, in the form [*schema*] of a myth, that these bodies are gods and that the divine encloses the whole of nature. The rest of the tradition has been added later in mythical form . . . ; they say that these gods are in the form of men or like some of the other animals . . . But if one were to separate the first point from these additions and take it alone—that they thought the first substances to be gods, one must regard this as an inspired utterance, and reflect that, while probably each art and each science has often been developed as far as possible and has again perished, these opinions, with others, have been preserved until the present like relics [*leipsana*] of the ancient treasure.

Aristotle, being a true Greek, cannot conceive of progress in our sense. Time proceeds for him in cycles of flowering and decay. But this absence of modern preconceptions had left his mind open to an ancient certainty. This certainty is what shines through the mist of ages and through a language dimly understood. It was attention to the events of heaven which shaped men's minds before recorded history; but since there was as yet no writing, these thoughts have receded, as astrophysicists would say, over the "event horizon." They can survive only through fragments of tale and myth because these made up the only technical language of those times.

Yet an enormous intellectual achievement is presupposed in this organization of heaven, in naming the constellations and in tracing the paths of the planets. Lofty and intricate theories grew to account for the motions of the cosmos. One would wonder about this obsessive concern with the stars and their motion, were it not the case that those early thinkers thought they had located the gods which rule the universe and with it also the destiny of the soul down here and after death.

In modern language, they had found the essential invariants where Being is. In paying respect to those forefathers, Aristotle shows himself clearly aware that his philosophical quest started with them.

One should pay attention to the cosmological information contained in ancient myth, information of chaos, struggle and violence.

They are not mere projections of a troubled consciousness: They are attempts to portray the forces which seem to have taken part in the shaping of the cosmos. Monsters, Titans, giants locked in battle with the gods and trying to scale Olympus are functions and components of the order that is finally established.

A distinction is immediately clear. The fixed stars are the essence of Being, their assembly stands for the hidden counsels and the unspoken laws that rule the Whole. The planets, seen as gods, represent the Forces and the Will: all the forces there are, each of them seen as one aspect of heavenly power, each of them one aspect of the ruthless necessity and precision expressed by heaven. One might also say that while the fixed stars represent the kingly power, silent and unmoving, the planets are the executive power.

Are they in total harmony? This is the dream that the contemplative mind has expressed again and again, that Kepler tried to fix by writing down the notes of his "Harmony of the Spheres," and that was consecrated in the "turning over" of the sky. This is the faith expressed by ancient thinkers in a Great Year, in which all the motions brought back all the planets to the same original configuration. But the computations created doubt very early and with it anxiety. Only rarely is there an explicit technical statement of those views. Here is one from the Egyptian Book of the Dead, Osiris speaking:

> "Hail, Thot! What is it that hath happened to the divine children of Nut? They have done battle, they have upheld strife, they have made slaughter, they have caused trouble: in truth, in all their doing the mighty have worked against the weak. Grant, O might of Thot, that that which the God Atum hath decreed (may be accomplished)! And thou regardest not evil nor art thou provoked to anger when they *bring their years to confusion and throng in and push to disturb their months;* for in all that they have done unto thee, they have *worked iniquity in secret!*"[1]

Thot is the god of science and wisdom; as for Atum, he precedes, so to speak, the divine hierarchy. Described only in metaphysical terms, he is the mysterious entity from which the All sprang: his

---

[1] Chapter 175, 1–8, W. Budge trans. The italics are ours.

name might be Beginning-and-End. He is thus the Presence and the secret Counsel whom one feels tempted to equate with the starry sky itself. His decree must be of immutable perfection. But here it appears that there are forces which have worked iniquity in secret. They appear everywhere, these forces, and regularly they are denounced as "overbearing," or "iniquitous," or both. But these "forces" are not iniquitous right from the beginning: they turn out to be, they *become* overbearing in the course of time. Time alone turns the Titans, who once ruled the Golden Age, into "workers of iniquity" (compare appendix #12). The idea of measure stated or implied will show the basic crime of these "sinners": it is the overreaching, overstepping of the ordained degree, and this is meant literally.[2] Says the *Mahabharata* about the Indian Titans, the Asura: "assuredly were the Asura originally just, good and charitable, knew the Dharma and sacrificed, and were possessed of many other virtues . . . But afterwards as they multiplied in number, they became proud, vain, quarrelsome . . . they made confusion in everything. Thereupon in the course of time . . ." they were doomed.[3]

Thus severe consequences must be expected when Gen. vi.1 commences with the formula, "when men began to multiply on the face of the earth . . ." And sure enough, ten verses later, Gen. vi.11, the time for grave decisions has come: "And God said to Noah, 'I have determined to make an end of all flesh!' " More outspoken is the 18th chapter of the Book of Enoch, where an Angel acts as Enoch's guide through the celestial landscape. In showing him the quarters destined for iniquitous personalities, the Angel tells Enoch: "These stars which roll around over the fire are those who, at rising time, overstepped the orders of God: *they did not rise at their ap-*

---

[2] It is only the careless manner in which we usually deal with precise terms that blocks the understanding: e.g., Greek *moira*, also written *moros*, is translated as "fate," "destiny," sometimes as "doom"; *moira* is one degree of the 360° of the circle; when we keep this in mind we understand better such lines as Od. *1.* 34–35, where Aegisthus is accused twice of having done deeds "*hyper moron,*" beyond degree. How could one overstep one's destiny? How could one be overmeasured against fate? This would invalidate the very concept of "destiny."

[3] V. Fausböll, *Indian Mythology according to the Mahabharata* (1902), pp. 40f.

pointed time. And He was wroth with them, and He bound them for 10,000 years until the time when their sin shall be fulfilled."[4]

Yet one must beware of simplifications. The wording, "assuredly were the Asura originally just, good and charitable," goes for the Titans, too, the forces of the first age of the world. But seen through the "eyeglasses" of the preceding state of things, Titans, Asura and their like had committed atrocities first. And so did Saturn, the "originator of times," and in the drastic measure he took to accomplish the "separation of the parents of the world," which stands for the falling apart of the axes of equator and ecliptic. Before this separation time did not exist. These "united parents"—heartlessly called "chaos" by Macrobius—resented the breaking up of the original eternity by the forces which worked iniquity in secret.[5] These forces as they appear in the *Enuma elish*, the so-called Babylonian Creation Epic, are the children of Apsu and Tiamat and they crowded in between their parents. "They disturbed Tiamat as they surged back and forth; yea, they troubled the mood of Tiamat. Apsu could not lessen their clamor . . . Unsavory were their ways, they were overbearing."[6]

Not having "multiplied" yet, this first generation of the world established the Golden Age under the rule of Him of many names —Enki, Yima, Freyr and many more. "But these sons whom he begot himself, great Heaven [*megas Ouranos*] used to call Titans [Strainers] in reproach, for he said that they strained and did presumptuously a fearful deed, and that vengeance for it would come afterwards," as Hesiod has it (*Theogony* 207–10).[7] And so it would, after their "multiplication," when they overstrained the

[4] E. Kautzsch, ed., *Die Apokryphen und Pseudoepigraphen des Alten Testaments* (1900), vol. 2, pp. 249f.

[5] There is no complete unanimity among mythographers, though; in Hesiod's *Theogony*, Gaia "rejoiced greatly in spirit" (173) when Kronos promised to do away with Father Ouranos according to Gaia's very own plan and advice.

[6] EE Tabl. *1.22–28* (E. Speiser trans.), ANET, p. 61.

[7] This translation by H. G. Evelyn-White (LCL) pays no regard to a "pun," a rather essential one, indeed. Hesiod makes use, side by side in these few lines, of both radicals from which "Titan" was supposed to have been derived: *titainō*, "to strain," and *tisis*, "vengeance."

measure. And it was bound to happen again when future generations would construct "forbidden ways to the sky,"[8] or build a tower which happened to be too high. The one secure measure, the "golden rope" of the solar year,[9] is stretched beyond repair. The equinoctial sun had been gradually pushed out of its Golden Age "sign," it had started on the way to new conditions, new configurations. This is the frightful event, the unexpiable crime that was ascribed to the Children of Heaven. They had nudged the sun out of place, and now it was on the move, the universe was out of kilter and nothing, nothing—days, months or years, the rising or setting of stars—was going to fall into its rightful place any more. The equinoctial point had nudged and nuzzled its way forward, in the very same way as a car with automatic gearshift will nuzzle its way forward unless we put it in neutral—and there was no way of putting the equinox in neutral. The infernal pushing and squeezing of the Children of Heaven had separated the parents, and now the time machine had been set rolling forever, bringing forth at every new age "a new heaven and a new earth," in the words of Scripture. As Hesiod says, the world had entered now the second stage, that of the giants, who were to wage a decisive battle with the restraining forces before their downfall.

The vision of a whole world-age with its downfall is given by the *Edda*. It comes in the very first poem, the Song of the Sibyl, the Völuspa, in which the prophetess Vala embraces past and future in adequately strange and obscure language. At the beginning of the Age of the Aesir, the gods gather in council, and give names to sun and moon, days and nights and seasons. They order the years and assign to the stars their places. On Idavollr (the "whirl-field"; *ida* = eddy), they establish their seat "in the Golden Age" and play checkers with golden pieces, and all is happiness until "the three awful maidens" come (this is another mystery).[10] But once before,

---

[8] *Claudianus* 26.69–71, speaking of the Aloads, who piled Ossa upon Olympus.

[9] See e.g., RV 5.85.5: "This great feat of the famous Asurian Varuna I shall proclaim who, standing in the air, using the Sun as an inch scale, measured the earth."

[10] The three maidens from Jötunheimr are not the Norns, this much can be safely said, but should be Gulveig the "thrice born," whom the Aesir killed "thrice, and still she is living" (Völuspa 8): one more "iniquity" asking for vengeance.

it is hinted, there has been a "world war" between Aesir and Vanir, which was terminated by a sharing of power. In a vision in which past and future blend in a flash, Vala sees the outcome and announces it to the "high and low children of Heimdal," that is, to all men. She asks them to open their eyes, to understand what the gods had to know: the breaking of the peace, the murder of Thjassi, Odin himself abetting the crime and nailing Thjassi's eyes to heaven. With this a curtain is lifted briefly over a phase of the past. For Thjassi belongs to the powers that preceded the Aesir. In Greek terms, the Titans came before the gods. The main Vana or Titanic powers (in Rydberg's thoughtful reconstruction) are the three brothers, Thjassi/Volund, Orvandil/Eigil, and Slagfin: the Maker, the Archer, and the Musician. This finally locates Orvandil the Archer, the father of Amlethus. He is one of the three "sons of Ivalde," just as their counterparts in the Finnish epic are the "sons of Kaleva."[11] And Ivalde, like Kaleva, is barely mentioned, never described, at least not under the name Ivalde: there is a glimpse at him under his other name, Wate. Like Kaleva, he is a meaningful void. But all this is of the past. The Sibyl's vision is projected toward the onrushing end. True, Loke has been chained in Hell since he brought about the death of Balder, the great Fenrir wolf is still fettered with chains, once cunningly devised by Loke himself, and they are made up of such unsubstantial things as the footfall of a cat, the roots of a rock, the breath of a fish, the spittle of a bird.[12]

Now the powers of the Abyss are beginning to rise, the world is coming apart. At this point Heimdal comes to the fore. He is the Warner of Asgard, the guardian of the Bridge between heaven and earth, the "Whitest of the Aesir," but his role, his freedom of action, is severely limited. He has many gifts—he can hear grass grow, he can see a hundred miles away—but these powers seem to

[11] Strange to say, the three brothers, Volund, Eigil and Slagfin, are called "synir Finnakonungs," i.e., "sons of a Finnish king" (J. Grimm, TM, p. 380).
[12] Again, strange to say, this very kind of "un-substance"— including the milk of Mother Eagle, and the tears of the fledglings—had to be provided for by Tibetan Bogda Gesser Khan, who also snared the sun.

remain ineffectual. He owns the Gjallarhorn, the great battle horn of the gods; he is the only one able to sound it, but he will blow it only once, when he summons the gods and heroes of Asgard to their last fight.

Nordic speculation down to Richard Wagner has dwelt with gloomy satisfaction on Ragnarok,[13] the Twilight of the Gods, which will destroy the world. There is the prediction in the Song of the Sibyl, and also in Snorri's *Gylfaginning:* when the great dog Garm barks in front of the Gnipa cave, when the Fenrir wolf breaks his fetters and comes from "the mouth of the river,"[14] his jaws stretching from heaven to earth, and is joined by the Midgard Serpent, then Heimdal will blow the Gjallarhorn, the sound of which reaches through all the worlds: the battle is on. But it is written that the forces of order will go down fighting to atone for the initial wrong done by the gods. The world will be lost, good and bad together. Naglfar, the ship of the dead, built with the nail parings of the living, will sail through the dark waters and bring the enemy to the fray. Then, adds Snorri:

> The heavens are suddenly rent in twain, and out ride in shining squadrons Muspel's sons, and Surt with his flaming sword, at the head of the fylkings.[15]

[13] For the etymology of *ragnarok*, see Cleasby-Vigfusson, *An Icelandic-English Dictionary*, in which *regin* (whence *ragna*) is defined as "the gods as the makers and rulers of the universe"; *rök* as "reason, ground, origin" or "a wonder, sign, marvel"; and *ragna rök* as "the history of the gods and the world, but especially with reference to the last act, the last judgment." The word *rökr*, a possible alternate to *rök*, is defined as "the twilight . . . seldom of the morning twilight," and "the mythological phrase, *ragna rökr*, the twilight of gods, which occurs in the prose Edda (by Snorri), and has since been received into modern works, is no doubt merely a corruption from *rök*, a word quite different from *rökr*."

Taking into consideration that the whole war between the Pandavas and the Kauravas, as told in the *Mahabharata*, takes place in the "twilight" between Dvapara and Kali Yuga, there is no cogent reason to dismiss Snorri's *ragna rökr* as a "corruption." But then, the experts also condemned Snorri's comparison between Ragnarok and the Fall of Troy: the logical outcome of their conviction that "poetry" is some kind of *creatio ex nihilo*, whence the one question never raised is whether the poets might not be dealing with hard scientific facts.

[14] *Lokasenna* 41; see also V. Rydberg, *Teutonic Mythology* (1907), p. 563.
[15] *Gylf.* 51.

All-engulfing flames come out with Surt "the Black," who kills Freyr, the Lord of the Mill. Snorri makes Surt "Lord of Gimle" and likewise the king of eternal bliss "at the southern end of the sky."[16] He must be some timeless force which brings destructive fire to the world; but of this later.

Hitherto all has been luridly and catastrophically and murkily confused as it should be. But the character of Heimdal raises a number of sharp questions. He has appeared upon the scene as "the son of nine mothers"; to be the son of several mothers is a rare distinction even in mythology, and one which Heimdal shares only with Agni in the *Rigveda*,[17] and with Agni's son Skanda in the *Mahabharata*. Skanda (literally "the jumping one" or "the hopping one") is the planet Mars, also called *Kartikeya*, inasmuch as he was borne by the Krittika, the Pleiades. The *Mahabharata*[18] insists on *six* as the number of the Pleiades as well as of the mothers of Skanda and gives a very broad and wild description of the birth and the installation of Kartikeya "by the assembled gods . . . as their generalissimo," which is shattering, somehow, driving home how little one understands as yet.[19]

The nine mothers of Heimdal bring to mind inevitably the nine goddesses who turn the mill. The suspicion is not unfounded. Two of these "mothers," Gjalp and Greip, seem to appear with changed

[16] *Gylf.* 17;, cf. R. B. Anderson, *The Younger Edda* (1880), p. 249. That Surt is Lord of Gimle is a particularly important statement; it will not be found in the current translations of Snorri, but only in the *Uppsala Codex*: "there are many good abodes and many bad; best it is to be in Gimle with Surt" (Rydberg, p. 651).

[17] RV *10.45.2* points to nine births, or mothers; *1.141.2* tells of the seven mothers of Agni's second birth. Most frequently, however, Agni has three "mothers," corresponding to his three birthplaces: in the sky, on the earth, in the waters.

[18] Mbh. *9.44–46* (Roy trans. vol. 7, pp. 130–43). It should be emphasized, aloud and strongly, that in Babylonian astronomy Mars is the *only* planetary representative of the Pleiades. See P. F. Gössmann, *Planetarium Babylonicum* (1950), p. 279: "In der Planetenvertretung kommt für die Plejaden nur Mars in Frage."

[19] The least which can be said, assuredly: Mars was "installed" during a more or less close conjunction of all planets; in Mbh. *9* 45 (p. 133) it is stressed that the powerful gods assembled "all poured water upon Skanda, even as the gods had poured water on the head of Varuna, the lord of waters, for investing him with dominion." And this "investiture" took place at the beginning of the Krita Yuga, the Golden Age.

names or generations as Fenja and Menja.[20] Rydberg claims Heimdal to be the son of Mundilfoeri. The story is then astronomical. Where does it lead? Thanks to the clues provided by Jacob Grimm, Rydberg and O. S. Reuter, and thanks to many hints hidden in the *Rigveda, Atharva Veda* and at other unexpected places, one can offer a probable conclusion: Heimdal stands for the world axis, the *skambha*. His head is the "measurer" (*mjotudr*) of the same measures that the Sibyl claims to understand: "Nine worlds I know, nine spaces of the measure-tree which is beyond (*fyr*) the earth." "Measure-tree" is the translation of *mjotvidr*,[21] which so-called poetic versions usually render as "world tree." The word *fyr* appears here again; it connotes priority; in this verse 2 of Völuspa it is translated as "below" in most of the cases. The question "who measures what?" would require an extensive analysis; here, with no need for so many details, it is important only to learn that Heimdal is honored by a second name, *Hallinskidi* (appendix #16). This name is said to mean a bent, bowed or slanted stake or post. To be bent or inclined befits the world axis and all that belongs to it, with the one exception of the observer who stands exactly at the terrestrial North Pole. Why not call it "oblique" or slanting right away?[22] Whether bent or oblique, Grimm rightly says that it

[20] For the names of these mothers, see *Hyndluljod* 38; for Gjalp and Greip, daughters of the giant Geirroed, see Snorri's *Skaldskaparmal* 2, and *Thorsdrapa*, broadly discussed by Rydberg (pp. 932–52), who established Greip as the mother of the "Sons of Ivalde." R. Much claims the identity of Geirroed with Surt ("Der germanische Himmelsgott," in *Ablandlungen zur germanische Philologie* [1898], p. 221). The turning up of a plurality of mothers in the ancient North, and in India (see also J. Pokorny, "Ein neun-monatiges Jahr im Keltischen," OLZ *21* [1918], pp. 130–33) might induce the experts eventually to reopen the trial of those perfectly nonsensical seven or nine, even fourteen, "motherwombs" which haunt the Babylonian account of the creation of man. (Cf. E. Ebeling, *Tod und Leben* [1931], pp. 172–77; E. A. Speiser (trans.), "Akkadian Myths and Epics," ANET, pp. 99f.; W. von Soden, Or. *26*, pp. 309ff.)

[21] O. S. Reuter, *Germanische Himmelskunde* (1934), pp. 236, 319. As concerns *mjotudr* (measurer) and its connection with Sanskrit *matar* and with *meter, mensar*, etc., see Grimm, TM, pp. 22, 1290. Reuter (p. 236) quotes *Lex. Poet. Boreale* 408, where *mjotudr* = fate.

[22] We have more of this mythological species of oblique posts or trees—e.g., the Rigvedic "sacrificial post"—and even Bears are not afraid to inhabit the one or the other. See F. G. Speck and J. Moses, *The Celestial Bear Comes Down to Earth: The Bear Sacrifice Ceremony of the Munsee-Mohican in Canada* (1945).

is "worthy of remark that Hallinskidi and Heimdal are quoted among the names for the ram."[23] Heimdal is the "watcher" of the much-trodden Bridge of the gods which finally breaks down at Ragnarok; his "head" measures the crossroads of ecliptic and equator at the vernal equinox in Aries,[24] a constellation which is called "head" also by Cleomedes,[25] and countless astromedical illustrations show the Ram ruling the head (Pisces the feet). Accordingly, one might say that the Sibyl addresses herself to "the high and low children of Aries."

Recalling Rigvedic Agni, son of seven to nine mothers like Heimdal, and remembering what has been said of "fire," that it means a great circle connecting the celestial poles, the scheme becomes more understandable. Heimdal stands for the equinoctial colure which "accompanies" the slowly turning, wholly abstract and invisible axis along the surface of the sphere. It will emerge presently that "axis" always means the whole "frame" of a world-age, given by the equinoctial and solstitial colures.[26] More understandable also becomes another epithet of Heimdal, namely, *Vindler*, of which Rydberg states (p. 595): "The name is a subform of *vindill* and comes from *vinda*, to twist or turn, wind, to turn anything around rapidly. As the epithet 'the turner' is given to that god who brought friction-fire (bore-fire) to man, and who is himself the personification of this fire, then it must be synonymous with 'the borer.' "

[23] TM, p. 234. Rydberg (p. 593) spells it: "In the old Norse poetry *Vedr* (wether, ram) Heimdal and the Heimdal epithet hallinskidi, are synonymous."

[24] A. Ohlmarks, *Heimdalls Horn und Odins Auge* (1937), p. 144, makes the god a he-goat. That would not be bad, either, if he is right, since Capella, alpha Aurigae, "capricious" all over, whether male or female, has the name "asar bardagi = Fight of the Aesir" (Reuter, p. 279). Of Auriga-Erichthonios we shall hear more in the future.

[25] Instead of "head" (*kephalos*), Nonnos calls Aries *mesomphalos*, "midnavel," of Olympus.

[26] It should be remarked that Snorri's identification (*Gylf.* 13) of the bridge Bifroest with the rainbow made scholars rush to rescue a definitely regular phenomenon from the hazardous existence which is allotted to a rainbow; they voted for the Milky Way instead. With this we are not likely to agree. See A. Ohlmarks, "Stellt die mythische Bifroest den Regenbogen oder die Milchstrasse dar?" *Medd. Lunds Astron. Observ.* (1941), ser. ii, no. 110, and Reuter, p. 284, quoting additional literature.

The Sibyl's prophecy does not end with the catastrophes, but it moves from the tragic to the lydic mode, to sing of the dawning of the new age:

Now do I see
the Earth anew
Rise all green
from the waves again . . .
Then fields unsowed
bear ripened fruit
All ills grow better.

Even if that generation of gods has perished, the younger ones remain: Balder and Hoder, also the two sons of Thor, and Vidar the son of Odin. The House of the Wise Vanir is not affected as a whole, even if Freyr fell in battle. As the Vanir belong to a past age, this crisis apparently does not concern them. There is in fact a certain perversely nightmarish or neurotic unreality about the tragedy as a whole. The Wolf's fetters were made of nothing but he was able to snap them only when the time came, when Odin and the Sun had to be devoured. The next instant, young Vidar kills the monster simply by thrusting his shoe down his throat (he has one shoe only, just like Jason). It is guilt and the ensuing chaos, more than actual forces, which dragged down the Establishment once the appointed time came, as decreed by fate and sounded on the Gjallarhorn.

What happens after (or happened, or will happen sometime, for this myth is written in the future tense), is told in the Völuspa, but it is also amplified in Snorri's *Gylfaginning* (53), a tale of a strange encounter of King Gylfi with the Aesir themselves, disguised as men, who do not reveal their identity but are willing to answer questions: "What happens when the whole world has burned up, the gods are dead, and all of mankind is gone? You have said earlier, that each human being would go on living in this or that world." So it is, goes the answer, there are several worlds for the good and the bad. Then Gylfi asks :"Shall any gods be alive, and shall there be something of earth and heaven?" And the answer is:

"The earth rises up from the sea again, and is green and beautiful and things grow without sowing. Vidar and Vali are alive, for neither the sea nor the flames of Surt have hurt them and they dwell on the Eddyfield, where once stood Asgard. There come also the sons of Thor, Modi and Magni, and bring along his hammer. There come also Balder and Hoder from the other world. All sit down and converse together. They rehearse their runes and talk of events of old days. Then they find in the grass the golden tablets that the Aesir once played with. Two children of men will also be found safe from the great flames of Surt. Their names, Lif and Lifthrasir, and they feed on the morning dew and from this human pair will come a great population which will fill the earth. And strange to say, the sun, before being devoured by Fenrir, will have borne a daughter, no less beautiful and going the same ways as her mother."

Then, all at once, concludes Snorri's tale wryly, a thunderous cracking was heard from all sides, and when the King looked again, he found himself on the open plain and the great hall had vanished.

The times and tenses are deliberately scrambled, but the statements, even if elliptical, are pregnant with ancient meaning. The rediscovery of the pieces of the game lying around in the grass, already told in the Völuspa, becomes clearer if one thinks of the *Rigveda,* where the gods themselves are said to go around like *ayas,* that is, casts of dice.[27] It becomes more understandable still when one considers that the name of the Indian world-ages (Yuga) has been taken from the idiom of dicing.[28] But both data could be dismissed as unrevealing were it forgotten that in several kinds of "proto-chess"—to use an expression of J. Needham—board games and dicing were combined: the number of eyes thrown by the dice determined the figure which was to be moved.[29] That this very rule was also valid for *tafl,* the board game mentioned in the Völuspa, has been shown by A. G. van Hamel.[30] Thus, the dice forced the hands of the chess player—a game called "planetary battles" by the Indians, and in 16th-century Europe still termed "Celestial War, or

[27] RV *10.116.9;* in *10.34.8,* the dice are called *vrata,* i.e., an organized "gang" under a king; the king is Rudra.
[28] Krita, Treta, Dvapara, Kali, this last one being the worst cast (which the Greeks termed "dog"). See H. Lüders, *Das Würfelspiel im Alten Indien* (1907), pp. 41, 63f.
[29] H. Lüders, p. 69; see also S. Culin, *Chess and Playing Cards* (1898), p. 857.
[30] "The Game of the Gods," *Arkiv für Nordisk Filologi 50* (1934), p. 230.

Astrologer's Game,"[31] whereas the Chinese chessboard shows the Milky Way dividing the two camps. Which goes to show that the Icelanders knew what they were talking about.

Finally, there is one remarkable and disturbing coincidence from the same direction. It is known that in the final battle of the gods, the massed legions on the side of "order" are the dead warriors, the "Einherier" who once fell in combat on earth and who have been transferred by the Valkyries to reside with Odin in Valhalla—a theme much rehearsed in heroic poetry. On the last day, they issue forth to battle in martial array. Says the *Grimnismal* (23): "Five hundred gates and forty more—are in the mighty building of Wal-halla—eight hundred 'Einherier' come out of each one gate—on the time they go out on defence against the Wolf."

That makes 432,000 in all, a number of significance from of old.

This number must have had a very ancient meaning, for it is also the number of syllables in the *Rigveda*. But it goes back to the basic figure 10,800, the number of stanzas in the *Rigveda* (40 syllables to a stanza) which, together with 108, occurs insistently in Indian tradition. 10,800 is also the number which has been given by Heraclitus for the duration of the Aiōn, according to Censorinus (*De die natali* 18), whereas Berossos made the Babylonian Great Year to last 432,000 years. Again, 10,800 is the number of bricks of the Indian fire-altar (Agnicayana).[32]

"To quibble away such a coincidence," remarks Schröder, "or to ascribe it to chance, is in my opinion to drive skepticism beyond its limits."[33] Shall one add Angkor to the list? It has five gates, and to each of them leads a road, bridging over that water ditch which surrounds the whole place. Each of these roads is bordered by a row of huge stone figures, 108 per avenue, 54 on each side, altogether 540 statues of Deva and Asura, and each row carries a huge Naga

---

[31] A. Bernhardi, "Vier Könige," BA *19* (1936), pp. 171f. See J. Needham, *Science and Civilization in China*, vol. 4, Pt. I: *Physics* (1962), p. 325, about a book on chess published in 1571 under the title *Uranomachia seu Astrologorum Ludus*.

[32] See J. Filliozat, "L'Inde et les échanges scientifiques dans l'antiquité," *Cahiers d'histoire mondiale 1* (1953), pp. 358f.

[33] F. R. Schröder, *Altgermanische Kulturprobleme* (1929), pp. 8of.

Horus and Seth in the act of drilling or churning. Horus has the head of a falcon; the head of Seth-Typhon shows the peculiar mixture of dog and ass which are characteristic of the so-called "Seth-beast." This feature is continuously mislabeled the "uniting of the two countries," whether Horus and Seth serve the churn or, as is more often the case, the so-called "Nile-Gods."

The "incomparably mighty churn" of the Sea of Milk, as described in the *Mahabharata* and *Ramayana*. The heads of the deities on the right are the Asura, with unmistakable "Typhonian" characteristics. They stand for the same power as the Titans, the Turanians, and the people of Untamo, in short, the "family" of the bad uncle, among whom Seth is the oldest representative, pitted against Horus, the avenger of his father Osiris.

The simplified version of the Amritamanthana (or Churning of the Milky Ocean) still shows Mount Mandara used as a pivot or churning stick, resting on the tortoise. And here, also, the head on the right has "Typhonian" features.

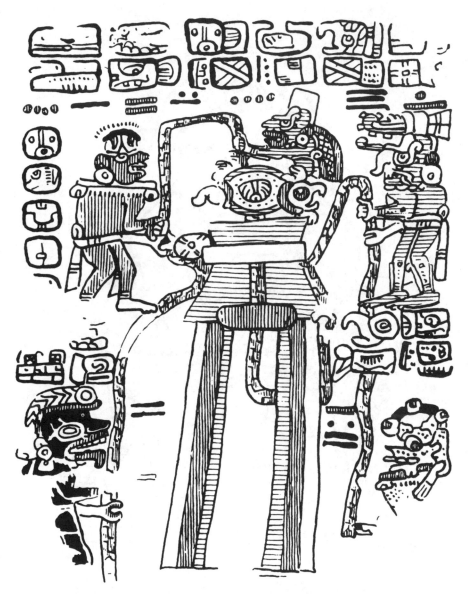

The Maya Codex Tro-Cortesianus presents the same event in a different "projec-tion." The illustration is harder to decode—as all Maya pictures are—but the rope, the tortoise, and the churn (indicating an hourglass?) can be made out, and "kin," the sign of the sun, glides along the serpent-rope.

serpent with nine heads. Only, they do not "carry" that serpent, they are shown to "pull" it, which indicates that these 540 statues are churning the Milky Ocean, represented (poorly, indeed) by the water ditch,[34] using Mount Mandara as a churning staff, and Vasuki, the prince of the Nagas, as their drilling rope. (Just to prevent misunderstanding: Vasuki had been asked before, and had agreeably consented, and so had Vishnu's tortoise avatar, who was going to serve as the fixed base for that "incomparably mighty churn," and even the Milky Ocean itself had made it clear that it was willing to be churned.) The whole of Angkor thus turns out to be a colossal model set up for "alternative motion" with true Hindu fantasy and incongruousness to counter the idea of a continuous one-way Precession from west to east.

Now there is a last paragraph in the *Gylfaginning*, which is usually considered an afterword, and its authorship is in doubt, for it is supposed that Snorri's *Edda* was completed by Olaf Hvitaskald (d. 1259), Snorri's nephew. In any case, this addition is somewhat out of the previous context, but it reinforces it:

> The Aesir now sat down to talk, and held their counsel, and remembered all the tales that were told to Gylfi. They gave the very same names that had been named before to the men and places that were there. This they did for the reason that, when a long time had elapsed, men should not doubt that those to whom the same names were given, were all identical. There was one who is called Thor, and he is Asa-Thor, the old. He is Oeku-Thor (Chariot-Thor) and to him are ascribed the great deeds by Hektor in Troy.

As for the rebirth of the world, another "Twilight" comes to mind. It is in the *Kumulipo*, a Polynesian cosmogonic myth from Hawaii. "Although we have the source of all things from chaos, it is a chaos which is simply the wreck and ruin of an earlier world."[35]

> *Now turns the swinging of time over on the burnt-out world*
> *Back goes the great turning of things upwards again*
> *As yet sunless the time of shrouded light;*

[34] R. von Heine-Geldern, "Weltbild und Bauform in Südostasien," in *Wiener Beiträge zur Kunst- und Kulturgeschicte 4* (1930), pp. 41f.
[35] R. B. Dixon, *Oceanic Mythology* (1910), p. 15.

Unsteady, as in dim moon-shimmer,
From out Makalii's night-dark veil of cloud
Thrills, shadow-like, the prefiguration of the world to be.[36]

So sang an Oceanian Empedocles long ago. The poem was drawn from very old royal tradition, just as Virgil had drawn his from the story of the Gens Julia, for the true original line of Hawaiian kings was supposed to come from Kane, the Demiurge God of the Pacific.

[36] A. Bastian, *Die Heilige Sage der Polynesier* (1881), pp. 69–121. Along with Roland B. Dixon, who translated the last three lines above, we have relied on the German of Bastian, who was an outstanding authority on Polynesian culture and language. Modern experts have their own way. M. Beckwith (*Hawaiian Mythology* [1940], p. 58) translates these lines thus: "At the time when the earth became hot/ At the time when the heavens turned about/At the time when the earth was darkened/To cause the moon to shine/The time of the rise of the Pleiades."

As concerns *Makalii* (Maori: *Matariki;* Micronesian and Melanesian dialects spell it *Makarika,* and the like), it is the name for the Pleiades, although more often we come across the phrase "the net of Makalii" (the correct form: Huihui-o-Matariki, i.e., the cluster of M.). The "person" Makalii, to whom this net belongs, as well as a second one (see p. 175) which we have reason to take for the Hyades, remains in the dark. See E. Tregear, *The Maori-Polynesian Comparative Dictionary* (1891) s.v. Matariki; N. B. Emerson, *Unwritten Literature of Hawaii* (1909), p. 17; M. W. Makemson, *The Morning Star Rises: An Account of Polynesian Astronomy* (1941), nos. 327, 380; Beckwith, p. 368; K. P. Emory, *Tuamotuan Religious Structures and Ceremonies* (1947), p. 61. For the Hyades and Pleiades as "celestial hunting nets" of the Chinese sphere, see G. Schlegel, *L'Uranographie Chinoise* (1875; repr. 1967), pp. 365–70.

CHAPTER XI

Samson Under Many Skies

> Why was my breeding ordered and prescribed
> As of a person separate to God,
> Designed for great exploits, if I must die
> Betrayed, captived, and both my eyes put out . . .
> O dark dark dark, amid the blaze of noon,
> Irrevocably dark, total eclipse
> Without all hope of day!
> O first-created beam, and thou great Word,
> "Let there be light, and light was over all"
> Why am I thus bereaved thy prime decree?
>
> *Samson Agonistes*

THE STORY OF Samson stands out in the Bible as a grand tissue of absurdities. Sunday school pupils must long have been puzzled about his weapon for killing Philistines. But there is much more to puzzle about (Judges xv):

15. And he found a new jawbone of an ass, and put forth his hand and took it, and slew a thousand men therewith.

16. And Samson said, with the jawbone of an ass, heaps upon heaps, with the jaw of an ass have I slain a thousand men.

17. And it came to pass, when he had made an end of speaking, that he cast away the jawbone out of his hand, and called that place Ramathle'hi.

18. And he was sore athirst, and called on the Lord, and said, Thou hast given this great deliverance into the hand of thy servant: and now shall I die for thirst, and fall into the hand of the uncircumcised?

19. But God clave an hollow place that was in the jaw, and there came water thereout; and when he had drunk, his spirit came again, and he revived: wherefore he called the name thereof En-hak'ko-re, which is in Le'hi unto this day.

20. And he judged Israel in the days of the Philistines twenty years.

The passage has been bowdlerized in the Revised Version to make it more plausible, but verse 18 is an unshakable reminder that this was not an ordinary bone, or even "the place" of it as suggested recently. For that jaw is in heaven. It was the name given by the Babylonians to the Hyades, which were placed in Taurus as the "Jaw of the Bull." If we remember the classic tag "the rainy Hyades" it is because Hyades meant "watery." In the Babylonian creation epic, which antedates Samson, Marduk uses the Hyades as a boomeranglike weapon to destroy the brood of heavenly monsters. The whole story takes place among the gods. It is known, too, that Indra's powerful weapon, Vajra, the Thunderbolt made of the bones of horse-headed Dadhyank, was not of this earth (see appendix #19).

The story is so universal that it must be seen as spanning the globe. In South America, where bulls were still unknown, the Arawaks, the Tupi, the Quechua of Ecuador spoke of the "jaw of the tapir," which was connected with the great god, Hunrakán, the hurricane, who certainly knows how to slay his thousands. In our sky, the name of the celestial Samson is Orion, the mighty hunter, alias Nimrod. He remains such even in China as "War Lord Tsan," the huntmaster of the autumn hunt, but the Hyades are changed there into a net for catching birds. In Cambodia, Orion himself became a trap for tigers; in Borneo, tigers not being available, pigs have to substitute; and in Polynesia, deprived of every kind of big game, Orion is found in the shape of a huge snare for birds. It is this snare that Maui, creator-hero and trickster, used to catch the Sun-bird; but having captured it, he proceeded to beat it up, and with what?—the jawbone of Muri Ranga Whenua, his own respected grandmother.

If one brings Samson—the biblical Shimshon—back to earth, he becomes a preposterous character, or rather, no character at all, except for his manic violence and his sudden passions. It comes as a shock, after reading that chaotic and whimsical life, to find: "And he judged Israel twenty years." For if anyone was bereft of judgment, it was this berserker. As Frazer remarks, one doubts whether he particularly adorned the bench. Yet there is a mysterious importance to his person. On him was piled a hoard of classic fairy tales, like "the man whose soul was placed elsewhere" (the external soul), and the insistent motif of fatal betrayal by women, the motif of Herakles and Llew Llaw Gyffes. More than that, he is an incongruous montage of nonhuman functions which could no longer be put together intelligibly, and were crowded together with cinematographic haste. Even his feats as a young Herakles, tearing a lion apart, change over in a flash to the generation of bees from a carcass, recalling the time-honored *bougonia* of the fourth book of Virgil's *Georgics*.

Of the many nonsense feats there are some which take particular relief from the context. Samson was displeased (Judges xiv–xv) because the wife of his heart, a Philistine, had given away to the children of her people the meaning of his riddle on the lion: "Out of the eater came forth meat, and out of the strong came forth sweetness," so that he was held to pay forfeit for his last bet.

xiv.19. And the Spirit of the Lord came upon him, and he went down to Ashkelon, and slew thirty men of them, and took their spoil, and gave change of garments unto them which expounded the riddle. And his anger was kindled, and he went up to his father's house.

20. But Samson's wife was given to his companion, whom he had used as his friend.

xv.1. But it came to pass within a while after, in the time of wheat harvest, that Samson visited his wife with a kid; and he said, I will go in to my wife into the chamber. But her father would not suffer him to go in.

2. And her father said, I verily thought that you hadst utterly hated her; therefore I gave her to your companion. Is not her younger sister fairer than she? Take her, I pray thee, instead of her to your companion.

3. And Samson said concerning them, Now shall I be more blameless than the Philistines, though I do them a displeasure.

4. And Samson went and caught three hundred foxes, and took firebrands, and turned tail to tail, and put a firebrand in the midst between two tails.

5. And when he had set the brands on fire, he let them go into the standing corn of the Philistines, and burnt up both the shocks, and also the standing corn, with the vineyards and olives.

6. Then the Philistines said, Who hath done this? And they answered, Samson, the son in law of the Timnite, because he had taken

his wife, and given her to his companion. And the Philistines came up, and burnt her and her father with fire.

7. And Samson said unto them, Though ye have done this, yet will I be avenged of you, and after that I will cease.

8. And he smote them hip and thigh with a great slaughter: and he went down and dwelt in the top of the rock Etam.

Leaving the great Shimshon there sitting in the top of the rock, a brief interlude before he goes out again on his own wayward, rash and splenetic way to provoke his enemies, one is moved to reflection.

To catch and corral three hundred foxes, and tie them in pairs by the tail, just to work off a spite, seems more the daydream of a juvenile delinquent or a Paul Bunyan or a "Starke Hans" than the feat of a warrior. It is as if Scripture had remembered that he had to stand out as a great hunter, but had misplaced the occasion of his hunts. After all, lions are not to be found behind every hedge-row, and foxes might do, if only to annoy. But we know from Ovid (*Fasti 4*, 631ff.) that in April, at the feast of Ceres, foxes with burning fur were chased through the Circus. This might be the real context. The modern "fertility rite" explanations are so futile that it might be more to the point to be reminded of the three hundred elite "dogs" that Gideon recruited for his band, and which still stand unexplained. One should also consider a more important occasion to which attention has been drawn by Felix Liebrecht: the "Sada-Festival," during which animals were kindled and chased, burning, through the whole Iranian countryside. This, however, would lead back to Firdausi's Book of Kings, and beyond that to the whole problem of Kynosoura, that cannot be tackled at this point because it calls for an examination of all that was implied by the starting of celestial fires.

But the main theme of the story will appear more clearly if it is transposed in an utterly different narrative convention, the adventures of Susanowo the Japanese god. They are found in the Japanese Scriptures, in this case the *Nihongi*, compiled about the 8th century A.D. but going back to unknown times. They are the full equivalent of what the Bible was in our recent past, and even more, for "this body of legend, folklore to us but credible history to the people of the archipelago, is tangled in the roots of everything Japanese." The quotation is from Post Wheeler, who prepared the latest edition of the Japanese mythical corpus. To quote him further: "In no other land do we find a people's sacred legend so interknit with the individual's daily thoughts and life. Its episodes peer at us from every nook and byway. The primeval myth of the slaughter of the Eight-Forked-Serpent by the deity Brave-Swift-Impetuous-Male, brother of Bright-Shiner the Sun-Goddess, is pictured on Japan's paper currency. I have seen it produced au grand sérieux at Tokyo's Imperial Theatre, in the same week as one of Ibsen's tragedies and a Viennese light opera."[1]

Most of Hebrew mythology wears the hempen homespun of peasants and patriarchs from Palestine. Japanese myth bears the mark of an already refined perverse feudal world, back of which there is the baroque elegance and fantasy of late Chinese culture. With this premise, here is the story of the Japanese Samson, Susanowo, whose name means Brave-Swift-Impetuous-Male. No better set of attributes for Mars; he is also officially a god, since his sister Amaterasu, the sun-goddess, is still today the worshiped ancestor of the Imperial dynasty; the courtly precedences are neatly established. The hero need no longer masquerade as a boor from the tribe of Dan who raged in Ashkelon and destroyed himself in Gaza.

Now Susanowo was banished from the sky for having thrown the hind part of his backward-flayed piebald stallion in the weaving hall of his sister Amaterasu. These sudden discourteous gestures seem to be part of the code: Enkidu had thus thrown the hind quarter of the Bull from Heaven in the face of Ishtar, but here there is the additional code feature (it is code) of the backward-flayed animal. Susanowo's gesture caused the Sun-lady to withdraw in anger into a cave: the world was plunged into darkness. The 80,000 gods assembled in the Milky Way to take counsel, and at last came upon a device to coax the Sun out of the cave and end the great blackout. It was a low-comedy trick, part of the stock-in-

[1] P. Wheeler, *The Sacred Scriptures of the Japanese* (1952), pp. vf.

trade that is used to coax Rā in Egypt, Demeter in Greece (the so-called Demeter Agelastos or Unlaughing Demeter) and Skadi in the North—obviously another code device.[2]

Now light was restored to the world, but on earth the hero-god moving out of the darkness had nowhere to lay his head; he wandered around and succeeded in killing the "Eight-Forked-Serpent," thus saving a damsel.

Afterwards he arranged "The Drawing of the Lands," and the sowing of more land, giving the islands the shape which they have now. Finally, Brave-Swift-Impetuous-Male, having traveled about the limits of the sky and the earth, even to the Sky-Upright-Limiting-Wall, dwelt on Mount Bear-Moor and finally went to the Lower World, also called the Nether-Distant-Land.

To this his place came a Jason, namely the Kami (Divine Prince) Great-Land-Master, looking for some helpful device against his brothers, "the 80 Kami" who had succeeded in killing him several times (Sky-Producer revived him). Before reaching the house, he married Susanowo's daughter, Princess-Forward, and this Medea was to support him faithfully, so that he survived the different "stations" which Susanowo had prepared for him[3] as proper guest rooms: the fire, the snake-house, the centipede-and-wasp-house (Dostoyevsky's Svidrigailov must have been a great seer):

> Then Brave-Swift-Impetuous-Male, having shot a humming arrow into the midst of a great grass-moor, sent him to fetch it, and when he had entered the moor, set fire to it on all sides. But when Great-Land-Master found no place of exit, there came a mouse which said, "The inside is hollowly hollow; the outside is narrowly narrow." Even as it spoke thus, he trod on the spot, and falling into the hollow, hid himself until the fire had burned over, when the mouse brought him the humming arrow in its mouth, and the arrow's feathers were brought in like manner by its young ones.

[2] The obscene dance of old Baubo, also called Iambe in Eleusis, parallels the equally unsavory comic act of Loke in the *Edda*. The point in all cases is that the deities must be made *to laugh* (cf. also appendix #36).

[3] For a comparison of the sequence of troublesome caves, holes, or "houses" that heroes of the Old World as well as of the New World have to pass through, see L. Frobenius, *Das Zeitalter des Sonnengottes* (1904), pp. 371f.

Now his wife, Princess-Forward, weeping, made preparation for the funeral, and her father, deeming Great-Land-Master dead, went out and took stand on the moor, but he found his guest standing there, who brought the arrow and gave it to him. Then the great Kami Susanowo took him into the palace and into a great-spaced room, where he made Great-Land-Master pick the lice from his head, among which were many centipedes. His wife, however, gave him aphananth berries and red earth, and he chewed up the berries and spat them out with the red earth which he held in his mouth, so that the great Kami, believing him to be chewing and spitting out the centipedes, began to feel a liking for him in his heart and fell asleep.

Then Great-Land-Master bound Brave-Swift-Impetuous-Male's hair fast to the palace rafters, and blocking up the door with a five-hundred-man-lift rock, took his wife Princess-Forward on his back, possessed himself of the Kami's great life-preserving sword, his bow-and-arrows, and his Sky-speaking lute, and fled. But the Sky-speaking lute smote against a tree so that the earth resounded, and the great Kami [Susanowo] started from sleep at the sound and pulled down the palace.

While he was freeing his hair from the rafters, however, Great-Land-Master fled a long way; so pursuing after him to the Level-Pass-of-the-Land-of-Night, and gazing on him from afar, Brave-Swift-Impetuous-Male called out to him, saying, "With the great, life-preserving sword and the bow-and-arrow which you carry, pursue your low-born brethren till they crouch on the hill-slopes and are swept into the river currents! And do you, fellow! make good your name of Great-Land-Master, and your name of Spirit-of-the-Living-Land, and making my daughter Princess-Forward your chief wife, make strong the pillars of your palace at the foot of Mount Inquiry in the lowest rock bottom, and rear its crossbeams to the Plain-of-the-High-Sky, and dwell there!"

Then, bearing the great sword and book, Great-Land-Master pursued and scattered the eighty Kami, saying, "They shall not be permitted within the circle of the blue fence of mountains." He pursued them till they crouched on every hill-slope, he pursued them till they were swept into every river, and then he began to rule the Land. (Therefore the place where he overtook them was called Come-Overtake.)[4]

Later on, the "Genesis" part of the *Nihongi* will be shown to meet the requirements of archaic theory very exactly. Even incidents that seem like minor embellishments, the little mouse in her

4 Wheeler, pp. 44f.

burrow, are really recurrent elements in the ancient fugue. Because it is necessary to deal with one theme at a time, much of the tale of Susanowo appears wildly arbitrary although no more so than that of Samson. Also the narrative is confusingly interwoven with other classic plots, recognizably those of Theseus and the Argonauts. And yet there Susanowo is, a maker of darkness at noon, Samson strength-in-hair, who "went away with the pin of the beam, and with the web," walking off with rafters and rocks and gates and posts, pulling down a palace (his own, for a change), smiting and scattering low-born workers of iniquity "not to be permitted again within the circle of the blue fence." But the *Nihongi* shows the ampler scheme in which the old order is smashed and the new foundation of an order is undertaken: "make strong the pillars of your palace at the foot of Mount Inquiry in the lowest rock-bottom, and rear its crossbeams to the Plain-of-the-High Sky, and dwell there."

The god has not only judged and apportioned, he has also established and sowed for the future in his capacity as the new king of the Underworld; he has gone to sleep in his Ogygia, and appointed his successor as ruler of the new age. Further, the Great-Land-Master had to procure something in the Nether-Distant-Land (in Japan the dead go down there by land with countless windings, whereas the whirlpool in the ocean is good only for transporting there the "sinful dirt"). He had been sent there to get "counsel" from Susanowo (who identified him at the first glance as: "This is the Kami Ugly-Male-of-the-Reed-Plains"), he eventually got it, and added to it the precious life-preserving sword which Susanowo had found in the tail of the Eight-Forked Dragon, and the "bow-and-arrows," and his Orphic Sky-speaking lute, not to forget Princess-Forward. A complicated affair. But the Great-Land-Master undeniably plays a Jupiter role against Susanowo's Mars, the more so, as his beloved Princess-Forward turns out to be extremely jealous.

Now, after this Far Eastern interlude, Samson's own tragedy can be seen in better focus (Judges xvi):

19. And [Delilah] made him sleep upon her knees; and she called for a man, and she caused him to shave off the seven locks of his head; and she began to afflict him, and his strength went from him.

20. And she said, The Philistines be upon thee, Samson. And he awoke out of his sleep, and said, I will go out as at other times before, and shake myself. And he wist not that the Lord was departed from him.

21. But the Philistines took him, and put out his eyes, and brought him down to Gaza, and bound him with fetters of brass; and he did grind in the prison house. [Appendix #17]

22. Howbeit the hair of his head began to grow again after he was shaven.

23. Then the lords of the Philistines gathered them together for to offer a great sacrifice unto Dagon their god, and to rejoice: for they said, Our god hath delivered Samson our enemy into our hand.

24. And when the people saw him, they praised their god: for they said, Our god hath delivered into our hands our enemy, and the destroyer of our country, which slew many of us.

25. And it came to pass, when their hearts were merry, that they said, Call for Samson, that he may make us sport. And they called for Samson out of the prison house; and he made them sport; and they set him between the pillars.

26. And Samson said unto the lad that held him by the hand, Suffer me that I may feel the pillars whereupon the house standeth, that I may lean upon them.

27. Now the house was full of men and women; and all the lords of the Philistines were there; and there were upon the roof about three thousand men and women, that beheld while Samson made sport.

28. And Samson called unto the Lord, and said, O Lord God, remember me, pray thee, and strengthen me, I pray thee, only this once, O God, that I may be at once avenged of the Philistines for my two eyes.

29. And Samson took hold of the two middle pillars upon which the house stood, and on which it was borne up, of the one with his right hand, and of the other with his left.

30. And Samson said, Let me die with the Philistines. And he bowed himself with all his might; and the house fell upon the lords, and upon all the people that were therein. So the dead which he slew at his death were more than they which he slew in his life.

Such is the great story, and it has gone through innumerable variations.

The general design of the tragedy is obviously faulty, more even than most Bible narratives which are superbly indifferent to such considerations. If Samson had been bred as "a person separate to God," by the care of the Lord "who sought an occasion against the Philistines," he does not compare with chiefs like Joshua and Gideon. He remains, mythically speaking, a misguided missile. Most great feats of the mythistorical past would not have rated the attention of news media, but Samson's achievements make so little sense, even on the micro-scale of Palestine power politics, that Milton finds it hard to justify the ways of God to man. Certain "central" events like the fall of royal houses, whether in Greece or Babylon or Denmark, are capable of a truer and deeper reverberation. That is why great motifs like "darkness at noon" and "pulling down the edifice" combine into a larger theme, obviously cosmic, which is here obscured. The *Nihongi* is truer to this larger style.

In the arabesque of interlaced motifs, one can mark those where the theme of "pulling down the structure" is in evidence. The powerful Maori hero Whakatau, bent on vengeance,

> laid hold of the end of the rope which had passed round the posts of the house, and, rushing out, pulled it with all his strength, and straightaway the house fell down, crushing all within it, so that the whole tribe perished, and Whakatau set it on fire.[5]

This is familiar. At least one such event comes down dimly from history. It happened to the earliest meetinghouse of the Pythagorean sect, and it is set down as a sober account of the outcome of a political conflict, but the legend of Pythagoras was so artfully constructed in early times out of prefabricated materials that doubt is allowable. The essence of true myth is to masquerade behind seemingly objective and everyday details borrowed from known circumstances. However that may be, in many other stories the destruction of the building is linked with a *net*. Saxo's Amlethus does not pull down pillars; he reappears at the banquet set by the king for his own supposed funeral, like Great-Land-Master himself. He throws the knotted carpet net prepared by his mother over the drunken crowd and burns down the hall. In Japan the parallel does not go farther than that but it has its own relevance nevertheless. It suggests the fall of the House of Atreus. The net thrown by

[5] See Sir George Grey, *Polynesian Mythology* (1956; 1st ed. 1855), pp. 97f.

Clytemnestra over the king struggling in his bath cannot have come in by chance. But this is an uncertain lead as yet.

The Sacred Book of the ancient Maya Quiche, the famous *Popol Vuh* (the Book of Counsel) tells of Zipacna, son of Vucub-Caquix (=Seven Arara). He sees 400 youths dragging a huge log that they want as a ridgepole for their house. Zipacna alone carries the tree without effort to the spot where a hole has been dug for the post to support the ridgepole. The youths, jealous and afraid, try to kill Zipacna by crushing him in the hole, but he escapes and brings down the house on their heads. They are removed to the sky, in a "group," and the Pleiades are called after them (appendix #18).

Then there is a true avenger-of-his-father, the Tuamotuan Tahaki, who, after long travels, arrives in the dark at the house of the goblin band who tortured his father. He conjures upon them "the intense cold of Havaiki" (the other world) which puts them to sleep.

> Then Tahaki gathered up the net given to him by Kuhi, and carried it to the door of the long house. He set fire to the house. When the goblin myriads shouted out together "Where is the door?" Tahaki called out: "Here it is." They thought it was one of their own band who had called out, and so they rushed headlong into the net, and Tahaki burned them up in the fire.[6]

What the net could be is known from the story of Kaulu. This adventurous hero, wanting to destroy a she-cannibal, first flew up to Makalii the great god, and asked for his nets, the Pleiades and the Hyades, into which he entangled the evil one before he burned down her house.[7] It is clear who was the owner of the nets up there. The Pleiades are in the right hand of Orion on the Farnese Globe,[8] and they used to be called the "lagobolion" (hare net). The Hyades were for big game.[9]

At the end of this far-ranging exploration, it is fair now to ask, who could Samson have been? Clearly a god, and a planetary

---

[6] J. F. Stimson, *The Legends of Maui and Tahaki* (1934), pp. 51, 66.
[7] A. Fornander, *Hawaiian Antiquities* (1916–1920), vol. 4, pp. 350f.; vol. 5, p. 368.
[8] R. Eisler, *Orpheus the Fisher* (1921), pp. 25f.
[9] G. Schlegel, *L'Uranographie Chinoise* (1967), pp. 351–58, 365–70.

Power, for such were the gods of old. As Brave-Swift-Impetuous-Male, as the Nazirite Strong One, he has all the countersigns that belong to Mars, and to none other. Clearly, while trying to draw the concluding episode of the investigation of Amlethus-Kronos, King of the Cosmic Mill, something else has come into view, the new and formidable personage of Mars—or Ares as the Greeks called him. He will come back more than once. Yet there is no question but that the name of Samson comes up quite spontaneously in connection with the Sampo, the original quern. It was clearly and unequivocally within the Amlethus design. At this point, the intrusion of this new planetary Power must be recognized. Even Susanowo substitutes for Kronos in his very reign of the Underworld. It would have been desirable to present the Powers separately, and each in his own shape, as will be done farther on. But the many-threaded tale has its own rules, and this exemplifies an important one. There are no Powers more diverse than Saturn and Mars; yet this is not the only time they will appear as a confusing and unexplained doublet of the two.

One of the motifs, destruction, is often associated with the Amlethus figure. The other belongs more specifically to Mars. There is a peculiar blind aspect to Mars, insisted on in both Harranian and Mexican myths. It is even echoed in Virgil: "*caeco Marte*." But it does not stand only for blind fury. It must be sought in the Nether World, which will come soon. Meanwhile, here is the first presentation of the double figure of Mars and Kronos. In Mexico, it stands out dreadfully in the grotesque forms of the Black and the Red Tezcatlipoca. There is a certain phase in the Great Tale, obviously, in which the wrecking powers of Mars unleashed make up a fatal compound with the avenging implacable design of Saturn. Shakespeare has, with his preternatural insight, alluded to both when he made Hamlet warn the raging Laertes before their final encounter:

> Though I am not by nature rash and splenetic
> Yet there is in me something dangerous
> Which let thy wisdom fear . . .

But obviously there is more, and what emerges here lifts the veil of a fundamental archaic design. The real actors on the stage of the universe are very few, if their adventures are many. The most "ancient treasure"—in Aristotle's word—that was left to us by our predecessors of the High and Far-Off Times was the idea that the gods are really stars, and that there are no others. The forces reside in the starry heavens, and all the stories, characters and adventures narrated by mythology concentrate on the active powers among the stars, who are the planets. A prodigious assignment it may seem for those few planets to account for all those stories and also to run the affairs of the whole universe. What, abstractly, might be for modern men the various motions of those pointers over the dial became, in times without writing, where all was entrusted to images and memory, the Great Game played over the aeons, a never-ending tale of positions and relations, starting from an assigned Time Zero, a complex web of encounters, drama, mating and conflict.

Lucian of Samosata, that most delightful writer of antiquity, the inventor of modern "science fiction," who knew how to be light and ironic on serious subjects without frivolity, and was fully aware of the "ancient treasure," remarked once that the ludicrous story of Hephaistos the Lame surprising his wife Aphrodite in bed with Mars, and pinning down the couple with a net to exhibit their shame to the other gods, was not an idle fancy, but must have referred to a conjunction of Mars and Venus, and it is fair to add, a conjunction in the Pleiades.

This little comedy may serve to show the design, which turns out to be constant: the constellations were seen as the setting, or the dominating influences, or even only the garments at the appointed time by the Powers in various disguises on their way through their heavenly adventures.

No one could deny, in the case of the Amlethus-Samson epiphany, that this fierce power, or momentary combination of powers, wears here the figure of Orion the blind giant, called also Nimrod the Hunter, brandishing the Hyades, working the Mill of the

Stars, like Talos, the bronze giant of Crete. For the feature which clinches the case has been named. Orion was blind, the only blind figure of constellation myth. He was said to have regained his sight eventually, as befits an eternal personage. But this is how legend portrays him, wading through the rushing flood of the whirlpool at his feet (where he will appear again), guided by the eyes of little Tom Thumb sitting on his shoulder, whose name, Kedalion, suggests a low-comedy occupation. But who are we to impose Mrs. Grundy on the assembly of heaven?

CHAPTER XII

Socrates' Last Tale

Al suo aspetto tal dentro mi fei
Qual si fe' Glauco nel gustar dell'erba
Che il fe' consorto in mar degli altri dei.

DANTE, *Paradiso* 1.67

WHAT A MAN has to say in the last hours of his life deserves attention. Most especially if that man be Socrates, awaiting execution in his jail and conversing with Pythagorean friends. He has already left the world behind, has made his philosophical will and is now quietly communing with his own truth. This is the close of the *Phaedo* (107D–115A), and it is expressed in the form of a myth. Strangely enough, innumerable commentators have not taken the trouble to scrutinize it, and have been content to extract from it some pious generalities about the rewards of the soul. Yet it is a thoughtful and elaborate statement, attributed to an authority whom Socrates (or Plato) prefers not to name. It is clothed in a strange physical garb. It is worth accepting Plato's suggestion to take it with due attention. Socrates is quietly moving into the other world, he is a denizen of it already, and his words stand, as it were, for a rite of passage:

"The story goes that when a man dies his guardian deity, to whose lot it fell to watch over the man while he was alive, undertakes to conduct him to some place where those who gather must submit their cases to judgment before journeying to the other world; and this they do with the guide to whom the task has been assigned of taking them there. When they have there met with their appropriate fates and waited the appropriate time, another guide brings them back here again, after many long cycles of time. The journey, then, is not as Aeschylus' Telephys describes it: he says that a single track leads to the other world, but I don't think that it is 'single' or 'one' at all. If it were, there would be no need of guides; no one would lose the way, if there were only one road. As it is, there seem to be many partings of the way and places where three roads meet. I say this, judging by the sacrifices and rites that are performed here. The orderly and wise soul follows on its way and is not ignorant of its surroundings; but that which yearns for the body, as I said before, after its long period of passionate excitement concerning the body and the visible region, departs only after much struggling and suffering, taken by force, with great difficulty, by the appropriate deity. When it arrives where the others are, the unpurified soul, guilty of some act for which atonement has not been made, tainted with wicked murder or the commission of some other crime which is akin to this and work of a kindred soul, is shunned and avoided by everyone, and no one will be its fellow-traveller or guide, but all by itself it wanders, the victim of every kind of doubt and distraction, until certain periods of time have elapsed, and when they are completed, it is carried perforce to its appropriate habitation. But that soul which has spent its life in a pure and temperate fashion finds companions and divine guides, and each dwells in the place that is suited to it. There are many wonderful places in the world, and the world itself is not of such a kind or so small as is supposed by those who generally discourse about it; of that a certain person has convinced me."

"How do you mean, Socrates?" asked Simmias. "I too have heard a great deal about the world, but not the doctrine that has found favour with you. I would much like to hear about it."

"Well, I don't think it requires the skill of a Glaucus[1] to relate my theory; but to prove that it is true would be a task, I think, too difficult for the skill of Glaucus. In the first place I would probably not even be capable of proving it, and then again, even if I did know how to, I don't think my lifetime would be long enough for me to give the explanation. There is, however, no reason why I should not tell you about the shape of the earth as I believe it to be, and its various regions."

"That will certainly do," said Simmias.

"I am satisfied," he said, "in the first place that if it is spherical and in the middle of the universe, it has no need of air or any other force of that sort to make it impossible for it to fall; it is sufficient by itself to maintain the symmetry of the universe and the equipoise of the

---

[1] Whoever this (unidentified) Glaucus is, he has nothing to do with the Glaucus of Anthedon mentioned in the epigraph, a fisherman who on eating a certain plant was overtaken by a transmutation and threw himself into the sea where he became a marine god.

earth itself. A thing which is in equipoise and placed in the midst of something symmetrical will not be able to incline more or less towards any particular direction; being in equilibrium, it will remain motionless. This is the first point," he said, "of which I am convinced!"[2]

"And quite rightly so," said Simmias.

"And again, I am sure that it is very big," he said, "and that we who live between the Phasis river and the pillars of Hercules inhabit only a small part of it, living round the coast of the sea like ants or frogs by a pond, while many others live elsewhere, in many similar regions. All over the earth there are many hollows of all sorts of shape and size, into which the water and mist and air have collected. The earth itself is a pure thing lying in the midst of the pure heavens, in which are the stars; and most of those who generally discourse about such things call these heavens the 'ether.' They say that these things I have mentioned are the precipitation of the 'ether' and flow continually into the hollows of the earth. We do not realize that we are living in the earth's hollows, and suppose that we are living up above on the top of the earth — just as if someone living in the middle of the sea-bed were to suppose that he was living on the top of the sea, and then, noticing the sun and the stars through the water, were to imagine that the sea was sky; through sluggishness and weakness he might never have reached the top of the sea, nor by working his way up and popping up out of the sea into this region have observed how much purer and more beautiful it is than theirs; nor even heard about it from anyone who had seen it. That is exactly what has happened to us: we live in a hollow in the earth, but suppose that we are living on top of it; and we call the air sky, as though this were the sky, and the stars moved across it. But the truth of the matter is just the same — through weakness and sluggishness we are not able to pass through to the limit of the air. If anyone could climb to the air's surface, or grow wings and fly up, then, as here the fishes of the sea pop their heads up and see our world, so he would pop his head up and catch sight of that upper region; and if his nature were such that he could bear the sight, he would come to realize that *that* was the real sky and the real light and the real earth. This earth of ours, and the stones, and all the region here is corrupted and corroded, just as the things in the sea are corroded by the brine; and in the sea nothing worth mentioning grows, and practically nothing is perfect—there are just caves and sand and indescribable mud and mire, wherever there is

[2] Thus far, this is Anaximander and his Principle of Sufficient Reason. But we cannot draw further conclusions: Socrates is, here, deep in his own myth already, and far beyond Ionian physics which, in his opinion, ought not to be taken seriously.

earth too, and there is nothing in any way comparable with the beautiful things of our world; but those things in the upper world, in their turn, would be seen far to surpass the things of our world. If it is a good thing to tell a story then you should listen, Simmias, and hear what the regions on the earth beneath the sky are really like."

"We should certainly very much like to hear this story, Socrates," said Simmias.

"In the first place, then, my friend, the true earth is said to appear to anyone looking at it from above like those balls which are made of twelve pieces of leather, variegated, a patchwork of colours, of which the colours that we know here—those that our painters use—are samples, as it were. There the whole earth is made of such colours, and of colours much brighter and purer than these: part of it is purple, of wondrous beauty, and part again golden, and all that part which is white is whiter than the whiteness of chalk or snow; and it is made up of all the other colours likewise, and of even more numerous and more beautiful colours than those that we have seen. Indeed these very hollows of the earth, full of water and of air, are said to present a kind of colour as they glitter amid the variety of all the other colours, so that the whole appears as one continuous variegated picture. And in this colourful world the same may be said of the things that grow up—trees and flowers and all the fruits; and in the same way again the smoothness and transparency and colours of the stars are more beautiful than in *our* world. Our little stones, these highly prized ones, sards and jaspers and emeralds and so on, are but fragments of those there; there, they say, *everything* is like this, or even more beautiful than these stones that we possess. The reason is that the stones there are pure, and not corroded or corrupted, as ours are, by rust and brine, as a result of all that has collected here, bringing ugliness and diseases to stones and to soil, and to animals and to plants besides. The earth itself, they say, is ornamented with all things, and moreover with gold and silver and all things of that sort. They are exposed to view on the surface, many in number and large, all over the earth, so that the earth is a sight for the blessed to behold. There are many living creatures upon it, including men; some live inland, some live round about the borders of the air as we do on the coasts of the sea, while others again live on islands encompassed by air near the mainland. In a word, what the water and the sea are to us, for our purposes, the air is to them; and what the air is to us, the 'ether' is to them. Their climate is such that they are free from illnesses, and live much longer than the inhabitants of our world, and surpass us in sight and hearing and wisdom and so on, by as much as the pureness of air surpasses that of water, and the pureness of 'ether' surpasses that of air. Moreover they have groves and temples sacred to the gods, in which the gods really dwell, and utterances and prophesies and visions of the gods; and other such means of intercourse are for them direct and face to face. And they see the sun and moon and stars as they really are, and their blessedness in other respects is no less than in these.

"This is the nature of the earth as a whole, and of the regions round about it, and in the earth, in the cavities all over its surface, are many regions, some deeper and wider than that in which we live, others deeper but with a narrower opening than ours, while others again are shallower than this one and broader. All of these are connected with each other by underground passages, some narrower, some wider—bored through in many different places; and they have channels along which much water flows, from one region to another, as into mixing-bowls; and they have, too, enormous ever-flowing underground rivers and enormous hot and cold springs, and a great deal of fire, and huge rivers of fire, and many rivers also of wet mud, some clearer, some denser, like the rivers of mud that flow before the lava in Sicily, and the lava itself; and they fill the several regions into which, at any given time, they happen to be flowing. They are all set in motion, upwards and downwards, by a sort of pulsation within the earth. The existence of this pulsation is due to something like this: one of the chasms of the earth is not only the biggest of them all, but is bored *right through* the earth—the one that Homer meant, when he said that it is 'very far off, where is the deepest abyss of all below the earth.' Homer elsewhere—and many other poets besides—have called this Tartarus. Now into this chasm all the rivers flow together, and then they all flow back out again; and their natures are determined by the sort of earth through which they flow. The reason why all these streams flow out of here and flow in is this, that this fluid has no bottom or resting place: it simply pulsates and surges upwards and downwards, and the air and the wind round about it does the same; they follow with it, whenever it rushes to the far side of the earth, and again whenever it rushes back to this side, and as the breath that men breathe is always exhaled and inhaled in succession, so the wind pulsates in unison with the fluid, creating terrible, unimaginable blasts as it enters and as it comes out. Whenever the water withdraws to what we call the lower region, the streams flow into the regions on the farther side of the earth and fill them, like irrigating canals; and whenever it leaves those parts and rushes back here, it fills the streams here afresh, and they when filled flow through their several channels and through the earth, and as each set of streams arrives at the particular regions to which its passages lead, it creates seas and marshes and rivers and springs; and then, sinking back again down into the earth, some encircling larger and more numerous regions, others fewer and smaller, these streams issue back into Tartarus again

—some of them at a point much lower down than that from which they were emitted, others only a little lower, but all flow in *below* the place from which they poured forth. Some flow into the same part of Tartarus from which they sprang, some into the part on the opposite side; and others again go right round in a circle, coiling themselves round the earth several times like snakes, before descending as low as possible and falling back again.

"It is possible to descend in either direction as far as the centre, but not beyond, for the ground on either side begins to slope *upwards* in the face of *both* sets of streams.

"There are many large streams of every sort, but among these many there are four that I would mention in particular. The largest, the one which flows all round in a circle furthest from the centre, is that which is called Oceanus; over against this, and flowing in the opposite direction, is Acheron, which flows through many desert places and finally, as it flows under the earth, reaches the Acherusian lake, where the souls of most of the dead arrive and spend certain appointed periods before being sent back again to the generations of living creatures. The third of these rivers issues forth between these two, and near the place where it issues forth it falls into a vast region burning with a great fire, and forms a marsh that is larger than our sea, boiling with water and mud. Thence it makes its way, turbulent and muddy, and as it coils its way round inside the earth it arrives, among other places, at the borders of the Acherusian lake, but it does not mix with the water of the lake; and having coiled round many times beneath the earth, it flows back at a lower point in Tartarus. This is the river they call Pyriphlegethon, and volcanoes belch forth lava from it in various parts of the world. Over against this, again, the fourth river flows out, into a region that is terrible and wild, all of a steely blue-grey colour, called the Stygian region; and the marsh which the river forms as it flows in is called the Styx. After issuing into this marsh and receiving terrible powers in its waters, it sinks down into the earth, and coiling itself round proceeds in the opposite direction to that of Pyriphlegethon, and then meets it coming from the opposite way at the Acherusian lake. The water of this river likewise mixes with no other, but itself goes round in a circle and then flows back into Tartarus opposite to Pyriphlegethon; and the name of this river, according to the poets, is Cocytus.

"Such is the nature of the world; and when the dead reach the region to which their divine guides severally take them, they first stand trial, those who have lived nobly and piously, as well as those who have not. And those who are found to have lived neither particularly well nor particularly badly journey to Acheron, and embarking on such vessels as are provided for them arrive in them at the lake. There they dwell and are purified; paying due penalties, they are absolved from any sins that they have committed, and receive rewards for their good deeds, each according to his merits. Those who are judged incurable because of the enormity of their crimes, having committed many heinous acts of sacrilege or many treacherous and abominable murders or crimes of that magnitude, are hurled by their fitting destiny into Tartarus, whence they never more emerge. Those who are judged to be guilty of crimes that are curable but nevertheless great —those, for example, who having done some act of violence to father or mother in anger live the rest of their lives repenting of their wickedness, or who have killed someone in other circumstances of a similar nature—must fall into Tartarus; but when they have fallen in and stayed there a year, the wave casts them forth—the murderers along Cocytus, those who have struck their fathers or mothers along Pyriphlegethon; and when they are being carried past the Acherusian lake, they shout and cry out to those whom they have murdered or outraged, and calling upon them beg and implore them to let them come out into the lake, and to receive them; and if they can prevail upon them, they come out and cease from their woes, but if not, they are carried again into Tartarus, and from there once more into the rivers, and they do not stop suffering this until they can prevail upon those whom they have wronged, for such is the sentence that the judges have pronounced upon them. Lastly, those who are found to have lived exceptionally good lives are released from these regions within the earth and allowed to depart from them as from a prison, and they reach the pure dwelling place up above and live on the surface of the earth; and of these, those who have sufficiently purified themselves by means of philosophy dwell free from the body for all time to come, and arrive at habitations even fairer than these, habitations that it is not easy to describe; and there is not time to make the attempt now. But for these reasons, Simmias, which we have discussed, we should do all in our power to achieve some measure of virtue and of wisdom during our lives, for great is the reward, and great the hope.

"No man of sense should affirm decisively that all this is exactly as I have described it. But that the nature of our souls and of their habitations is either as I have described or very similar, since the soul is shown to be immortal—that, I think, is a very proper belief to hold, and such as a man should risk: for the risk is well worth while. And one should repeat these things over and over again to oneself, like a charm, which is precisely the reason why I have spent so long in expounding the story now.

"For these reasons, then, a man should have no fears about his soul, if throughout his life he has rejected bodily pleasures and bodily adornments, as being alien to it and doing more harm than good, and has concentrated on the pleasures of learning, and having adorned his soul with adornments that are not alien to it, but appropriate—temperance and justice and courage and freedom and truth—continues to wait, thus prepared, for the time to come for him to journey to the other world. As for you, Simmias and Cebes and all you others, you will make your several journeys later, at an appointed time; but in my case, as a character in a tragedy might put it, Destiny is already summoning me; and it is almost time for me to go to the bath. I think it is better to have a bath before drinking the poison, and not to give the women the trouble of washing a corpse."[3]

The end has an invincible beauty, calm and serene, already shimmering with immortality, and yet preserving that light skeptical irony which makes "a man of sense" in this world. It puts the seal of confidence on what might otherwise be really an incantation that one repeats to himself in his last moments.

Readers who are insensitive to this magic will be tempted to dismiss the story as so much poetic nonsense. If Socrates, or rather Plato, is really talking of a system of rivers within the earth, then he obviously does not understand the first thing about hydraulics, and he has only let his fancy run wild. But looking again at the setting, one begins to wonder if he is referring at all to the earth as we understand it. He mentions a certain place where *we* live, and it looks like a marsh in a hollow or maybe like the bottom of a lake, full of rocks, and caverns, and sand, "and an endless slough of mud." The "true earth," which is like a ball of twelve colored pieces, is above us, and one may think instinctively that Plato refers to the upper limits of the stratosphere, but of course he has never heard of that. He is dealing with "another" world above us, and although there are some fantasies of lovely landscapes and animals and gems, it is in the "aether" as the Greeks understood it. It is above us, and centered like "our" place, whatever that is, on the center of the universe. There, the celestial bodies have become clear to the mind, and the gods are visible and present already. If they have "temples and houses in which they really dwell," these look very much like the houses of the zodiac. Although some features are scrambled for keeping up an impression of the wondrous, one

[3] R. S. Bluck trans. (1955), pp. 128–39.

suspects that this is heaven pure and simple. Then comes the unequivocal geometric countersign.

That world is a dodecahedron. This is what the sphere of twelve pieces stands for: there is the same simile in the *Timaeus* (55c), and then it is said further that the Demiurge had the twelve faces decorated with figures (*diazographōn*) which certainly stand for the signs of the zodiac. A. E. Taylor insisted rather prosily that one cannot suppose the zodiacal band uniformly distributed on a spherical surface, and suggested that Plato (and Plutarch after him) had a dodecagon in mind and they did not know what they were talking about. This is an unsafe way of dealing with Plato, and Professor Taylor's *suffisance* soon led him to grief. Yet Plutarch had warned him: the dodecahedron "seems to resemble both the Zodiac and the year."

> Is their opinion true who think that he ascribed a dodecahedron to the globe, when he says that God made use of its bases and the obtuseness of its angles, avoiding all rectitude, it is flexible, and by circumtension, like globes made of twelve skins, it becomes circular and comprehensive. For it has twenty solid angles, each of which is contained by three obtuse planes, and each of these contains one and the fifth part of a right angle. Now it is made up of twelve equilateral and equangular quinquangles (or pentagons), each of which consists of thirty of the first scalene triangles. Therefore it seems to resemble both the Zodiac and the year, it being divided into the same number of parts as these.[4]

In other words, it *is* stereometrically the number 12, also the number 30, the number 360 ("the elements which are produced when each pentagon is divided into 5 isosceles triangles and each of the latter into 6 scalene triangles")—the golden section itself. This is what it means to think like a Pythagorean.

Plato did not worry about future professional critics very much. He only provided a delectable image, and left them to puzzle it out. But what stands firm is the terminology. After the Demiurge had used the first four perfect bodies for the elements, says the *Timaeus*, he had the dodecahedron left over, and he used it for the

---

[4] *Quaestiones Platonicae* 5.1, 1003c (R. Brown trans.), in *Plutarch's Morals*, ed. W. W. Goodwin (1870), vol. 5, p. 433.

frame of the whole. There is no need to go into the reasons, geometrical and numerological, which fitted the "sphere of twelve pentagons," as it was called, for the role. What counts here: it was the whole, the *cosmos*, that was meant. Plato had stood by the original Pythagorean tradition, which called *cosmos* the order of the sun, moon and planets with what it comprised. As a free-roving soul, you can look at it "from above." (Archimedes in the *Sand-reckoner* still uses the term *cosmos* loosely in that sense, at least by way of a concession to old usage.)

To conclude: the "true earth" was nothing but the Pythagorean cosmos, and the rivers that flowed from its surface to the center and back can hardly be imagined as strictly terrestrial: although with that curious archaic intrication of earth and heaven which has become familiar and which makes great rivers flow from heaven to earth, it is not surprising to find oneself dealing with "real" fiery currents like Pyriphlegethon connected with volcanic fire. But where is Styx? Hardly down here, with its landscape of blue. And the immense storm-swept abyss of Tartaros is not a cavern under the ground, it belongs somewhere in "outer" space.

This is all the world of the dead, from the surface down and throughout. It localizes as poorly as the nether world of the *Republic*. The winding rivers which carry the dead and which go back on their tracks are suggestive more of astronomy than of hydraulics. The "seesaw" swinging of the earth (N.B.: it has to be the "true earth") might well be the swinging of the ecliptic and the sky with the seasons. There is no need now to go into the confusing earthy or infernal details of the description except to note that Numenius of Apamea, an important exegete of Plato, comes out flatly with the contention that the other world rivers and Tartaros itself are the "region of planets." But Proclus, an even more important and learned exegete, comes out flatly against Numenius.[5] Enough is known, indeed more than enough of the welter of oriental traditions on the Rivers of Heaven with their bewildering mixture of astronomical and biological imagery, which culminated in Anaximander's idea of the "Boundless Flow," the *Apeiron*, to see

[5] See F. Buffière, *Les Mythes d'Homère et la Pensée Grecque* (1956), p. 444.

whence early Greece got its lore. It can be left alone here. But Socrates is citing an Orphic version, whence his restraint in naming his authorities, and its strange entities, such as Okeanos and Chronos, deserve attention. What is meant here is not Kronos, Saturn, but really Chronos, Time. As concerns Okeanos, even Jane Harrison, who could hardly be accused of a tendency to search for the gods somewhere else than on the surface or in the interior of the earth, had to admit: "Okeanos is much more than Ocean and of other birth."[6] In her eyes he is "a daimon of the upper air." An important concession which may lead a long way.

We bypass for the moment the imposing work of Eisler, *Weltenmantel und Himmelszelt* (1910), an inexhaustible lode but one which provides more information than guidance. Onians' *Origins of European Thought* offers a more recent appraisal.[7] He compares Okeanos to Acheloüs, the primal river of water that "was conceived as a serpent with human head and horns." He goes on:

> The procreation element in any body was the psyche, which appeared in the form of a serpent. Okeanos was, as may now be seen, the primeval psyche and this would be conceived as a serpent in relation to procreative liquid . . . Thus we may see, for Homer, who refers allusively to the conception shared by his contemporaries, the universe had the form of an egg girt about by "Okeanos, who is the generation of All" . . . We can perhaps also better understand . . . why in this Orphic version [Frgs. 54, 57, 58 Kern] the serpent was called Chronos and why, when asked what Chronos was, Pythagoras answered that it was the psyche of the universe. According to Pherekydes it was from the seed of Chronos that fire and air and water were produced.

The great Orphic entity was Chronos Aiōn (the Iranian Zurvan akarana), commonly understood as "Time Unbounded," and in "Aiōn" Professor Onians sees "the procreative fluid with which the psyche was identified, the spinal marrow believed to take serpent form" and it may well be so, since these are timeless ideas which still live today in ophidic cults and in the "kundalini" of Indian Yoga. But Aiōn certainly meant "a period of time," and age, hence

[6] J. E. Harrison, *Themis* (1960), pp. 456f.

[7] P. B. Onians: *The Origins of European Thought about the Body, the Mind, the Soul, the World, Time, and Fate* (2d ed. 1953), pp. 249ff.

"world-age" and later "eternity," and there is no reason to think that the biological meaning must have been prior and dominant. It is known that for the Orphics Chronos was mated to Ananke, Necessity, which also, according to the Pythagoreans, surrounds the universe. Time and Necessity circling the universe, this is a fairly clear and fundamental conception; it is linked with heavenly motions independently from biology, and it leads directly to Plato's idea of time as "the moving image of eternity."

It would be helpful if historians of archaic thought would first present straight data, without pressing and squeezing their material into a shape that reflects their preconceived conclusion, that biological images must come first in "primitive" psychology, like all that is concerned with generation.

If one wants psychology, one can go back to Socrates in a very different phase of his life, where he is really talking psychology in the *Theaetetus* (152E): "When Homer sings the wonder of 'Ocean whence sprang the Gods and Mother Tethys,' does he not mean that all things are the offspring of flux and motion?" The question arises, would the ocean be an image of flux except for the tides? But Socrates' Aegean had no tides. The image comes to him from Hesiod's description of Okeanos (*Theogony* 790ff.): "With nine swirling streams he winds about the earth and the sea's wide back, and then falls into the main; but the tenth flows out from a rock, a sore trouble to the Gods." That dreaded tenth is the river of Styx. Jane Harrison was right. Okeanos is "of another birth" than our Ocean.

The authority of Berger can reconstruct the image.[8] The attributes of Okeanos in the literature are "deep-flowing," "flowing-back-on-itself," "untiring," "placidly flowing," "without billows." These images, remarks Berger, suggest silence, regularity, depth, stillness, rotation—what belongs really to the starry heaven. Later the name was transferred to another more earthbound concept: the actual sea which was supposed to surround the land on all sides. But the explicit distinction, often repeated, from the "main" shows that this was never the original idea. If Okeanos is a "silver-swirl-

[8] E. H. Berger, *Mythische Kosmographie der Griechen* (1904), pp. 1ff.

ing" river with many branches which obviously never were on sea or land, then the main is not the sea either, *pontos* or *thalassa*, it has to be the Waters Above. The Okeanos of myth preserves these imposing characters of remoteness and silence. He was the one who could remain by himself when Zeus commanded attendance in Olympus by all the gods. It was he who sent his daughters to lament over the chained outcast Prometheus, and offered his powerful mediation on his behalf. He is the Father of Rivers; he dimly appears in tradition, indeed, as the original god of heaven in the past. He stands in an Orphic hymn[9] as "beloved end of the earth, ruler of the pole," and in that famous ancient lexicon, the *Etymologicum magnum*, his name is seen to derive from "heaven."

[9] 83.7 (ed. Quandt, p. 55): terma philon gaiēs, archē polou.

CHAPTER XIII

Of Time and the Rivers

Di, quibus imperium est animarum, umbraeque silentes
Et Chaos et Phlegethon, loca tacentia late
Sit mihi fas audita loqui . . .

VIRGIL, Aeneid VI.264

SOCRATES' INIMITABLE HABIT of discussing serious things while telling an improbable story makes it very much worth while to take a closer look at his strange system of rivers.

It appears again in Virgil, almost as a set piece. The *Aeneid* is noble court poetry, and was not intended to say much about the fate of souls; one cannot expect from it the grave explicit Pythagorean indications of Cicero's *Dream of Scipio*. But while retaining conventional imagery and the official literary grand style which befitted a glorification of the Roman Empire, it repays attention to its hints, for Virgil was not only a subtle but a very learned poet. Thus, while Aeneas' ingress into Hades begins with a clangorous overture of dark woods, specters, somber caves and awesome nocturnal rites, which betoken a real descent into Erebus below the earth, he soon finds himself in a much vaguer landscape. *Ibant obscuri sola sub nocte per umbram* . . . "On they went dimly, beneath the lonely night amid the gloom, through the empty halls of Dis and his unsubstantial realm, even as under the grudging light of an inconstant moon lies a path in the forest."

The beauty of the lines disguises the fact that the voyage really is not through subterranean caverns crowded with the countless dead, but through great stretches of emptiness suggesting night space, and once the party has crossed the rivers and passed the gates of Elysium thanks to the magic of the Golden Bough, they are in a serene land "whence, in the world above, the full flood of Eridanus rolls amid the forest." Now Eridanus is and was in heaven—surely not, in this context, on the Lombard plain. And here also "an ampler aether clothes the meads with roseate light, and they know their own sun, and stars of their own." There is no mention here of the "pallid plains of asphodel" of Homeric convention. Those hovering souls, "peoples and tribes unnumbered," are clearly on the "true earth in heaven," for it is also stated that many of them await the time of being born or reborn on earth in true Pythagorean fashion. And there is more than an Orphic hint in the words of father Anchises: "Fiery is the vigour and divine the source of those life-seeds, so far as harmful bodies clog them not . . ." But when they have lived, and died, "it must needs be that many a taint, long linked in growth, should in wondrous use become deeply ingrained. Therefore, they are schooled with penalties, for some the stain of guilt is washed away under swirling floods or burned out in fire. Each of us suffers his own spirit." Some remain in the beyond and become pure soul; some, after a thousand years (this comes from Plato) are washed in Lethe and then sent to life and new trials.

This is exactly Socrates' belief. The words "above" and "below" are carefully equivocal, here as there, to respect popular atavistic beliefs or state religion, but this *is* Plato's other world.

When Dante took up Virgil's wisdom, his strong Christian preconceptions compelled him to locate the world of ultimate punishment "physically below." But his Purgatory is again above, under the open sky, and there is no question but that most, if not quite all, of Virgil's world is a Purgatory and definitely "up above" too. Socrates' strange descriptions have remained alive.

But Virgil offers even more than this. In the *Georgics* (*1.242f.*) it is said: "One pole is ever high above us, while the other, beneath our feet, is seen of black Styx and the shades infernal" (*sub pedibus Styx atra videt Manesque profundi*). What can it mean, except that

Styx flows in sight of the other pole? The circle which began with Hesiod is now closed.[1]

Great poets seem to understand each other, and to use information usually withheld from the public; Dante carries on where the *Aeneid* left off. As the wanderers, Dante and the shade of Virgil as his guide, make their way through the upper reaches of Hell (*Inferno* VII.102) they come across a little river which bubbles out of the rock. "Its water was dark more than grey-blue"; it is Styx, and as they go along it they come to the black Stygian marsh, where are immersed the souls of those who hated "life in the gentle light of the sun" and spent it in gloom and spite. Then they have to confront the walls of the fiery city of Dis, the ramparts of Inner Hell, guarded by legions of devils, by the Furies with the dreadful Gorgon herself. It takes the intervention of a Heavenly Messenger to spring the barred gates with the touch of his wand (a variant of Aeneas' Golden Bough) to admit the wanderers into the City of Perdition. As they proceed along the inner circle, there is a river of boiling red water, which eventually will turn into a waterfall plunging toward the bottom of the abyss (*baratro* = Tartaros). At this point Virgil remarks (XIV.85): "Of all that I have shown you since we came through the gate that is closed to none, there is nothing you have seen as notable as this stream, whose vapors screen us from the rain of fire." Those are weighty words after all that they have gone through; then comes the explanation, a rather farfetched one: "In the midst of the sea," Virgil begins, "there lies a ruined country which is called Crete, under whose king [i.e., Saturn] the world was without vice." There, at the heart of Mount Ida where Zeus was born of Rhea, there is a vast cavern in which sits a great statue. Dante is going back there to an ancient tradition

---

[1] The symmetry of both polar zones is clearly in the poet's mind. "Five zones comprise the heavens; whereof one is ever glowing with the flashing sun, ever scorched by his flames. Round this, at the world's ends, two stretch darkling to right and left, set fast in ice and black storms. Between them and the middle zone, two by grace of the Gods have been vouchsafed to feeble mortals; and a path is cut between the two [the ecliptic], wherein the slanting array of the Signs may turn" (*Georgics* 1.233–38).

to be found in Pliny, that an earthquake broke open a cavern in the mountain, where a huge statue was found, of which not much was said, except that it was 46 cubits high; but Dante supplies the description from a famous vision of Daniel, when the prophet was asked by King Nebuchadnezzar to tell him what he had seen in a frightening dream that he could not remember. Daniel asked God to reveal to him the dream:

> "Thou, O king, sawest, and beheld a great image. This great image, whose size was immense, stood before thee; and the form thereof was terrible. This image's head was of fine gold, his breast and his arms of silver, his belly and his thighs of bronze. His legs of iron, his feet part of iron, and part of clay.
>
> Thou sawest till that a stone was cut out without hands which smote the image upon his feet that were of iron and clay, and brake them to pieces . . . and the stone that brake the image became a great mountain and filled the whole earth."

At this point Dante takes leave of Daniel, and with that insouciance which marks him even when speaking of Holy Prophets, whom he treats as his equals, he dismisses the royal shenanigans in Babylon. His instinct tells him that the vision must really deal with older and loftier subjects, with the cosmos itself. Hence he proceeds to complete the vision on his own. The four metals stand for the four ages of man, and each of them except the gold (symbol of the Age of Innocence) is rent by a weeping crack from whence issue the rivers which carry the sins of mankind to the Nether World. They are Acheron, Styx and Phlegethon. We have noted that he describes the original flow of Styx as dark gray-blue, or steel-blue (*perso*), just as written in Hesiod and Socrates that he had never read. It may have come to him by way of Servius or Macrobius, no matter; what is remarkable is the strictness with which he preserves the dimly understood tradition of the lapis lazuli landscape of Styx, which will be seen to extend all over the world. As far as Phlegethon goes, the course of the stream follows quite exactly what Socrates had to say about Pyriphlegethon, the "flaming river." We have seen in the *Phaidon* a low-placed fiery region traversed by a stream of lava, which even sends off real fire to the surface of the earth. Whereas some interpreters thought it flowed through the interior of our earth, others transferred Pyriphlegethon, as well as the other rivers, into the human soul,[2] but there is little doubt that it was originally, as Dieterich has claimed,[3] a stream of fiery light in heaven, as Eridanus was. In any case, the flaming torrent, as the *Aeneid* calls it, goes down in spirals carefully traced in Dante's topography, until it cascades down with the other rivers to the icy lake of Cocytus, "where there is no more descent," for it is the center, the Tartaros where Lucifer himself is frozen in the ice. (Dante has been respectful of the Christian tradition which makes the universe, so to speak, diabolocentric.) But why does he say that the fiery river is so particularly "notable"?

G. Rabuse[4] has solved this puzzle in a careful analytical study of Dante's three worlds. First, he has found by way of a little-known manuscript of late antiquity, the so-called "Third Vatican Mythographer," that the circular territory occupied by the Red River in Hell was meant "by certain writers" to be the exact counterpart of the circle of Mars in the skies "because they make the heavens to begin in the Nether World" (3.6.4).[5] So Numenius was not wrong after all. The rivers are planetary. Dante subscribed to the doctrine and worked it out with a wealth of parallel features. Mars to him was important because, centrally placed in the planetary system, he held the greatest force for good or evil *in action*. As the central note in the scale, he can *also* become the harmonizing force. Both Hermetic tradition and Dante himself are very explicit about it. Is he the planetary Power that stands for Apollo? That requires future investigation.

[2] Cf. Macrobius, *Commentary on the Dream of Scipio* 1.10.11 (Stahl trans., p. 128): "Similarly, they thought that Phlegethon was merely the fires of our wraths and passions, that Acheron was the chagrin we experienced over having said or done something, . . . that Cocytus was anything that moved us to lamentation or tears, and that Styx was anything that plunged human minds into the abyss of mutual hatred."

[3] A. Dieterich, *Nekyia* (1893), p. 27.

[4] *Der kosmische Aufbau der Jenseitsreiche Dantes* (1958), pp. 58–66, 88–95.

[5] See *Scriptores Rerum Mythicarum Latini*, ed. G. H. Bode (1968; 1st ed. 1934) vol. *1*, p. 176: Eundem Phlegethontem nonnulli, qui a caelo infernum incipere autumant, Martis circulum dicunt sicut et Campos Elysios . . . circulum Jovis esse contendunt.

In the sky of Mars in his Paradise Dante placed the sign of the Cross ("I come to bring not peace but a sword"), a symbol of reckless valor and utter sacrifice, exemplified by his own ancestor the Crusader with whom he passionately identified. In the circle of Mars in Hell he placed, albeit reluctantly, most of the great characters he really admired, from Farinata, Emperor Frederick II, his Chancellor Pier della Vigna, to Brunetto, Capaneus and many proud conquerors. In truth, even Ulysses belongs in it, clothed in the "ancient flame," the symbol of his "ire" more than of his deceit. Virtues appear down there with the sign *minus;* they stand as fiery refusal, "blind greed and mad anger" which punish themselves: but their possessors are nonetheless, on the whole, noble, as, in the *Nihongi,* Brave-Swift-Impetuous-Male, the force of action par excellence. The meek may inherit the earth, but of the Kingdom of Heaven it has been written: *violenti rapiunt illud.* Christ stands in Dante as the Heliand, the conquering hero, the judge of the living and the dead: *rex tremendae majestatis.*

However that may be, the equivalence of above and below, of the rivers with the planets, remains established. By artifice Dante brings in at this point the figure of the Colossus of Crete, built out of archaic mythical material. By identifying the rivers with the world-ages, he emphasizes the identity of the rivers with Time: not here the Time that brings into being, but that of passing away—the Time that takes along with it the "sinful dirt," the load of errors of life as it is lived.

Men's minds in the 13th century were still very much alive to the archaic structure. But over and above this, by way of the Circle of Mars, an unexpected insight appears. Through the solemn Christian architecture of the poem, through the subtle logical organization, beyond the "veil of strange verses" and the intention they cloaked, there is a glimpse of what the author cared for more than he would say, of the man Alighieri's own existential choice. Poets cannot guard their own truth. Ulysses setting out toward the southwest in a last desperate attempt foreordained to failure by the order of things, trying to reach the "world denied to mortals," swallowed by the whirlpool in sight of his goal, *that* is the symbol. It is revealed not by the poet's conscious thinking, but by the power of the lines themselves, so utterly remote, like light coming from a "quasi-stellar object." To be sure, the Greek stayed lost in Hell for his ruthless resourcefulness in life as much as for his impiety: he was branded by Virgil as "dire and fierce"; the sentence was accepted. But he was the one who had willed to the last, even against God, to conquer experience and knowledge. His Luciferian loftiness remains in our memory more than the supreme harmony of the choirs of heaven.

To pursue this hazardous inquiry the first source is Homer, "the teacher of Hellas." The voyage of Odysseus to Hades is the first such expedition in Greek literature. It is undertaken by the weary hero to consult the shade of Teiresias about his future. The advice he eventually gets is startlingly outside the frame of his adventures and of the *Odyssey* itself (*10.508ff.*). It will be necessary to come back to this strange prophecy. But as far as the voyage itself goes, Circe gives the hero these sailing instructions:

> "Set your mast, hoist your sail, and sit tight: the North Wind will take you along. When you have crossed over the ocean, you will see a low shore, and the groves of Persephoneia, tall poplars and fruit-wasting willows; there beach your ship beside deep-eddying Okeanos, and go on yourself to the dank house of Hades.
>
> There into Acheron, the river of pain, two streams flow, Pyriphlege-thon blazing with fire, and Cocytos resounding with lamentation, which is a branch of the hateful water of Styx: a rock is there, by which the two roaring streams unite. Draw near to this, brave man, and be careful to do what I bid you. Dig a pit about one cubit's length along and across, and pour into it a drink-offering for all souls . . ."

Many centuries later, a remarkable commentary on this passage was made by Krates of Pergamon, a mathematician and mythographer of the Alexandrian period. It has been preserved by Strabo:[6] Odysseus coming from Circe's island, sailing to Hades and coming back, "must have used the part of the Ocean which goes from the hibernal tropic [of Capricorn] to the South Pole, and Circe helped

[6] *1.1.7.* Referring to *Odyssey 11.639–12.2.* See H. J. Mette, *Sphairophoiia* (1936), pp. 75, 250.

with sending the North Wind." This is puzzling geography, but astronomically it makes sense, and Krates seems to have had good reasons of his own to make the South Pole the objective.

The next information comes from Hesiod in his *Theogony* (775–814), and very obscure it is. After having heard of the "echoing halls" of Hades and Persephone, he says:

"And there dwells the goddess loathed by the deathless gods, terrible Styx, eldest daughter of backflowing Ocean. She lives apart from the gods in her glorious house vaulted over with great rocks and propped up to heaven all around with silver pillars. Rarely does the daughter of Thaumas, swift-footed Iris, come to her with a message over the sea's wide back.

"But when strife and quarrel arise among the deathless gods, and when any one of them who live in the house of Olympus lies, then Zeus sends Iris to bring in a golden jug the great oath of the gods from far away, the famous cold water which trickles down from a high and beetling rock.

"Far under the wide-pathed earth a branch of Oceanus flows through the dark night out of the holy stream, and a tenth part of his water is allotted to her. With nine silver-swirling streams he winds about the earth and the sea's wide back, and then falls into the main; but the tenth flows out from a rock, a sore trouble to the gods. For whoever of the deathless gods that hold the peaks of snowy Olympus pours a libation of her water and is forsworn, lies breathless until a full year is completed, and never comes near to taste ambrosia and nectar, but lies spiritless and voiceless on a strewn bed: and a heavy trance [*coma*] covers him.

"But when he has spent a long year in his sickness, another penance and a harder follows after the first. For nine years he is cut off from the eternal gods and never joins their councils or their feasts, nine full years. But in the tenth year he comes again to join the assemblies of the deathless gods who live in the house of Olympus. Such an oath, then, did the gods appoint the eternal and primeval water of Styx to be: and it spouts through a rugged place.

"And there, all in their order, are the sources and limits of the dark earth and misty Tartarus and the unfruitful sea [*pontos*] and starry heaven, loathsome and dank, which even the gods abhor. And there are shining gates and an immoveable threshold of bronze having unending roots and it is grown of itself. And beyond, away from all the gods, are the Titans, beyond gloomy Chaos."

This is Hesiod's version of the "Foundations of the Abyss." Its very details make confusion worse confounded, as befits the subject. The difficult word *ogygion*, translated often with "primeval," seems to designate things vaguely beyond time and place; one might say, the hidden treasure at the end of the rainbow. It was also the name for the resting place of Kronos, where he awaited the time of his return. But the paradoxical piling up of sources, limits, "unending roots" of earth, sea, heaven, and Tartaros too, remove any thought of a location at the earth's core, such as the cryptic words were popularly felt to convey. This "deeper than the deep" must have been "beyond the other side of the earth," and for reasons of symmetry, opposite to our pole. The shining gates and the immovable threshold of bronze are said elsewhere in the text to be the gates of Night and Day. Two centuries later, Parmenides, taking up Hesiod's allegorical language, speaks again of those gates of Night and Day.[7] But his image becomes clearer, as befits his invincibly geometrical imagination. The gates are "high up in the aether," leading to the abode of the Goddess of Truth and Necessity, and in his case too they must be at the Pole for explicit reasons of symmetry. We once tentatively suggested the North Pole, but many concurrent clues would indicate now the other one, the unknown, the Utterly Inaccessible. Hesiod says that Styx is a branch of Okeanos in heaven, "under the wide-pathed earth"; its dreaded goddess lives in a house "propped up to heaven all around with silver pillars," the water drips from a high rock. It can be reached by Iris coming with her rainbow "from snowy Olympus in the north." This *ogygion* region, that the gods abhor, has to be both under and beyond the earth; this should mean something like "on the other side of heaven." Homer never spoke of "above" and "below" in the strict sense. He simply made Odysseus land on a flat shore far away.

But what of the dreadful Styx which seems to be the core of the mystery? A river of death, even to gods, who can at least expect to come out of their coma at the appointed time. It is inimical to all

[7] G. de Santillana, *Prologue to Parmenides*, U. of Cincinnati, Semple Lecture, 1964. Reprinted in *Reflections on Man and Ideas* (1968), p. 82.

matter: it cracks glass, metal, stone, any container. Only a horse's hoof is proof against it, says the legend.[8] It adds that to men that water is inescapably lethal—except for one day of the year, which no one knows, when it becomes a water of immortality. This leads finally to the tragic ambiguity which gives drama to the tale of Gilgamesh and Alexander.

It is clear by now that the rivers are understood to be Time—the time of heaven. But images have their own logic. Where are the sources? The Colossus of Crete is Dante's own invention. Before him, there were many other accounts of the cracks from which flow the world-ages. Kai Khusrau, the Iranian Amlethus, was persecuted by a murderous uncle, established a Golden Age and then moved off in melancholy into the Great Beyond. The bad uncle, Afrasiyab, in his desperate efforts to seize the holy legitimacy, the "Glory" (*Hvarna*), had turned himself into a creature of the deep waters and plunged into the mystic Lake Vurukasha, diving after the "Glory." Three times he dove, but every time "this glory escaped, this glory went away": and at every try, it escaped through an outlet which led to a river to the Beyond. The name of the first outlet was Hausravah, the original Avestan name of Kai Khusrau. This should make the epoch and design tolerably plain.

An equally ancient story of three outlets comes from Hawaii. It appears in Judge Fornander's invaluable *Account* compiled a century ago, when the tradition was still alive. The "living waters" belong to Kane, the world-creating Demiurge or craftsman god. These waters are to be found in an invisible divine country, Pali-uli (= blue mountain), where Kane, Ku, and Lono created the first man, Kumu honua ("earth-rooted") or alternatively, the living waters are on the "flying island of Kane" (the Greek Hephaistos lived also on a floating island). Fornander describes the spring of this "living water" as

> beautifully transparent and clear. Its banks are splendid. It had three outlets: one for Kane, one for Ku, one for Lono; and through these outlets the fish entered the pond. If the fish of this pond were thrown

---

[8] Pausanias *8*.18.4–6; cf. J. G. Frazer, *Pausanias' Description of Greece 4*, pp. 248–56; also O. Waser, Roscher *4*, cols. 1574, 1576. Pausanias leaves it open whether or not Alexander was killed by means of Stygian water, as was fabled.

on the ground or on the fire, they did not die; and if a man had been killed and was after-wards sprinkled over with this water he did soon come to life again.[9]

An extraordinary theme has been set, that of the "revived fish" which will later show itself as central in Mid-Eastern myth, from Gilgamesh to Glaukos to Alexander himself. And then there are again the three outlets. These may help individualize the notion of Kane's "spring of life," which might otherwise sound as commonplace to folklorists as the Fountain of Youth. But something really startling can be found in good sound Pythagorean tradition. Plutarch in his essay "Why oracles no longer give answer" tells us (422E) that Petron, a Pythagorean of the early Italian school, a contemporary and friend to the great doctor Alcmaeon (c. 550 B.C.) theorized that there must be many worlds—183 of them. More about these 183 worlds was reported by Kleombrotos, one of the persons taking part in the conversation about the obsolescence of oracles, who had received his information from a mysterious "man" who used to meet human beings only once every year near the Persian Gulf, spending "the other days of his life in association with roving nymphs and demigods" (421A). According to Kleombrotos, he placed these worlds on an equilateral triangle, sixty to each side, and one extra at each corner. No further reason is given, but

> they were so ordered that one always touched another in a circle, like those who dance in a ring. The plain within the triangle is . . . the foundation and *common altar* to all these worlds, which is called the *Plain of Truth*, in which lie the designs, moulds, ideas, and invariable examples of all things which were, or ever shall be; and about there is *Eternity, whence flowed Time, as from a river, into the worlds.* Moreover, that the souls of men, if they have lived well in this world, do see these ideas *once in ten thousand years;* and that the most holy mystical ceremonies which are performed here are not more than a dream of this sacred vision.[10]

What is this? A mythical prefiguration of Plato's metaphysics? And why this *triangular* "Plain of Truth," which turns out again

[9] A. Fornander, *An Account of the Polynesian Race, Its Origin and Migrations* (1878), vol. *1*, pp. 72f. Cf. *Fornander Collection of Hawaiian Antiquities and Folk-Lore*, Mem. BPB Mus. *6* (1920), pp. 77f.

[10] Plutarch, *De defectu oraculorum*, ch. 22, 422BC.

to be a lake of Living Water? Pythagoreans did not care to explain. Nor did Plutarch.[11] But here is at least one original way of linking Eternity with the flow of Time. When it came to geometric fantasy, no one could outbid the Pythagoreans.

[11] Proclus (comm. on Plato's *Timaeus* 138B, ed. Diehl, BT, vol. *1*, p. 454) claimed this to be a "barbarous opinion" (doxē barbarikē). He shows no particular interest in the triangular plain of truth, alias our "lake" with its outlets, but he has more to say about the 180 "subordinate" and the 3 "leading" worlds (hēgemonas) at the angles, and how to interpret them. To which Festugière, in his (highly welcome and marvelous) translation of Proclus' commentary, remarks (vol. *2*, p. 336, n. 1): "On notera que Proclus donne à la fois moins et plus que Plutarque. A-t-il lu ces élucubrations pythagoriciennes elles-mêmes?"

CHAPTER XIV

The Whirlpool

Tre volte il fe' girar con tutte l'acque
alla quarta voltar la poppa in suso
e la prora ire in giu, com'altrui piacque
Infin che'l mar fu sopra noi richiuso.

DANTE, *Inferno*

DANTE KEPT to the tradition of the whirlpool as a significant end for great figures, even if here it comes ordained by Providence. Ulysses has sailed in his "mad venture" beyond the limits of the world, and once he has crossed the ocean he sees a mountain looming far away, "hazy with the distance, and so high I had never seen any." It is the Mount of Purgatory, forbidden to mortals.

"We rejoiced, and soon it turned to tears, for from the new land a whirl was born, which smote our ship from the side. Three times it caused it to revolve with all the waters, on the fourth to lift its stern on high, and the prow to go down, as Someone willed, until the sea had closed over us." The "many thoughted" Ulysses is on his way to immortality, even if it has to be Hell.

The engulfing whirlpool belongs to the stock-in-trade of ancient fable. It appears in the Odyssey as Charybdis in the straits of Messina—and again, in other cultures, in the Indian Ocean and in the Pacific. It is found there too, curiously enough, with the overhanging fig tree to whose boughs the hero can cling as the ship goes down, whether it be Satyavrata in India, or Kae in Tonga. Like Sindbad's magnetic mountain, it goes on in mariners' yarns through the centuries. But the persistence of detail rules out free invention. Such stories have belonged to the cosmographical literature since antiquity. Medieval writers, and after them Athanasius Kircher, located the *gurges mirabilis*, the wondrous eddy, somewhere off the coast of Norway, or of Great Britain. It was the Maelstrom, plus probably a memory of Pentland Firth.[1] It was generally in the direction north-northwest, just as Saturn's island, Ogygia, had been vaguely placed "beyond" the British Isles by the Greeks.

On further search this juxtaposition seems to be the result of the usual confusion between uranography and geography. There is frequently a "gap" in the northwest ("Nine-Yin" for the Chinese) of the heavens and inasmuch as the skeleton map of earth was derived from that of the sky, the gap was pinned down here as the Maelstrom, or Ogygia. Both notions are far from obvious, as are the localizations, and it is even more remarkable that they should be frequently joined.

For the Norse (see chapter VI) the whirlpool came into being from the unhinging of the Grotte Mill: the Maelstrom comes of the hole in the sunken millstone. This comes from Snorri. The older verses by Snaebjörn which described Hamlet's Mill stated that the nine maids of the island mill who in past ages ground Amlodhi's meal now drive a "host-cruel skerry-quern." That this skerry-quern means the whirlpool, and not simply the northern ocean, is backed up through some more lines which Gollancz ascribes to Snaebjörn; not that they were of crystal clarity, but again mill and whirlpool are connected:

> The island-mill pours out the blood of the flood goddess's sisters [i.e., the waves of the sea], so that [it] bursts from the feller of the land: whirlpool begins strong.[2]

No localization is indicated here, whereas the Finns point to directions which are less vague than they sound. Their statement that the Sampo has three roots—one in heaven, one in the earth, the third in the water eddy—has a definite meaning, as will be shown.

---

[1] See for Ireland, W. Stokes, "The Prose Tales in the Rennes Dindsenchas," RC *16* (1895), no. 145: "A great whirlpool there is between Ireland and Scotland on the North. It is the meeting of many seas [from NSEW]—it resembles an open caldron which casts the draught down [and] up, and its roaring is heard like far-off thunder . . ."

[2] I. Gollancz, *Hamlet in Iceland* (1898), pp. xvii.

But then also, Vainamoinen driving with his copper boat into the "maw of the Maelstrom" is said to sail to "the depths of the sea," to the "lowest bowels of the earth," to the "lowest regions of the heavens." Earth and heaven—a significant contraposition. As concerns the whereabouts of the whirlpool, one reads:

> *Before the gates of Pohjola,*
> *Below the threshold of color-covered Pohjola,*
> *There the pines roll with their roots,*
> *The pines fall crown first into the gullet of the whirlpool.*[3]

Then in Teutonic tradition, one finds in Adam of Bremen (11th century):

> Certain Frisian noblemen made a voyage past Norway up to the farthest limits of the Arctic Ocean, got into a darkness which the eyes can scarcely penetrate, were exposed to a maelstroem which threatened to drag them down to Chaos, but finally came quite unexpectedly out of darkness and cold to an island which, surrounded as by a wall of high rocks, contains subterranean caverns, wherein giants lie concealed. At the entrances of the underground dwellings lay a great number of tubs and vessels of gold and other metals which "to mortals seem rare and valuable." As much as the adventurers could carry of these treasures they took with them and hastened to their ships. But the giants, represented by great dogs, rushed after them. One of the Frisians was overtaken and torn into pieces before the eyes of the others. The others succeeded, thanks to our Lord and Saint Willehad, in getting safely on board their ships.[4]

The Latin text (Rydberg, p. 422) uses the classical familiar name of Euripus. The Euripus, which has already come up in the *Phaedo*, was really a channel between Euboea and the mainland, in which the conflict of tides reverses the current as much as seven times a day, with ensuing dangerous eddies—actually a case of standing waves rather than a true whirl.[5]

[3] M. Haavio, *Vainamoinen, Eternal Sage* (1952), pp. 191–98.

[4] V. Rydberg, *Teutonic Mythology* (1907), p. 320.

[5] We meet the name again at a rather unexpected place, in the Roman circus or hippodrome, as we know from J. Laurentius Lydus (*De Mensibus 1.12.*), who states that the center of the circus was called Euripos; that in the middle of the stadium was a pyramid, belonging to the Sun; that by the Sun's pyramid were three altars, of Saturn, Jupiter, Mars, and below the pyramid, altars of Venus, Mercury and the

And here the unstable Euripus of the Ocean, which flows back to the beginnings of its mysterious source, dragged with irresistible force the unhappy sailors, thinking by now only of death, towards Chaos. This is said to be the maw of the abyss, that unknown depth in which, it is understood, the ebb and flow of the whole sea is absorbed and then thrown up again, which is the cause of the tides.

This is reflection of what had been a popular idea of antiquity. But here comes a version of the same story in North America.[6] It concerns the canoe adventure of two Cherokees at the mouth of Suck Creek. One of them was seized by a fish, and never seen again. The other was

taken round and round to the very lowest center of the whirlpool, when another circle caught him and bore him outward. He told afterwards that when he reached the narrowest circle of the maelstroem the water seemed to open below and he could look down as through the roof beam of a house, and there on the bottom of the river he had seen a great company, who looked up and beckoned to him to join them, but as they put up their hands to seize him the swift current caught him and took him out of their reach.

It is almost as if the Cherokees have retained the better memory, when they talk of foreign regions, inhabited by "a great company" —which might equally well be the dead, or giants with their dogs— there, where in "the narrowest circle of the maelstroem the water seemed to open below." It will be interesting to see whether or not this impression is justifiable.[7]

Snorri, who has preserved the Song of Grotte for us, does not actually name the whirlpool in it, but there is only one at hand,

---

Moon, and that there were not more than seven circuits (*kykloi*) around the pyramid, because the planets were only seven. (See also F. M. Cornford's chapter on the origin of the Olympic games in J. Harrison's *Themis* (1962), p. 228; G. Higgins' *Anacalypsis* (1927), vol. 2, pp. 377ff.) This brings to mind (although not called Euripus, obviously, but "the god's place of skulls") the Central American Ball Court which had a round hole in its center, termed by Tezozomoc "the enigmatic significance of the ball court," and from this hole a lake spread out before Uitzilopochtli was born. See W. Krickeberg, "Der mittelamerikanische Ballspielplatz und seine religiöse Symbolik," *Paideuma 3* (1948), pp. 135ff., 155, 162.

[6] J. Mooney, *Myths of the Cherokee* (1900), p. 340.

[7] See illustrations (p. 60) showing Mount Meru in the shape of an hourglass.

namely the "Hvergelmer" in Hel's abode of the dead, from and to which "all waters find their way."[8] Says Rydberg:

It appears that the mythology conceived Hvergelmer as a vast reservoir, the mother fountain of all the waters of the world. In the front rank are mentioned a number of subterranean rivers which rise in Hvergelmer, and seek their courses thence in various directions. But the waters of earth and heaven also come from this immense fountain, and after completing their circuits they return thither.

The myth about Hvergelmer and its subterranean connection with the ocean gave our ancestors the explanation of ebb and flood tide. High up in the northern channels the bottom of the ocean opened itself in a hollow tunnel, which led down to the "kettle-roarer," "the one roaring in his basin" (hverr = kettle; galm = Anglo-Saxon gealm = a roaring). When the waters of the ocean poured through this tunnel down into the Hades-well there was ebb-tide, when it returned water from its superabundance then was flood-tide.

Between the death-kingdom and the ocean there was, therefore, one connecting link, perhaps several. Most of the people who drowned did not remain with Ran. Aegir's wife, Ran, received them hospitably, according to the Icelandic sagas of the middle ages. She had a hall in the bottom of the sea, where they were welcomed and offered . . . seat and bed. Her realm was only an ante-chamber to the realms of death.[9]

There are several features of the *Phaedo* here, but they will turn up again in Gilgamesh. This is not to deny that Hvergelmer, and other whirlpools, explain the tides, as indicated previously. (Perhaps it will be possible to find out what tides "mean" on the celestial level.) But it is clear that the Maelstrom as the cause of the tides does not account for the surrounding features, not even for the few mentioned by Rydberg—for instance, the wife of the Sea-god Aegir who receives kindly the souls of drowned seafarers in her antechamber at the bottom of the sea—nor the circumstance that the Frisian adventurers, sucked into the Maelstrom, suddenly find themselves on a bright island filled with gold, where giants lie concealed

[8] *Grimnismal* 26; cf. Snorri, *Gylf.* 15.

[9] Rydberg, pp. 414, 421f. Cf. the notions about the nun Saint Gertrude, patron of travelers, particula ly on sea voyages, who acted also as patron saint of inns "and finally it was claimed that she was the hostess of a public house, where the souls spent the first night after death" (M. Hako, *Das Wiesel in der europäischen Volksüberlieferung*, FFC *167* [1956], p. 119).

in the mountain caves. This island begins to look very much like Ogygia I, where Kronos/Saturn sleeps in a golden mountain cave, whereas the reception hall of Ran—her husband Aegir was famous for his beer brewing, and his hall it was, where Loke offended all his fellow gods as reported in the *Lokasenna*—would suggest rather Ogygia II, the island of Calypso, sister of Prometheus, called *Omphalos Thalasses*, the Navel of the Sea. Calypso was the daughter of Atlas, "who knew the depths of the whole sea." She, Calypso, has been authoritatively compared[10] to the divine barmaid Siduri, who dwells by the deep sea and will be found later on in the tale of Gilgamesh.

Mythology, meaning proper poetic fable, has been of great assistance but it can help no further. The golden island of Kronos, the tree-girt island of Calypso, remain unlocatable, notwithstanding the efforts of Homeric scholars. Through careful analysis of navigational data, one of them (Berard) has placed Calypso in the island of Perejil near Gibraltar, another (Bradfield) in Malta, others even off Africa. Presumably it should not be too far from Sicily, since Ulysses reaches it riding on the mast of his ship, right after having escaped from Charybdis in the straits of Messina, in the setting that Homer describes so plausibly. It appears throughout time in many places.[11] Some data in Homer look like exact geography, as Circe's Island with its temple of Feronia, or the Land of the Laistrygones, which should be the bay of Bonifacio. But most elements from past myth, like Charybdis or the Planktai, are illusionistic. They throw the whole geography into a cocked hat, as do the Argonauts themselves.

Without trying to fathom Ogygia, or Ogygos, the adjective "Ogygian"—which has been used as a label for the Waters of Styx —has also assumed the connotation of "antediluvian." As for Hvergelmer, "roaring kettle," it is the "navel of the waters" but it is certainly "way down," as is the strange "Bierstube" of Aegir. And when it is found, as it soon will be, that Utnapishtim (the builder

---

[10] See chapter XXII, "The Adventure and the Quest."
[11] The last learned attempt to locate it—by H. H. and A. Wolf, *Der Weg des Odysseus* (1968)—proves as illusionistic as the previous ones.

of the Ark, who can be reached only by the road leading through the bar of the divine Siduri and hence also, one would say, through the inn of beer-brewing Aegir) lives forever at the "confluence of the rivers," this might have charmed Socrates with his idea of confluences, but it will not make things much clearer.

Yet there are some footholds to climb back from the abyss. It is known (chapter XII) that Socrates and the poets really referred to heaven "seen from the other side."

It has been shown that the way through the "navel of the waters" was taken by Vainamoinen, and we shall see (chapter XIX) that the same goes for Kronos-Phaethon, and other powerful personalities as well, who reached the Land of Sleep where time has ceased. One can anticipate that the meaning will be ultimately astronomical. Hence, backing out of fable, one can turn again for assistance to the Royal Science.

That there is a whirlpool in the sky is well known; it is most probably the essential one, and it is precisely placed. It is a group of stars so named (*zalos*) at the foot of Orion, close to Rigel (beta Orionis, Rigel being the Arabic word for "foot"), the degree of which was called "death," according to Hermes Trismegistos,[12] whereas the Maori claim outright that Rigel marked the way to Hades (Castor indicating the primordial homeland). Antiochus the astrologer enumerates the whirl among the stars rising with Taurus. Franz Boll takes sharp exception to the adequacy of his description, but he concludes that the *zalos* must, indeed, be Eridanus "which flows from the foot of Orion."[13] Now Eridanus, the watery grave of Phaethon—Athanasius Kircher's star map of the southern hemisphere still shows Phaethon's mortal frame lying in the stream— was seen as a starry river leading to the other world. The initial frame stands, this time traced in the sky. And here comes a crucial confirmation. That mysterious place, *pī nārāti*, literally the "mouth of the rivers," meaning, however, the "confluence" of the rivers, was traditionally identified by the Babylonians with Eridu. But the

---

[12] *Vocatur mors.* W. Gundel, *Neue Astrologische Texte des Hermes Trismegistos* (1936), pp. 196f., 216f.

[13] *Sphaera* (1903), pp. 57, 164–67.

archaeological site of Eridu is nowhere near the confluence of the Two Rivers of Mesopotamia. It is between the Tigris and Euphrates, which flow separately into the Red Sea, and placed rather high up. The proposed explanation, that it was the expanding of alluvial land which removed Eridu from the joint "mouth" of the rivers, did not contribute much to an understanding of the mythical topos of *pi narāti*, and some perplexed philologist supposed in despair that those same archaic people who had built up such impressive waterworks had never known which way the waters flow and had believed, instead, that the two rivers had their source in the Persian Gulf.

This particular predicament was solved by W. F. Albright, who exchanged "mouth" and "source";[14] he left us stranded "high and dry"—a very typical mythical situation, by the way—in the Armenian mountains around the "source." And though he stressed, rightly, that Eridu–*pi narāti* could not mean geography, he banished it straightaway into the interior of the planet.

The "source" is as unrevealing as the "mouth" has been, and as every geographical localization is condemned to be. Eridu, Sumerian [mul]NUN[ki], is Canopus, alpha Carinae, the bright star near the South Pole, as has been established irrefragably by B. L. van der Waerden,[15] a distinguished contemporary historian of astronomy. That one or another part of Argo was meant had been calculated previously.[16] And that, finally, made sense of the imposing configuration of myths around Canopus on the one hand, and of the preponderance of the "confluence of the rivers" on the other hand. This unique topos will be dealt with later.

One point still remains a problem. The way of the dead to the other world had been thought to be the Milky Way, and that since the oldest days of high civilization. This image was still alive with the Pythagoreans. When and how did Eridanus come in? A reasonable supposition is that this was connected with the observed shift-

---

[14] "The Mouth of the Rivers," AJSL *35* (1919), pp. 161–95.

[15] "The Thirty-six Stars," JNES *8* (1949), p. 14. "The bright southern star Canopus was Ea's town Eridu (NUN[ki] [d]E–a)."

[16] See P. F. Gössmann, *Planetarium Babylonicum* (1950), 306.

ing of the equinoctial colure[17] due to the Precession. But the analysis of this intricate problem of rivers will come in the chapter on the Galaxy.

One thing meanwhile stands firm: the real, the original, way from the whirlpool lies in heaven. With this finding, one may plunge again into the bewildering jungle of "earthly" myths concerning the Waters from the Deep.

[17] The equinoctial colure is the great circle which passes through the celestial poles and the equinoctial points: the solstitial colure runs through both the celestial and ecliptic poles and through the solstitial points. Macrobius has it, strange to say, that "they are not believed to extend to the South Pole," whence *kolouros*, meaning "dock-tailed," "which are so called because they do not make complete circles" (*Comm. Somn. Scip. 1.15.14*). The translator, W. H. Stahl (p. 151), refers, among others, to Geminus 5.49–50. Geminus, however (5.49, Manitius, pp. 60–61), does not claim such obvious nonsense; he states the following: "*Kolouroi* they are called, because certain of their parts are *not visible* (dia to merē tina autōn atheōrēta ginesthai). Whereas the other circles become visible in their whole extension with the revolution of the cosmos, certain parts of the Colures remain invisible, 'docked' by the antarctical circle below the horizon."

CHAPTER XV

The Waters from the Deep

> The glacier knocks in the cup-
> board,
> The desert sighs in the bed,
> And the crack in the tea cup
> opens
> A lane to the land of the dead.
>
> W. H. AUDEN, *"As I Walked
> Out One Evening"*

T HERE IS A TRADITION from Borneo of a "whirlpool island" with a tree that allows a man to climb up into heaven and bring back useful seeds from the "land of the Pleiades."[1] The Polynesians have not made up their mind, apparently, concerning the exact localization of their whirlpool which serves in most cases as entrance to the abode of the dead; it is supposed to be found "at the end of the sky," and "at the edge of the Milky Way."[2]

On this side of the Atlantic the Cuna Indians also knew the basic scheme,[3] although they, too, failed to give the accepted localization: "God's very own whirlpool" (*tiolele piria*) was right beneath the Palluwalla tree, "Saltwater-Tree," and when the Sun-God, or the Tapir, a slightly disguised Quetzalcouatl, chopped down the tree, saltwater gushed forth to form the oceans of the world.

[1] A. Maass, "Sternkunde und Sterndeuterei im Malaiischen Archipel" (1924), in *Tijdschrift Indische Taal-, Land-, en Volkenkunde 64*, p. 388.

[2] M. W. Makemson (*The Morning Star Rises: An Account of Polynesian Astronomy* [1941], no. 160) suggests Sagittarius. For Samoa, see A. Kraemer, *Die Samoa-Inseln* (1902), vol. *1*, p. 369. For Mangaia, see P. Buck, *Mangaian Society* (1934), p. 198; and R. W. Williamson, *Religious and Cosmic Beliefs of Central Polynesia* (1924), vol. 2, p. 251.

[3] C. E. Keeler, *Secrets of the Cuna Earthmother* (1960), pp. 67ff., 78f.

There are three elements here, which combine into a curious tangle: (a) the whirlpool represents, or is, *the* connection of the world of the living with the world of the dead; (b) a tree grows close to it, frequently a life-giving or -saving tree; (c) the whirl came into being because a tree was chopped down or uprooted, or a mill's axle unhinged, and the like. This basic scheme works into many variants and features in many parts of the world, and it provides a very real paradox or conundrum: it is as if the particular waters hidden below tree, pillar, or mill's axle waited only for the moment when someone should remove that plug—tree, pillar, or mill's axle—to play tricks.

This is no newfangled notion. Alfred Jeremias remarks casually, "The opening of the navel brings the deluge. When David wanted to remove the navel stone in Jerusalem, a flood was going to start [see below, p. 220]. In Hierapolis in Syria the altar of Xisuthros [= Utnapishtim] was shown in the cave where the flood dried up."[4]

The pattern reveals itself in the Indonesian Rama epic.[5] When Rama is building the huge dike to Lanka (Ceylon) the helpful monkeys throw mountain after mountain into the sea, but all of them vanish promptly. Enraged, Rama is going to shoot his magic arrow into the unobliging sea, when there arises a lady from the waters who warns him that right here was a hole in the ocean leading to the underworld, and who informs him that the water in that hole was called Water of Life.

Rama would seem to have won out with his threat since the dike was built. But the same story comes back in Greece when Herakles crosses the sea in order to steal the cattle of Geryon. Okeanos, represented here as a god, works up the waters into a tumult which are the waters of the original flood; Herakles threatens with his drawn bow, and calm is re-established.

Neither whirlpool nor confluence are mentioned in these cases, but they clearly extend to them. This gives great importance to the Catlo'ltq story from the American Northwest that is paradigmatic (see chapter XXII) of the maiden who shoots her arrow into the

4 HAOG, p. 156, n. 7 ("wo die Flut versiegte").
5 W. Stutterheim, *Rama-Legenden und Rama-Reliefs in Indonesien* (1925), p. 54.

"navel of the waters which was a vast whirlpool," thus winning fire. Some very fundamental idea must be lurking behind the story, and a pretty old one, since it was said of Ishtar that it is "she who stirs up the apsu before Ea."[6]

A strange pastime for the heavenly queen, but it seems to have been a rather celestial sport. The eighth Yasht of the *Avesta*,[7] dedicated to Sirius-Tishtriya, says of this star: "We worship the splendid, brilliant Tishtriya, which soars rapidly to Lake Vurukasha, like the arrow quick-as-lightning, which Urxsa the archer, the best archer among the Aryans, shot from Mount Aryioxsutha to Mount Huvanvant."[8] And what does Sirius do to this sea? It causes "Lake Vurukasha to surge up, to flood asunder, to spread out; at all shores surges Lake Vurukasha, the whole center surges up" (Yt. *8.31*; see also *5.4*). Whereas Pliny[9] wants to assure us that "the whole sea is conscious of the rise of that star, as is most clearly seen in the Dardanelles, for sea-weed and fishes float on the surface, and everything is turned up from the bottom." He also remarks that at

---

[6] "Descent of Ishtar to the Nether World," obv. l. 27, ANET, p. 107; see also W. F. Albright, "The Mouth of the Rivers," AJSL *35* (1919), p. 184.

[7] Yasht *8.6* and *8.37* (H. Lommel, *Die Yashts des Awesta* [1927]).

[8] See for the feat of this unpronounceable archer (Rkhsha) the report given by Al-Biruni, who spells him simply *Arish* (*The Chronology of Ancient Nations*, trans. E. Sachau [1879], p. 205). The background of the tale: Afrasiyab had promised to restore to Minôcihr a part of Erânshar (which had been conquered by him) as long and as broad as an arrow shot. Arish shot the arrow on the 13th day of the month Tîr-Mâh, after having announced: "I know that when I shoot with this bow and arrow I shall fall to pieces and my life will be gone." Accordingly, when he shot, he "fell asunder into pieces. By order of God the wind bore the arrow away from the mountain of Rûyân and brought it to the utmost frontier of Khurâsân between Farghâna and Tabaristân; there it hit the trunk of a nut-tree that was so large that there had never been a tree like it in the world. The distance between the place where the arrow was shot and that where it fell was 1,000 Farsakh." (See also S. H. Taqizadeh, *Old Iranian Calendars* [1938], p. 44.) Tîr or Tîra is the name for Mercury (see T. Hyde, *Veterum Persarum et Parthorum Religionis historia* [1760], p. 24: "Tîr, i.e., Sagitta . . ., quo etiam nomine appellatur Mercurius Planeta propter velociorem motum"), but it is also, along with Tishtriya, the name for Sirius (see A. Scherer, *Gestirnnamen bei den indogermanischen Völkern* [1953], pp. 113f.), and the 13th day of every month is dedicated to Sirius-Tishtriya (see Lommel, p. 5). We must leave it at that: Sirius-the-arrow has made more mythical "noise" than any other star, and also its connection with the ominous number 13 appears to be no Iranian monopoly.

[9] *9.58*. Cf. Aristotle, *Historia Animalium 8.15.599B–600.*

the rising of the Dog-Star the wine in the cellars begins to stir up and that the still waters move (2.107)—and the *Avesta* offers as explanation (Yt. *8.41*) that it is Tishtriya, indeed, "by whom count the waters, the still and the flowing ones, those in springs and in rivers, those in channels and in ponds."[10]

This is, however, no Iranian invention: the ritual text of the Babylonian New Year addresses Sirius as "[mul]KAK.SI.DI. who measures the depth of the Sea." [mul] is the prefix announcing the star, KAK.SI.DI means "arrow," and it is this particular arrow which is behind most of the bewildering tales of archery. The bow from which it is sent on its way is a constellation, built from stars of Argo and Canis Major, which is common to the spheres of Mesopotamia, Egypt and China.[11] And since the name Ishtar is shared by both Venus and Sirius, one may guess who "stirs up the apsu before Ea."

And here is what the "fire" accomplished, according to a Finnish rune of origin,[12] after it had been "cradled . . . over there on the navel of the sky, on the peak of the famous mountain," when it rushed straightaway through seven or nine skies and fell into the sea: "The spark . . . rolled . . . to the bottom of Lake Aloe, roaring it rushed to the bottom of the sea, down into the narrow depression (?). This Lake Aloe then, thrice in the summernight, rose foaming to the height of its firs, driven in fury beyond its banks. Thereupon again Lake Aloe thrice in the summernight dried up its waters to the bottom, its perch on the rocks, its pope [small fishes] on the skerries."

A violent spark this seems to have been; yet—is it not also said of the old Sage: "Vainamoinen in the mouth of the whirlpool boils like fire in water"?[13] Which goes to show that mythical "fire" means more than meets the eye. Actually, the enigmatical events in "Lake

---

[10] Trans. E. Herzfeld, *Zoroaster and His World* (1947), p. 587.

[11] There is strong circumstantial evidence of this bow and arrow in Mexico also: the bow of the Chichimeca, the Dog-people.

[12] K. Krohn, *Magische Ursprungsrunen der Finnen* (1924), pp. 115ff. See also F. Ohrt, *The Spark in the Water* (1926), pp. 3f.

[13] M. Haavio, *Vainamoinen, Eternal Sage* (1952), p. 196.

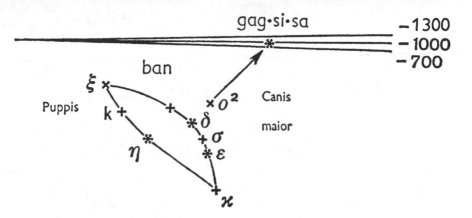

The Mesopotamian constellation of Bow and Arrow ([mul]BAN and [mul]KAK.SI.DI, or gag.si.sa), as reconstructed on the evidence of astronomical cuneiform texts; gag.si.sa/KAK.SI.DI is Sirius, the "Arrow-Star."

T'ien-lang ou Sirius

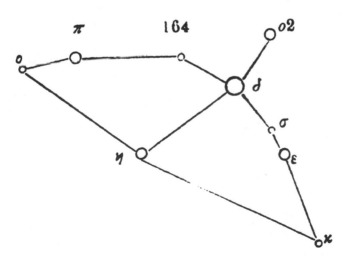

The Chinese constellation built up by the same stars. In China, however, the arrow is shorter; Sirius is not the tip of the arrow, but the target: the celestial jackal T'ien-lang.

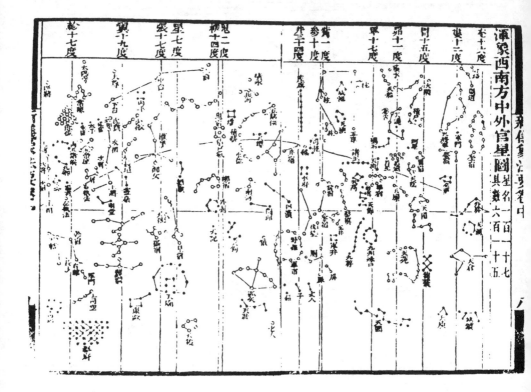

The star maps for the celestial globe in the Hsin I Hsiang Fa Yao of 1092 "Mercator's" projection. The constellation of the bow is seen near the center of the lower half.

Drawing the bow at Sirius, the celestial jackal, as it was done by the mythical emperors of Ancient China.

In the so-called "Round Zodiac" of Dendera (Roman Egypt), the goddess Satit is aiming her arrow from the same bow at the star on the head of the Sothiscow—Sirius again (on the right, lower half). The Egyptian conception is closer to that of the Chinese than to the Babylonian.

Aloe" cannot be severed from those occurring in Lake Vurukasha and the coming into being of the "three outlets," the first of which had the name Hausravah/Kai Khusrau (see chapter XIII, "Of Time and the Rivers," p. 201).

Before we move on to many motifs which will be shown as related to the same "eddy-field" or whirl, it is appropriate to quote in full a version of the fire and water story from the Indians of Guyana. This not only provides charming variations, but presents that rarest of deities, a creator power neither conceited nor touchy nor jealous nor quarrelsome nor eager to slap down unfortunates with "inborn sin," but a god aware that his powers are not really unlimited. He behaves modestly, sensibly and thoughtfully and is rewarded with heartfelt cooperation from his creatures, at least from all except for the usual lone exception.

> The Ackawois of British Guiana say that in the beginning of the world the great spirit Makonaima [or Makunaima; he is a twin-hero; the other is called Pia] created birds and beasts and set his son Sigu to rule over them. Moreover, he caused to spring from the earth a great and very wonderful tree, which bore a different kind of fruit on each of its branches, while round its trunk bananas, plantains, cassava, maize, and corn of all kinds grew in profusion; yams, too, clustered round its roots; and in short all the plants now cultivated on earth flourished in the greatest abundance on or about or under that marvellous tree.

> *In order to diffuse* the benefits of the tree all over the world, Sigu resolved to *cut it down* and plant slips and seeds of it everywhere, and this he did with *the help of all* the beasts and birds, all *except* the brown *monkey*, who, being both lazy and mischievous, refused to assist in the great work of transplantation. So to keep him out of mischief Sigu set the animal to fetch water from the stream in a basket of open-work, calculating that the task would occupy his misdirected energies for some time to come.

> In the meantime, proceeding with the labour of felling the miraculous tree, he discovered that the *stump was hollow and full of water* in which the fry of every sort of fresh-water *fish* was swimming about. The benevolent Sigu determined to stock all the rivers and lakes on earth with the fry on so liberal a scale that every sort of fish should swarm in every water.

But this generous intention was unexpectedly frustrated. For *the water in the cavity*, being *connected with the great reservoir* somewhere in the bowels of the earth, *began to overflow;* and to arrest the rising flood Sigu covered the stump with a *closely woven basket.* This had the desired effect. But unfortunately the brown monkey, tired of his fruitless task, stealthily returned, and his curiosity being aroused by the sight of the basket turned upside down, he imagined that it must conceal something good to eat. So he cautiously lifted it and peeped beneath, and *out poured the flood*, sweeping the monkey himself away and inundating the whole land. Gathering the rest of the animals together Sigu led them to the highest points of the country, where grew some tall coconut-palms. Up the tallest trees he caused the birds and climbing animals to ascend; and as for the animals that could not climb and were not amphibious, he shut them in a cave with a very narrow entrance, and having sealed up the mouth of it with wax he gave the animals inside a long thorn with which to pierce the wax and so ascertain when the water had subsided. After taking these measures for the preservation of the more helpless species, he and the rest of the creatures climbed up the palm-tree and ensconced themselves among the branches.

During the darkness and storm which followed, they all suffered intensely from cold and hunger; the rest bore their sufferings with stoical fortitude, but the red howling monkey uttered his anguish in such horrible yells that his throat swelled and has remained distended ever since; that, too, is the reason why to this day he has a sort of bony drum in his throat.

Meanwhile Sigu from time to time let fall seeds of the palm into the water to judge of its depth by the splash. As the water sank, the interval between the dropping of the seed and the splash in the water grew longer; and at last, instead of a splash the listening Sigu heard the dull thud of the seeds striking the soft earth. Then he knew that the flood had subsided, and he and the animals prepared to descend. But the trumpeter-bird was in such a hurry to get down that he flopped straight into an ant's nest, and the hungry insects fastened on his legs and gnawed them to the bone. That is why the trumpeter-bird has still such spindle shanks. The other creatures profited by this awful example and came down the tree cautiously and safely.

Sigu now rubbed two pieces of wood together to make fire, but just as he produced the first spark, he happened to look away, and the bush-turkey, mistaking the spark for a fire-fly, gobbled it up and flew off. The spark burned the greedy bird's gullet, and that is why turkeys have red wattles on their throats to this day.

The alligator was standing by at the time, doing no harm to anybody; but as he was for some reason an unpopular character, all the other animals accused him of having stolen and swallowed the spark. In order to recover the spark from the jaws of the alligator Sigu tore out the animal's tongue, and that is why alligators have no tongue to speak of down to this very day.[14]

There are many more stories over the world of a plug whose removal causes the flood: with the Agaria, an ironsmith tribe of Central India, it is the breaking of a nail of iron which causes their Golden Age town of Lohripur to be flooded.[15] According to the Mongolians, the Pole star is "a pillar from the firm standing of which depends the correct revolving of the world, or a stone which closes an opening: if the stone is pulled out, water pours out of the opening to submerge the earth."[16] In the Babylonian myth of Utnapishtim, "Nergal [the God of the Underworld] tears out the posts; forth comes Ninurta and causes the dikes to follow" (GE *11.101*f.). But the new thing to be faced is the appearance of the Ark in the flood, Noah's or another's.

The first ark was built by Utnapishtim in the Sumerian myth; one learns in different ways that it was a cube—a modest one, measuring 60 x 60 x 60 fathoms, which represents the unit in the sexagesimal system where 60 is written as 1. In another version, there is no ark, just a cubic stone, upon which rests a pillar which reaches from earth to heaven. The stone, cubic or not, is lying under a cedar, or an oak, ready to let loose a flood, without obvious reasons.

Confusing as it is, this seems to provide the new theme. In Jewish legends, it is told that "since the ark disappeared there was a stone in its place . . . which was called foundation stone." It was called foundation stone "because from it the world was founded [or started]." And it is said to lie above the Waters that are below the Holy of Holies.

[14] W. H. Brett, *The Indian Tribes of Guiana* (1868), pp. 378–84; Sir Everard F. im Thurn, *Among the Indians of Guiana* (1883), pp. 379–81 (quoted in J. G. Frazer, *Folklore in the Old Testament* [1918], vol. *1*, p. 265). The italics are ours.

[15] V. Elwin, *The Agaria* (1942), pp. 96ff.

[16] G. M. Potanin, quoted by W. Lüdtke, "Die Verehrung Tschingis-Chans bei den Ordos-Mongolen," ARW *25* (1927), p. 115.

This might look like a dream sequence, but it is buttressed by a very substantial tradition, taken up by the Jews but to be found also in Finno-Ugrian tradition.[17] The Jewish story then goes on:

When David was digging the foundations of the Temple, a shard was found at a depth of 1500 cubits. David was about to lift it when the shard exclaimed: "Thou canst not do it." "Why not?" asked David. "Because I rest upon the abyss." "Since when?" "Since the hour in which the voice of God was heard to utter the words from Sinai, 'I am the Lord, your God,' causing the world to quake and sink into the Abyss. I lie here to cover up the Abyss."

Nevertheless David lifted the shard, and the waters of the Abyss rose and threatened to flood the earth. Ahithophel was standing by and he thought to himself: "Now David shall meet with his death and I shall be king." Just then David said: "Whoever knows how to stem the tide of waters and fails to do it, will one day throttle himself."

Thereupon Ahithophel had the name of God inscribed upon the shard, and the shard thrown into the Abyss. The waters at once commenced to subside, but they sank to so great a depth that David feared the earth might lose her moisture, and he began to sing the fifteen "Songs of Ascents," to bring the waters up again.

The foundation stone here has become a shard and its name in tradition is *Eben Shetiyyah*, which is derived from a verb of many meanings:[18] "to be settled, satisfied; to drink; to fix the warp, to lay the foundations of," among which "to fix the warp" is the most revealing, and a reminder of the continuing importance of "frames." Within that "frame" there is a surging up and down of the waters below (as in the *Phaedo* myth) which suggests catastrophes unrecorded by history but indicated only by the highly colored terminology of cosmologists. Had they only known of a Cardan suspension, the world might have been conceived as more stable.

Hildegard Lewy's researches[19] on Eben Shetiyyah brought up a passage in the Annals of Assur-nasir-apli in which the new temple

---

[17] L. Ginzberg, *The Legends of the Jews* (1954), vol. 4, p. 96; cf. also vol. 1, p. 12; vol. 5, p. 14. We are indebted to Irvin N. Asher for the quotation, as well as for the ones from Jastrow that follow. Cf. V. J. Mansikka, "Der blaue Stein," FUF *11* (1911), p. 2.

[18] The verb is *shatan;* the meanings are given in Jastrow's dictionary.

[19] "Origin and Significance of the Mâgên Dâwîd," *Archiv Orientalni 18* (1950), Pt. 3, pp. 344ff.

of Ninurta at Kalhu is described as founded at the depth of 120 layers of bricks down "to the level of the waters," or, down to the water table. This comes back to the waters of the deep in their natural setting. But what people saw in them is something else again. If David and the Assyrian king dug down to subsoil water, so did the builders of the Ka'aba in Mecca. In the interior of that most holy of all shrines there is a well, across the opening of which had been placed, in pre-Islamic times, the statue of the god Hubal. Al-Biruni says that in the early Islamic period this was a real well, where pilgrims could quench their thirst at least at the time of the Arab pilgrimage. The statue of Hubal had been meant to stop the waters from rising. According to the legends, the same belief had once been current in Jerusalem. Hence the holy shard. But Mecca tells more. Hildegard Lewy points out that, in pre-Islamic days, the god Hubal was Saturn, and that the Holy Stone of the Ka'aba had the same role, for it was a cube, and hence originally Saturn. Kepler's polyhedron inscribed in the sphere of Saturn is only the last witness of an age-old tradition.

The humble little shard was brought in by pious legend to try to say that what counted was the power of the Holy Name. But the real thing was the cube: either as Utnapishtim's ark or, in other versions, as a stone upon which rests a pillar which reaches from earth to heaven. Even Christ is compared to "a cube-shaped mountain, upon which a tower is erected."[20] Hocart writes that "the Sinhalese frequently placed inside their topes a square stone representing Meru. If they placed in the center of a tope a stone representing the center of the world it must have been that they took the tope to represent the world"[21]—which goes without saying. But it is said otherwise that this stone, the foundation stone, lies under a great tree, and that from under the stone "a wave rose up to the sky."

This sounds like a late mixture, with no reasons given; the way to unscramble the original motifs is to take them separately.

[20] In the ninth simile of the "Pastor of Hermas," according to F. Kampers (*Vom Werdegange der abendländischen Kaisermystik* [1924], p. 53).

[21] *Kingship*, p. 179 (quoted by P. Mus, *Barabudur* [1935], p. 108, n. 1).

But first, some stock-taking is in order at this point. There are a number of figures to bring together. The brown monkey, father of mischief in Sigu's idyllic creation, is familiar under many disguises. He is the Serpent of Eden, the lone dissenter. He is Loke who persuaded the mistletoe not to weep over Balder's death, thus breaking the unanimity of creatures. Sigu himself, benevolent king of the Golden Age, is an unmistakably Saturnian figure, who dwelt among his creatures, and so is Iahwe, at least when he still "walked with Adam in the garden." A ruler who "means well" is a Saturnian character. No one but Saturn dwelt among men. Says an Orphic fragment: "Orpheus reminds us that Saturn dwelt openly on earth and among men."[22] Dionysius of Halicarnassus (*1.36.1*) writes: "Thus before the reign of Zeus, Kronos ruled on this very earth" to which Maximilian Mayer crisply annotates: "We find no mention anywhere of such an earthly sojourn on the part of Zeus."[23] In a similar way, Sandman Holmberg states with respect to Ptah, the Egyptian Saturn: "The idea of Ptah as an earthly king returns again and again in Egyptian texts," and also points to "the remarkable fact that Ptah is the only one of the Egyptian gods who is represented with a straight royal beard, instead of with a bent beard."[24]

The Saturnalia, from Rome to Mexico, commemorated just this aspect of Saturn's rule, with their general amnesties, masters serving slaves, etc., even if Saturn was not always directly mentioned. When this festival was due in China, so to speak "sub delta Geminorum"—more correctly, delta and the Gemini stars 61 and 56 of Flamsteed—"there was a banquet in which all hierarchic distinctions were set aside . . . The Sovereign invited his subjects through the 'Song of Stags.' "[25]

The cube was Saturn's figure, as Kepler showed in his *Mysterium Cosmographicum*; this is the reason for the insistence on cubic stones and cubic arks. Everywhere, the power who warns "Noah"

[22] *Orphicorum Fragmenta* (1963), frg. 139, p. 186, from Lactantius.
[23] M. Mayer, in Roscher s.v. Kronos, pp. 1458f.
[24] M. Sandman Holmberg, *The God Ptah* (1946), pp. 83, 85.
[25] G. Schlegel, *L'Uranographie Chinoise* (1875), p. 424.

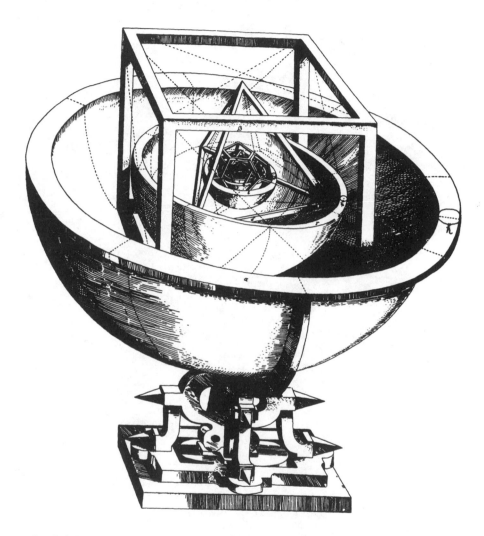

The Polyhedra inscribed into the planetary orbits. Kepler's drawing is a pure geometrical fancy, but it is meant to correspond to the actual relation between the radii of the planetary orbits. Most important here is the cube, fitted into the outermost sphere of Saturn.

and urges him to build his ark is Saturn, as Jehovah, as Enki, as Tane, etc. Sigu's basket stopper was obviously an inadequate version of the cube seen through the fantasy of basket-weaving natives. This leads to the conclusion that Noah's ark originally had a definite role in bringing the flood to an end. An interesting and unexpected conclusion for Bible experts.

One of the great motifs of myth is the wondrous tree so often described as reaching up to heaven. There are many of them—the Ash Yggdrasil in the *Edda*, the world-darkening oak of the *Kalevala*, Pherecydes' world-oak draped with the starry mantle, and the Tree of Life in Eden. That tree is often cut down, too. The other motif is the foundation stone, which sometimes becomes a cubic ark.

These motifs must first be traced through. After reading the beautiful story of Sigu's wonder tree, in whose stump are all the kinds of fish to populate the world, it needs patience to cope with the cubic stone which is found in the middle of the sea, under which dwells a mystic character whose guises vary from a miraculous fish, even a whale, to a "green fire," the "king of all fires," the "central fire," to the Devil himself. The chief source for him are Russian[26] and Finnish magic formulae, and these "superstitions" ("left-overs") are Stone Age fragments of flinty hardness embedded in the softer structure of historic overlay. Magic material withstands change, just because of its resistance to the erosion of common sense. As far as these magic formulae go, they became embedded in a Christian context as the particular populations underwent conversion, but they remain as witnesses for a very different understanding of the cosmos. For example, Finnish runes on the origin of water state that "all rivers come from the Jordan, into which all rivers flow," that "water has its origin in the eddy of the holy river—it is the bathing water of Jesus, the tears of God."[27] On the other hand, Scandinavian formulae stress the point that Christ "stopped up the Jordan" or "the Sea of Noah" (Mansikka, pp. 244f., 297, n. 1) which, in its turn, fits into the *Pastor of Hermas*, where Christ is compared to a "cube-shaped mountain" (see above, p. 221). From this it is not

[26] V. J. Mansikka, *Über russische Zauberformeln* (1909), pp. 184–87, 189, 192.
[27] Krohn, *Ursprungsrunen*, pp. 106f.

strange that the Cross becomes the "new tree," marking new cross-roads. One need not go as far as Russia for that. In the famous frescoes of Piero della Francesca in Arezzo there is "the discovery of the True Cross." It begins with the death of Adam, lying at the foot of the tree. The wood from the tree will later provide the material for the Cross. Later still, St. Helena, mother of Constantine, sees it in a dream and causes the wood to be dug up to become the holiest of relics. Piero illustrated nothing that was not in good medieval tradition. This is, one might say, sensitive ground.

CHAPTER XVI

The Stone and the Tree

In Xanadu did Kubla Khan
A stately pleasure dome decree
Where Alph, the sacred river, ran
Through caverns measureless to man
Down to a sunless sea.

COLERIDGE, *Kubla Khan*

THE GROUND, indeed, is not only sensitive but difficult and shifting as well. If the whirlpool turns up in the theory of the Cross, it is certainly without the consent of theologians. Yet the instances so far given are not isolated ones. It is necessary to deal with material which may appear suspicious to the trained historical reader, who is bound to be wary of *omne ignotum pro magnifico*. One should, therefore, preface this chapter with a small case history, which may show the infrangible tenacity of certain kinds of transmitted material, fragments of a sort official memory is prone to dismiss or neglect.

In the Gospel of Mark III.17, the "twins" James and John, the sons of Zebedee, are given by Jesus the name of Boanerges, which the Evangelist explains as meaning "Sons of Thunder."[1] This was long overlooked but eventually became the title of a work by a distinguished scholar, too soon forgotten, Rendel Harris. Here the Thunder Twins were shown to exist in cultures as different as Greece, Scandinavia and Peru. They call to mind the roles of Magni and Modi, not actually called twins, but successors of Thor, in Ragnarok. But to quote from Harris:

[1] Kai epethēken autois onoma Boanērges, ho estin hyioi brontēs.

We have shown that it does not necessarily follow that when the parenthood of the Thunder is recognised, it necessarily extends to both of the twins. The Dioscuri may be called unitedly, Sons of Zeus; but a closer investigation shows conclusively that there was a tendency in the early Greek cults to regard one twin as of divine parentage, and the other of human. Thus Castor is credited to Tyndareus, Pollux to Zeus . . . The extra child made the trouble, and was credited to an outside source. Only later will the difficulty of discrimination lead to the recognition of both as Sky-boys or Thunder-boys. An instance from a remote civilization will show that this is the right view to take.

For example, Arriaga, in his "Extirpation of Idolatry in Peru" tells us that "when two children are produced at one birth, which they call Chuchos or Curi, and in el Cuzco Taqui Hua-hua, they hold it for an impious and abominable occurrence, and they say, that *one of them is the child of the Lightning,* and require a severe penance, as if they had committed a great sin."

And it is interesting to note that when the Peruvians, of whom Arriaga speaks, became Christians, they replaced the name of Son of Thunder, given to one of the twins, by the name of Santiago, having learnt from their Spanish (missionary) teachers that St. James (Santiago) and St. John had been called Sons of Thunder by our Lord, a phrase which these Peruvian Indians seem to have understood, where the great commentators of the Christian Church had missed the meaning . . .

Another curious and somewhat similar transfer of the language of the Marcan story in the folk-lore of a people, distant both in time and place . . . will be found, even at the present day, amongst the Danes . . . Besides the conventional flint axes and celts, which commonly pass as thunder-missiles all over the world, the Danes regard the fossil sea-urchin as a thunderstone, and give it a peculiar name. Such stones are named in Salling, *sebedaei*-stones or *s'bedaei;* in North Salling they are called *sepadeje*-stones. In Norbaek, in the district of Viborg, the peasantry called them *Zebedee* stones! At Jebjerg, in the parish of Cerum, district of Randers, they called them *sebedei*-stones . . . The name that is given to these thunderstones is, therefore, very well established, and it seems certain that it is derived from the reference to the Sons of Zebedee in the Gospel as sons of thunder. *The Danish peasant, like the Peruvian savage, recognised at once what was meant by Boanerges,* and called his thunderstone after its patron saint.[2]

This might have given pause to later hyperscholars like Bultmann, before they proceeded to "de-mythologize" the Bible. One never knows what one treads underfoot. Conversely, it shows that

[2] R. Harris, *Boanerges* (1913), pp. 9ff.

some misunderstanding beyond the knowledge of the experts must be accounted for before one deals with the whole information. Thus, there is no intention to dismiss the abundant legends and runes dealing with the wood of the Cross. Lack of time, however, does not allow for a proper investigation,[3] and permits only some remarks on Finnish and Russian notions about the "Great Oak," which is the nearest "relative" of Sumerian trees. Says one of the Finnish runes: "Long oak, broad oak. What is the wood of its root? Gold is the wood of its root. The sky is the wood of the oak's summit. An enclosure within the sky. A wether in the enclosure. A granary on the horn of the wether."[4] The next version boldly puts "the granary upon the top of the cross." According to a further version, in the crown of the oak is a cradle with a little boy, who has an axe upon his shoulder. More stunning notions occur in a Russian Apocryph where Satanael planted the tree in the paradise intending to get out of it a weapon against Christ: "The branches of the tree spread over the whole paradise, and it also covered the Sun. Its summit touched the sky, and from its roots sprang fountains of milk and honey."[5]

This latter idea in its turn fits the medieval tradition according to which the rivers of Paradise gushed forth from under the Cross. There will be other bewildering "trees" in the chapter on Gilgamesh, but there also no attempt will be made to exhaust the huge and ambiguous evidence.

But with the caveats distilled from the Sons of Thunder, and similar instances, it is possible to deal with more outlandish data. First, there is in the *Atharva Veda*, a whole hymn dedicated to what may be called the world pillar (a highly multivalent pillar), called the *skambha* from which—see above, p. 111—the Finnish Sampo is derived. At this point only one verse will serve, in which the fiery monster of the deep is mentioned:[6]

---

[3] For a rich collection of material see F. Kampers, *Mittelalterliche Sagen vom Paradiese und vom Holze des Kreuzes Christi* (1897).

[4] K. Krohn, *Magische Ursprungsrunen der Finnen* (1924), p. 192.

[5] Krohn, p. 197.

[6] To prevent relentless experts from pointing to "fundamental" investigations which are, no doubt, unknown to us: the chapter on *yaksa* in Pischel and Geldner's *Vedische Studien* is not unknown to us; there are several momentous reasons why we prefer to stick to the "obsolete" submarine "monster."

AV *10.7.38*. A great monster [yaksa] in the midst of the creation, strode in penance on the back of the sea—in it are set whatever gods there are, like the branches of a tree roundabout the trunk.

Or, to take a testimony from "late" astrological sources, these statements given by the *Liber Hermetis Trismegisti* which became so famous in the Middle Ages, to the degrees of *Taurus* (Gundel, pp. 54f., 217ff.):

18–20° *oritur Navis et desuper Draco mortuus, vocatur Terra*
rises the Ship, and on it the dead Dragon, called Earth

21–23° *oritur qui detinet navem, Deus disponens universum mundum*
rises he who keeps (or detains) the ship, the God that orders the whole universe. [*Disponere* corresponds to Greek *kosmeō.*]

Whatever it is that rules "below" seems, indeed, a truly omnipotent entity: There are, after all, very few, if any, characters who are simply said to "order the whole universe."

This remarkable "kosmokrator" will be dealt with; the fiery creature deep down in the sea, however, has to be banished into an appendix. That it is relevant to the whole scheme can be seen from the fact that "Vainamoinen in the mouth of the whirlpool boils like fire in the water"[7] (appendix #19).

The words of Hermes-Three-Times-Great, cryptic as they sound, are part of the highly organized technical language of astrologers; we mean not those who cast people's fortunes for pay, but those who speculated on the traditional system of the world, and made use of whatever there was of astronomy, geography, mythology, holy texts of the laws of time and change, to build up an ambitious system. Abu Ma'shar and Michael Scotus were later dismissed as triflers, false prophets, and magicians, but Tycho and Kepler still held them in high esteem: they represented whatever there was of real science in the 13th century, and produced many daring thoughts. The *ignotum* may conceivably turn out to be *magnificum.*

[7] M. Haavio, *Vainamoinen, Eternal Sage* (1952), p. 196.

The few disconnected sayings quoted may be called lacking in sense and method. They will be shored up with more material. Actually, we had to sentence this chapter—once "swelling" enough to burst every seam—to the most meager of diets until it shriveled to its present state of emaciation and apparent lack of coherence. But first, one should understand what the latent geometrical design can imply, as it broke through, time and again, in the past chapters.

CHAPTER XVII

The Frame of the Cosmos

La mythologie, dans son origine,
est l'ouvrage de la science; la
science seule l'expliquera.

CHARLES DUPUIS

IN GREEK MYTH, the basic frame of the world is described in the famous Vision of Er in the 10th Book of the *Republic*. In it we find Er the Armenian, who was resurrected from the funeral pyre just before it was kindled, and who describes his travel through the other world (*10.615ff.*). He and the group of souls bound for re-birth whom he accompanies travel through the other world. They come to "a straight shaft of light, like a pillar, stretching from above throughout heaven and earth—and there, at the middle of the light, they saw stretching from heaven the extremities of its chains; for this light binds the heavens, holding together all the revolving firmament like the undergirths of a ship of war. And from the extremities stretched the Spindle of Necessity, by means of which all the circles revolve."

Cornford adds in a note: "It is disputed whether the bond hold-ing the Universe together is simply the straight axial shaft or a circular band of light, suggested by the Milky Way,[1] girdling the heaven of fixed stars."[2] Eisler understood it as the zodiac, strange to say.[3] Since those "undergirths" of the trireme did not go around the ship horizontally, but were meant to secure the mast (the

---

[1] Cf. O. Gruppe, *Griechische Mythologie und Religionsgeschichte* (1906), p. 1036, n. 1: "probably the Milky Way."

[2] Plato's *Republic* (Cornford trans.), p. 353.

[3] Eisler, *Weltenmantel und Himmelszelt* (1910), pp. 97ff.

"tree" of the ship) which points upwards, we stand, on principle, for the Galaxy, which, however, had to be "replaced" by invisible colures in later times.[4] But Er also talks of the adventures of the souls between incarnations, and in this context we might rely on the Milky Way. Surely the "model" is far from clear, even, on Cornford's concession, obviously intentionally so. And indeed, a few paragraphs later, there comes the complete planetarium with its "whorls," the "Spindle of Necessity" held by the goddess, by which sit the Fates as they unwind the threads of men's lives. The souls can listen to the Song of Lachesis, if they are still in the "meadow," but the chains and shaft or band are no longer in the picture. Plato refuses to be a correct geometrician of the Other World, just as he would not be sensible about the hydraulics of it. But previously in the *Phaedo*, Socrates had been ironic about the "truths" of science, and insisted that the truths of myth are of another order, and rebellious to ordinary consistency. It is here as if Plato had juxtaposed a number of revered mythical traditions (including the planetary harmony) without pretending to fit them into a proper order. And so his image of the "framework" of the cosmos is left inconclusive. But somehow the axis and the band and the chains stand together, and this, one concludes, was the original idea. The rotation of the polar axis must not be disjointed from the great circles which shift along with it in heaven. The framework is thought of as all one with the axis. This leads back to a Pythagorean authority whom Plato was supposed to have followed (Timon even viciously said: plagiarized) and whom Socrates often quotes with unfeigned respect. It is Philolaos, surely a creative astronomer of high rank, from whom there are only a few surviving fragments, and the authenticity of these has been rashly challenged by many modern philologists.[5] In fragment 12 of Philolaos, there is a brief definition of the cosmos, very much in the spirit of Plato's dode-

---

[4] Cf. also the discussion in J. L. E. Dreyer, *A History of Astronomy from Thales to Kepler* (1953), pp. 56ff. Concerning the "chains," which he translates "ligatures," Dreyer states: "The ligatures (*desmoi*) of the heavens are the solstitial and equinoctial colures intersecting in the poles, which points therefore may be called their extremities (*akra*)."

[5] G. de Santillana and W. Pitts, "Philolaos in Limbo," *ISIS 42* (1951), pp. 112–20; also in *Reflections on Men and Ideas* (1968), pp. 190–201.

cahedron quoted in chapter XII. "In the sphere there are five elements, those inside the sphere, fire, and water and earth and air,
and what is the hull of the sphere, the fifth."[6] Notwithstanding
Philolaos' graceless Doric, the statement is perfectly clear. The
"hull," (olkas) was the common name for freighters, built for bulk
cargo, broad in the beam. It is really more adequate than Plato's
slim trireme; and it is closer in shape to what both men meant
apparently: the dodecahedron, the "hull," i.e., the sphere, the
actual containing frame. It is clear from Plato that the "fifth" is the
sphere that he calls ether which contains the four earthly elements
but is wholly removed from them. Aristotle was to change it to the
crystalline heavenly "matter" that he needed for his system, but
it remained for him a "fifth essence." There has thus been twice
repeated the original "hull," the frame that has been sought. What
happened, and was noted in chapter VII, was that the etymology
of Sampo was discovered to be in the Sanskrit skambha.

The abstract idea of a simple earth axis, so natural today, was
by no means so logical to the ancients, who always thought of the
whole machinery of heaven moving around the earth, stable at the
center. One line always implied many others in a structure. So,
apparently one must accept the idea of the world frame as an implex
(as used here and later this word involves the necessary attributes
that are associated with a concept: e.g., the center and circumfer-
ence of a circle, the parallels and meridians implied by a sphere),
of which Grotte and Sampo were the rude models with their pon-
derous moving parts.

Like the axle of the mill, the tree, the skambha, also represents
the world axis. This instinctively suggests a straight, upright post,
but the word axis is a simplification of the real concept. There is the
invisible axis, of course, which is crowned by the North Nail, but
this image needs to be enriched by two more dimensions. The term
world axis is an abbreviation of language comparable to the visual
abbreviation achieved by projecting the reaches of the sky onto a
flat star map. It is best not to think of the axis in straight analytical
terms, one line at a time, but to consider it, and the frame to which

6 See H. Diels, Die Fragmente der Vorsokratiker (1951), vol. 1, pp. 412f.

it is connected, as one whole. This involves the use of multivalent terms and the recognition of a convergent involution of unusual meanings.

As *radius* automatically calls *circle* to mind, so *axis* must invoke the two determining great circles on the surface of the sphere, the equinoctial and solstitial colures. Pictured this way, the *axis* resembles a complete armillary sphere. It stands for the system of coordinates of the sphere and represents the frame of a world-age. Actually the frame defines a world-age. Because the polar axis and the colures form an indivisible whole, the entire frame is thrown out of kilter if one part is moved. When that happens, a new Pole star with appropriate colures of its own must replace the obsolete apparatus.

Thus the Sanskrit *skambha*, the world pillar, ancestor of the Finnish Sampo, is shown to be an integral element in the scheme of things. The hymn *10.7* of the *Atharva Veda* is dedicated to the *skambha*, and Whitney, its translator and commentator[7] sounds puzzled in his footnote to *10.7.2*: "Skambha, lit. 'prop, support, pillar,' strangely used in this hymn as frame of the universe or held personified as its soul." Here are two verses of it:

12. In whom earth, atmosphere, in whom sky is set, where fire, moon, sun, wind stand fixed, that Skambha tell . . .

35. The Skambha sustains both heaven-and-earth here; the skambha sustains the wide atmosphere, the skambha sustains the six wide directions; into the skambha entered this whole existence.

The good old Sampo sounds less pretentious, but it does have its three "roots," "one in heaven, one in the earth, one in the water-eddy."[8] To make a drawing of a pillarlike tree (let alone a mill), with its roots distributed in the manner indicated, would be quite a task. Notably it takes the "enormous bull of Pohja"—obviously a cosmic bull—to plow up these strange roots: the Finnish heroes by themselves had not been able to uproot the Sampo.

In the case of Yggdrasil, the World Ash, Rydberg tried his hardest to localize the three roots, to imagine and to draw them. Since he

[7] Harvard Oriental Series, vol. *8*, p. 590.
[8] K. Krohn, *Kalevalastudien 4. Sampo* (1927), p. 13.

looked with steadfast determination into the interior of our globe, the result was not overly convincing. One of the roots is said to belong to the Asa in heaven, and beneath it is the most sacred fountain of Urd. The second is to be found in the quarters of the frost-giants "where Ginnungagap formerly was," and where the well of Mimir now is. The third root belongs to Niflheim, the realm of the dead, and under this root is Hvergelmer, the Whirlpool (*Gylf*. 15).[9]

This precludes any terrestrial diagram. It looks as though the "axis," implicating the equinoctial and solstitial colures, runs through the "three worlds" which are, to state it roughly and most inaccurately, the following:

(a) the sky north of the Tropic of Cancer, i.e., the sky proper, domain of the gods

(b) the "inhabited world" of the zodiac between the Tropics, the domain of the "living"

(c) the sky south from the Tropic of Capricorn, alias the Sweet-Water Ocean, the realm of the dead.

The demarcation plane between solid earth and sea is represented by the celestial equator; hence half of the zodiac is under "water," the southern ecliptic, bordered by the equinoctial points. There are more refined subdivisions, to be sure, "zones" or "belts" or "climates," dividing the sphere from north to south and, most important, the "sky" as well as the waters of the south have a share in the "inhabited world" allotted to them.[10] This summary is an almost frivolous simplification, but for the time being it may be sufficient.

[9] We are aware that either Grotte "should" have three roots, or that Yggdrasil should be uprooted, and that the Finns do not tell how the maelstrom came into being. All of which can be explained; we wish, however, to avoid dragging more and more material into the case. Several ages of the world have passed away, and they do not perish all in the same manner; e.g., the Finns know of the destruction of Sampo and of the felling of the huge Oak.

[10] To clear up the exact range of the three worlds, it would be necessary to work out the whole history of the Babylonian "Ways of Anu, Enlil, and Ea" (cf. pp. 431f.), and how these "Ways" were adapted, changed, and defined anew by the many heirs of ancient oriental astronomy. And then we would not yet be wise to the precise whereabouts of Air, Saltwater, and other ambiguous items.

Meanwhile, it is necessary to explain again what this "earth" is that modern interpreters like to take for a pancake. The mythical earth *is*, in fact, a plane, but this plane is not *our* "earth" at all, neither our globe, nor a presupposed homocentrical earth. "Earth" is the implied plane through the four points of the year, marked by the equinoxes and solstices, in other words the ecliptic. And this is why this earth is very frequently said to be quadrangular. The four "corners," that is, the zodiacal constellations rising heliacally at both the equinoxes and solstices, parts of the "frame" *skambha*, are the points which determine an "earth." Every world-age has its own "earth." It is for this very reason that "ends of the world" are said to take place. A new "earth" arises, when another set of zodiacal constellations brought in by the Precession determines the year points.

Once the reader has made the adjustment needed to think of the frame instead of the "pillar" he will understand easily many queer scenes which would be strictly against nature—ideas about planets performing feats at places which are out of their range, as both the poles are. He will understand why a force planning to uproot (or to chop down) a tree, or to unhinge a mill, or merely pull out a plug, or a pin, does not have to go "up"—or "down"—all the way to the pole to do it. The force causes the same effect when it pulls out the nearest available part of the "frame" within the inhabited world.

Here are some examples of the manipulation of the frame, beginning with a most insignificant survival. Actually this is a useful approach, because the less meaningful the example, the more astonishing is the fact of its surviving. Turkmen tribes of southern Turkestan tell about a copper pillar marking the "navel of the earth," and they state that "only the nine-year-old hero Kara Pār is able to lift and to extract" it.[11] As goes without saying, nobody comments on the strange idea that someone should be eager to "extract the navel of the earth." When Young Arthur does it with Excalibur, the events have already been fitted into a more familiar frame and they provoke no questions.

[11] Radloff, quoted by W. E. Roscher, *Der Omphalosgedanke* (1918), pp. 1f.

In its grandiose style, the *Mahabharata* presents a similar prodigy as follows:

> It was Vishvāmitra who in anger created a second world and numerous stars beginning with Sravana . . . He can burn the three worlds by his splendour, can, by stamping (his foot), cause the Earth to quake. He can *sever the great Meru from the Earth* and hurl it to any distance. He can go round the 10 points of the Earth in a moment.[12]

Vishvāmitra is one of the seven stars of the Big Dipper, this at least has been found out. But each planet is represented by a star of the Wain, and vice versa, so this case does not look particularly helpful.[13]

A cosmic event of the first order can be easily overlooked when it hides modestly in a fairy tale. The following, taken from the Indian "Ocean of Stories," tells of Shiva: "When he drove his trident into the heart of Andhaka, the king of the Asuras, though he was only one, the dart which that monarch had infixed into the heart of the three worlds was, strange to say, extracted."[14]

A plot can also shrink to unrecognizable insignificance when it

[12] Mbh. *1.71*, Roy trans., vol. *1*, p. 171.

[13] The notion of "numerous [newly appointed] stars beginning with Sravana" *should* enlighten us. Sravana, "the Lame," is, in the generally accepted order, the twenty-first lunar mansion, alpha beta gamma Aquilae, also called by the name *Ashvatta*, which stands for a sacred fig tree but which means literally "below which the horses stand" (Scherer, *Gestirnnamen*, p. 158), and which invites a comparison with Old Norse *Yggdrasil*, meaning "the tree below which Odin's horse grazes" (Reuter, *Germanische Himmelskunde*, p. 236). Actually, the solstitial colure ran through alpha beta gamma Aquilae around 300 B.C., and long after the time when it used to pass through one or the other of the stars of the Big Dipper; the *equinoctial* colure, however, comes down very near eta Ursae Majoris. Considering that eta maintains the most cordial relations with Mars in occidental astrology, Vishvāmitra might be eta, and might represent Mars, and that would go well with the violent character of this Rishi. But even if we accept this for a working hypothesis, there remains the riddle of the "second world," i.e., "second" with respect to which "first" world? Although we have a hunch, we are not going to try to solve it here and now. Two pieces of information should be mentioned, however: (1) Mbh. *14.44* (Roy trans., vol. *12*, p. 83) states: "The constellations [= lunar mansions, *nakshatras*] have Sravana for their first"; (2) Sengupta (in Burgess' trans. of *Surya Siddhanta*, p. xxxiv) claims that "the time of the present redaction of the *Mahabharata*" was called "Sravanadi kala, i.e., the time when the winter solstitial colure passed through the *nakshatra* Sravana."

[14] N. M. Penzer, *The Ocean of Story* (1924), vol. *1*, p. 3.

comes disguised as history, but this next story at least has been pinned down to the proper historical character, and even has been checked by a serious military historian like Arrianus, who tells us the following:

> Alexander, then, reached Gordium, and was seized with an ardent desire to ascend to the acropolis, where was the palace of Gordius and his son Midas, and to look at Gordius' wagon and the knot of the chariot's yoke. There was a widespread tradition about this chariot around the countryside; Gordius, they said, was a poor man of the Phrygians of old, who tilled a scanty parcel of earth and had but two yoke of oxen: with one he ploughed, with the other he drove his wagon. Once, as he was ploughing, an eagle settled on the yoke and stayed, perched there, till it was time to loose the oxen; Gordius was astonished at the portent, and went off to consult the Telmissian prophets, who were skilled in the interpretation of prodigies, inheriting—women and children too—the prophetic gift. Approaching a Telmissian village, he met a girl drawing water and told her the story of the eagle: she, being also of the prophetic line, bade him return to the spot and sacrifice to Zeus the King. So then Gordius begged her to come along with him and assist in the sacrifice; and at the spot duly sacrificed as she directed, married the girl, and had a son called Midas.

> Midas was already a grown man, handsome and noble, when the Phrygians were in trouble with civil war; they received an oracle that a chariot would bring them a king and he would stop the war. True enough, while they were discussing this, there arrived Midas, with his parents, and drove, chariot and all, into the assembly. The Phrygians, interpreting the oracle, decided that he was the man whom the gods had told them would come in a chariot; they thereupon made him king, and he put an end to the civil war. The chariot of his father he set up in the acropolis as a thank-offering to Zeus the king for sending the eagle.

> Over and above this there was a story about the wagon, that anyone who should untie the knot of the yoke should be lord of Asia. This knot was of cornel bark, and you could see neither beginning nor end of it. Alexander, unable to find how to untie the knot, and not brooking to leave it tied, lest this might cause some disturbance in the vulgar, smote it with his sword, cut the knot, and exclaimed, "I have loosed it!"—so at least say some, but Aristobulus puts it that *he took out the pole pin*, a dowel driven right through the pole, holding the knot together, and so *removed the yoke from the pole*. I do not attempt to be precise how Alexander actually dealt with this knot. Anyway, he and his suite left the wagon with the impression

that the oracle about the loosed knot had been duly fulfilled. It is certain that there were that night thunderings and lightnings, which indicated this; so Alexander in thanksgiving offered sacrifice next day to whatever gods had sent the signs and certified the undoing of the knot.[15]

Without going now into the relevant comparative material it should be stressed that in those cases where "kings" are sitting in a wagon (Greek *hamaxa*), i.e., a four-wheeled truck, it is most of the time *Charles' Wain*.

Alexander was a true myth builder, or rather, a true myth-attracting magnet. He had a gift for attracting to his fabulous personality the manifold tradition that, once, had been coined for Gilgamesh.

But the time is not yet ripe either for Alexander or for Gilgamesh, nor for further statements about deities or heroes who could pull out pins, plugs and pillars. The next concern is with the decisive features of the mythical landscape and their possible localization, or their fixation in time. It is essential to know where and when the first whirlpool came into being once Grotte, Amlodhi's Mill, had been destroyed. This is, however, a misleading expression because our terminology is still much too imprecise. It would be better to say the first exit from, or entrance to, the whirlpool. It appears advisable to recapitulate the bits of information that have been gathered on the whirlpool as a whole:

The maelstrom, result of a broken mill, a chopped-down tree, and the like, "goes through the whole globe," according to the Finns. So does Tartaros, according to Socrates. To repeat it in Guthrie's words: "The earth in this myth of Socrates is spherical, and Tartaros, the bottomless pit, is represented in this mythical geography by a chasm which pierces the sphere right through from side to side."[16]

It is source and mouth of all waters.

It is *the* way, or one among others, to the realm of the dead.

Medieval geographers call it "Umbilicus Maris," Navel of the Sea, or "Euripus."

[15] *Anabasis of Alexander* 2.3.1–8 (Robson trans., LCL).
[16] *Orpheus and Greek Religion* (1952), p. 168.

Antiochus the astrologer calls Eridanus proper, or some abstract topos not far from Sirius, "zalos," i.e., whirlpool.

M. W. Makemson looks for the Polynesian whirlpool, said to be "at the end of the sky," "at the edge of the Galaxy," in Sagittarius.

A Dyak hero, climbing a tree in "Whirlpool-Island," lands himself in the Pleiades.

But generally, one looks for "it" in the more or less northwest–north-northwest direction, a direction where, equally vaguely, Kronos-Saturn is supposed to sleep in his golden cave notwithstanding the blunt statements (by Homer) that Kronos was hurled down into deepest Tartaros.

And from those "infernal" quarters, particularly from the (Ogygian) Stygian landscape, "one"—who else but the souls?—sees the celestial South Pole, invisible to us.

The reader might agree that this summary shows clearly the insufficiency of the general terminology accepted by the majority. The verbal confusion provokes sympathy for Numenius (see above, p. 188), and the Third Vatican Mythographer who took the rivers for planets, their planetary orbs respectively. We think that the whirlpool stands for the "ecliptical world" marked by the whirling planets, embracing everything which circles obliquely with respect to the polar axis and the equator—oblique by 23½ degrees, more or less, each planet having its own obliquity with respect to the others and to the sun's path, that is, the ecliptic proper. It has been mentioned earlier (p. 206, n. 5) that in the axis of the Roman circus was a *Euripus*, and altars of the three outer planets (Saturn, Jupiter, Mars), and the three inner planets (Venus, Mercury, Moon) on both sides of the pyramid of the sun, and that there were not more than seven circuits because the "planets are seven only."

The ecliptic as a whirl is only one aspect of the famous "implex." It must be kept in mind that being the seat of all planetary powers, it represented, so to speak, the "Establishment" itself. There is no better symbol of the thinking of those planet-struck Mesopotamian civilizations than the arrogant plan of the royal cities themselves, as it has been patiently reconstructed by generations of Orientalists and archaeologists. Nineveh proclaimed itself as the seat of stable order and power by its seven-times crenellated circle of walls, colored with the seven planetary colors, and so thick that chariots could run along the top. The planetary symbolism spread to India, as was seen in chapter VIII, and culminated in that prodigious cosmological diagram that is the temple of Barabudur in Java.[17] It is still evident in the innumerable *stupas* which dot the Indian countryside, whose superimposed crowns stand for the planetary heavens. And here we have the Establishment seen as a Way Up and Beyond, as Numenius would have seen immediately, the succession of spheres of transition for the soul, a quiet promise of transcendence which marks the Gnostic and Hinduistic scheme. The skeleton map will always lack one or the other dimension. The Whirl is then a way up or a way down? Heraclitus would say both ways are one and the same. You cannot put into a scheme everything at once.

This general conception of the whirlpool as the "ecliptical world" does not, of course, help to understand any single detail. Starting from the idea of the whirlpool as a way to the other world, one must look at the situation through the eyes of a soul meaning to go there. It has to move from the interior outwards, to "ascend" from the geocentric earth through the planetary spheres "up" to the fixed sphere, that is, right through the whole whirlpool, the ecliptical world. But in order to leave the ecliptical frame, there must be a station for changing trains at the equator. One would expect this station to be at the crossroads of ecliptical and equatorial coordinates at the equinoxes. But evidently, this was not the arrangement. A far older route was followed. It is true that it sometimes looks as though the transfer point were at the equinoxes. The astrological tradition that followed Teukros,[18] for example, provided a rich offering of celestial locations for Hades, the Acherusian lake, Charon the ferryman, etc., all of them under the chapter *Libra*. But this is a trap and one can only hope that many hapless souls have not been deceived. For these astrological texts mean the *sign*

[17] P. Mus, *Barabudur* (1935).
[18] F. Boll, *Sphaera* (1903), pp. 19, 28, 47, 246–51. Antiochus does not mention any of these star groups.

Libra, not the constellation. All "change stations" are found invariably in two regions: one in the South between Scorpius and Sagittarius, the other in the North between Gemini and Taurus; and this is valid through time and space, from Babylon to Nicaragua.[19] Why was it ever done in the first place? Because of the Galaxy, which has its crossroads with the ecliptic between Sagittarius and Scorpius in the South, and between Gemini and Taurus in the North.

[19] The notion is not even foreign to the cheering adventures of *Sun*, the Chinese Monkey (*Wou Tch'eng Ngen*, French trans. by Louis Avenol [1957]). One day, two "harponneurs des morts" get hold of him, claiming that he has arrived at the term of his destiny, and is ripe for the underworld. He escapes, of course. The translator remarks (vol. *1*, p. iii) that it is the constellation Nan Teou, the Southern Dipper, that decides everybody's death, and the orders are executed by these "harponneurs des morts." The Southern Dipper consists of the stars mu lambda phi sigma tau zeta Sagittarii (cf. G. Schlegel, *L'Uranographie Chinoise* [1875], pp. 172ff.; L. de Saussure, *Les Origines de l'Astronomie Chinoise* [1930], pp. 452f.).

Chapter XVIII

The Galaxy

Voie Lactée, soeur lumineuse
des blanches rivières de Canaan,
et des corps blancs de nos amoureuses,
nageurs morts suivrons nous d'ahan
ton cours vers d'autres nébuleuses.

APOLLINAIRE,
La Chanson du Mal-Aimé

Men's spirits were thought to dwell in the Milky Way between incarnations. This conception has been handed down as an Orphic and Pythagorean tradition[1] fitting into the frame of the migration of the soul. Macrobius, who has provided the broadest report on the matter, has it that souls ascend by way of Capricorn, and then, in order to be reborn, descend again through the "Gate of Cancer."[2] Macrobius talks of *signs;* the constellations rising at the solstices in his time (and still in ours) were Gemini and Sagittarius: the "Gate of Cancer" means Gemini. In fact, he states explicitly (*1.12.5*) that this "Gate" is "where the Zodiac and the Milky Way intersect." Far away, the Mangaians of old (Austral Islands, Polynesia), who kept the precessional clock running instead of switching over to "signs," claim that only at the evening of the solstitial days can spirits enter heaven, the inhabitants of the

---

[1] See F. Boll, *Aus der Offenbarung Johannes* (1914), pp. 32, 72 (the first accepted authority has been Herakleides of Pontos); W. Gundel, RE s.v. Galaxias; A. Bouché-Leclerq, *L'Astrologie Grecque* (1899), pp. 22f.; F. Cumont, *After Life in Roman Paganism* (1959), pp. 94, 104, 152f.

[2] *Commentary on the Dream of Scipio 1.12.1–8.*

northern parts of the island at one solstice, the dwellers in the south at the other.[3] This information, giving precisely fixed dates, is more valuable than general statements to the effect that the Polynesians regarded the Milky Way as "the road of souls as they pass to the spirit world."[4] In Polynesian myth, too, souls are not permitted to stay unless they have reached a stage of unstained perfection, which is not likely to occur frequently. Polynesian souls have to return into bodies again, sooner or later.[5]

Two instances of relevant American Indian notions are worth mentioning without discussion. The important thing is that the tradition is there, more or less intact. Among the Sumo in Honduras and Nicaragua their "Mother Scorpion . . . is regarded as dwelling at the end of the Milky Way, where she receives the souls of the dead, and from her, represented as a mother with many breasts, at which children take suck, come the souls of the newborn."[6] Whereas the Pawnee and Cherokee say:[7] "The souls of the dead are received by a star at the northern end of the Milky Way, where it bifurcates, and he directs the warriors upon the dim and difficult arm, women and those who die of old age upon the brighter and easier path. The souls then journey southwards. At the end of the celestial pathway they are received by the Spirit Star, and there they make their home." One can quietly add "for a while," or change it to "there they make their camping place." Hagar takes the "Spirit Star" to be Antares (alpha Scorpii). Whether or not it is precisely alpha, because the star marks the southern "end" of the Galaxy, the southern crossroads with the

---

[3] W. W. Gill, *Myths and Songs from the South Pacific* (1876), pp. 156ff., 185ff.

[4] E. Best, *The Astronomical Knowledge of the Maori* (1955), p. 45.

[5] Since so many earlier and recent "reporters at large" fail to inform us of traditions concerning reincarnation, we may mention that according to the Marquesans "all the souls of the dead, after having lived in one or the other place (i.e., Paradise or Hades) for a very long time, returned to animate other bodies" (R. W. Williamson, *Religious and Cosmic Beliefs of Central Polynesia* [1924], vol. 1, p. 208), which recalls the wording of the case as we know it from book X of Plato's *Republic*.

[6] H. B. Alexander, *Latin American Mythology* (1916), p. 185.

[7] S. Hagar, "Cherokee Star-Lore," in *Festschrift Boas* (1906), p. 363; H. B. Alexander, *North American Mythology*, p. 117.

ecliptic, it is at any rate a star of Sagittarius, or Scorpius.[8] That fits "Mother Scorpion" of Nicaragua and the "Old goddess with the scorpion tail" of the Maya as it also fits the Scorpion-goddess Selket-Serqet of ancient Egypt and the Ishara tam.tim of the Babylonians. Ishara of the sea, goddess of the constellation Scorpius, was also called "Lady of the Rivers" (compare appendix #30).

Considering the fact that the crossroads of ecliptic and Galaxy are crisis-resistant, that is, not concerned with the Precession, the reader may want to know why the Mangaians thought they could go to heaven only on the two solstitial days. Because, in order to "change trains" comfortably, the constellations that serve as "gates" to the Milky Way must "stand" upon the "earth," meaning that they must rise heliacally either at the equinoxes or at the solstices. The Galaxy is a very broad highway, but even so there must have been some bitter millennia when neither gate was directly available any longer, the one hanging in midair, the other having turned into a submarine entrance.

Sagittarius and Gemini still mark the solstices in the closing years of the Age of Pisces. Next comes Aquarius. The ancients, no doubt, would have considered the troubles of these our times, the overpopulation, the "working iniquity in secret," as an inevitable prelude to a new tilting, a new world-age.

But the coming of Pisces was long looked forward to, heralded as a blessed age. It was introduced by the thrice-repeated Great Conjunction of Saturn and Jupiter in Pisces in the year 6 B.C., the star of Bethlehem. Virgil announced the return of the Golden Age under the rule of Saturn, in his famous Fourth Eclogue: "Now the Virgin returns, the reign of Saturn returns, now a new generation descends from heaven on high. Only do thou, pure Lucina, smile on the birth of the child, under whom the iron brood shall first cease, and a golden race spring up throughout the world!" Although pro-

---

[8] This is no slip of the tongue; the zodiacal Sagittarius of Mesopotamian boundary stones had, indeed, the tail of a Scorpion: but we just must not be drowned in the abyss of details of comparative constellation lore, and least of all in those connected with Sagittarius, two-faced as he is, half royal, half dog.

moted to the rank of a "Christian honoris causa" on account of this poem, Virgil was no "prophet," nor was he the only one who expected the return of Kronos-Saturn.[9] "*Iam redit et Virgo, redeunt Saturnia regna.*" What does it mean? Where has Virgo been, supposedly, so that one expected the constellation "back"?

Aratus, in his renowned astronomical poem (95–136), told how Themis-Virgo, who had lived among humans peacefully, retired at the end of the Golden Age to the "hills," no longer mingling with the silver crowd that had started to populate the earth, and that she took up her heavenly abode near Bootes, when the Bronze Age began.[10] And there is Virgil announcing Virgo's return. This makes it easy to guess time and "place" of the Golden Age. One need only turn back the clock for one quarter "hour" of the Precession (about 6,000 years from Virgil), to find Virgo standing firmly at the summer solstitial corner of the abstract plane "earth." "Returning," that is moving on, Virgo would indicate the autumnal equinox at the time when Pisces took over the celestial government of the vernal equinox, at the new crossroads.

Once the Precession had been discovered, the Milky Way took on a new and decisive significance. For it was not only the most spectacular band of heaven, it was also a reference point from which the Precession could be imagined to have taken its start. This would have been when the vernal equinoctial sun left its position in Gemini in the Milky Way. When it was realized the sun *had* been there once, the idea occurred that the Milky Way might mark the abandoned track of the sun—a burnt-out area, as it were, a scar in heaven. Decisive notions have to be styled more carefully, however: so let us say that the Milky Way was a reference "point" from which the Precession could be termed to have taken its start, and that the idea which occurred was not that the Milky Way

[9] See, for example, A. A. Barb, "St. Zacharias the Prophet and Martyr," *Journal of the Warburg and Courtauld Institutes 11* (1948), pp. 54f., and "Der Heilige und die Schlangen," MAGW *82* (1953), p. 20.

[10] Cf. Al-Biruni, dealing with the Indian ages of the world, and quoting the above passages from Aratus with a scholion (*Alberuni's India*, trans. E. C. Sachau [1964], vol. *1*, pp. 383–85).

might mark the abandoned track of the sun, but that the Milky Way was an image of an abandoned track, a formula that offered rich possibilities for "telling" complicated celestial changes.

With this image and some additional galactic lore, it is now possible to concentrate on the formula by which the Milky Way became the way of the spirits of the dead, a road abandoned by the living. The abandoned path is probably the original form of the notions insistently built around a projected Time Zero. If the Precession was seen as the great clock of the Universe, the sun, as it shifted at the equinox, remained the measure of all measures, the "golden cord," as Socrates says in Plato's *Theaetetus* (153c). In fact, apart from the harmonic intervals, the sun was the only absolute measure provided by nature. The sun must be understood to be conducting the planetary fugues at any given moment as Plato also showed in the *Timaeus*. Thus, when the sun at his counting station moved on toward the Milky Way, the planets, too, were termed to hunt and run this way.

This does not make very sound geometrical sense, but it shows how an image can dominate men's minds and take on a life of its own. Yet the technical character of these images should not be forgotten, and it is to prevent this that the verbs "to term" and "to spell out" are used so often instead of the customary expression "to believe."

To the American Plains Indians, the Milky Way was the dusty track along which the Buffalo and the Horse once ran a race across the sky.[11] For the Fiote of the African Loango Coast the race was run by Sun and Moon.[12] The East African Turu took it for the "cattle track" of the brother of the creator,[13] which is very close to the Greek legend of Herakles moving the herd of Geryon.[14] The convergence of so many animal tracks along this heavenly way is, once again, not a pointless conjunction of fancies. The Arawak of Guyana call the Galaxy "the Tapir's way." This is confirmed in a

[11] J. Mooney, *Myths of the Cherokee*, 19th ARBAE 1897–98 (1900), p. 443.
[12] E. Pechuel-Loesche, *Volkskunde von Loango* (1907), p. 135.
[13] S. Lagercrantz, "The Milky Way in Africa," *Ethnos* (1952), p. 68.
[14] See W. Gundel, RE s.v. Galaxias.

tale of the Chiriguano and some groups of the Tupi-Guarani of South America. According to Lehman-Nitsche, these people speak of the Galaxy as "the way of the true father of the Tapir," a Tapir-deity which is itself invisible.[15] Now, if this hidden deity turns out to be Quetzalcouatl himself, ruler of the Golden Age town Tollan, no other than "Tixli cumatz," the tapir-serpent dwelling in the "middle of the sea's belly," as the Maya tribes of Yucatán describe him,[16] the allusions begin to focus. Finally, the actual scheme is found in that Cuna tradition described earlier: the Tapir chopped down the "Saltwater Tree," at the roots of which is God's whirl-pool, and when the tree fell, saltwater gushed out to form the oceans of the world.

Should the Tapir still seem to lack the appropriate dignity, some Asiatic testimonies should be added. The Persian Bundahishn calls the Galaxy the "Path of Kay-us," after the grandfather and co-regent of Kai Khusrau, the Iranian Hamlet.[17] Among the Altaic populations the Yakuts call the Milky Way the "tracks of God," and they say that, while creating the world, God wandered over the sky; more general in use seems to have been the term "Ski-tracks of God's son," whereas the Voguls spelled it out "Ski-tracks of the forest-man." And here the human tracks fade out, although the snowshoes remain. For the Tungus the Galaxy is "Snowshoe-tracks of the Bear." But whether the figure is the son of God, the forest-man, or the Bear, he hunted a stag along the Milky Way, tore it up and scattered its limbs in the sky right and left of the white path, and so Orion and Ursa Major were separated.[18] The "Foot of the Stag" reminded Holmberg immediately of the "Bull's Thigh" of ancient Egypt—Ursa Major. With his penetrating insight he might easily have gone on to recognize, in that potent thigh, the isolated "one-leg" of Texcatlipoca, Ursa Major again, in Mexico—the day-sign

---

[15] O. Zerries, "Sternbilder als Ausdruck jägerischer Geiteshaltung in Südamer-ika," *Paideuma* 5 (1951), pp. 220f.

[16] E. Seler, *Gesammelte Abhandlungen* (1961), vol. 4, p. 56.

[17] Bdh. V B 22, B. T. Anklesaria, *Zand-Akasih. Iranian or Greater Bundahishn* (1956), pp. 69, 71.

[18] U. Holmberg, *Die religiösen Vorstellungen der altaischen Völker* (1938), pp. 201f.

"Crocodile" (Cipactli) had bitten it off—the great Hunrakán (= 1 leg) of the Maya Quiche.[19]

There is an insistent association here, right below the surface, which is still revealed by the old Dutch name for the Galaxy, "Brunelstraat." Brunel, Bruns, Bruin (the Brown) is the familiar name of the bear in the romance of Reynard the Fox, and is as ancient as anything that can be traced.[20] It is a strange lot of characters that were made responsible for the Milky Way: gods and animals leaving the path that had been used at "creation" time.[21] But where did they go, the ones mentioned, and the many whom we have left out of consideration? It depends, so to speak, from where they took off. This is often hard to determine, but the subject of "tumbling down" will be dealt with next.

As for Virgo, who had left the "earth" at the end of the Golden Age, her whereabouts in the Silver Age could have been described

[19] Going farther south, he would have found there again the lining up of Ursa and Orion and the violent tearing up of celestial figures. Says W. E. Roth ("An Inquiry into the Animism and Folk-lore of the Guiana Indians," 30th ARBAE 1908–09 [1915], p. 262; cf. Zerries, pp. 220f.) of the Indians of Guiana: "All the legends relating to the constellations Taurus and Orion have something in common in the detail of an amputated arm or leg." And that goes for parts of Indonesia too. But then, Ursa Major is the thigh of a *Bull*, and the zodiacal Taurus is so badly amputated, there is barely a half of him left. More peculiar still, in later Egyptian times it occurs, if rarely, that Ursa is made a ram's thigh (see G. A. Wainwright, "A Pair of Constellations," in *Studies Presented to F. L. Griffith* [1932], p. 373); and on the round zodiac of Dendera (Roman period) we find a ram sitting on that celestial leg, representing Ursa, and it even looks back, as befits the traditional zodiacal Aries. We must leave it at that.

[20] The notion of the Milky Way as "Brunelstraat" seems to be present in ancient India: the *Atharva Veda 18.2.31* mentions a certain path or road called *rikshaka*. Riksha is the bear in both senses, i.e., the animal and Ursa Major (see H. Grassmann, *Wörterbuch zum Rig-Veda* [1915] s.v. Riksha). Whitney (in his translation of AV, p. 840) suggested *rikshaka* as a road "infested by bears (?)." A. Weber, however, proposed to identify *rikshaka* with the Milky Way ("Miszellen aus dem indogermanischen Familienleben," in *Festgruss Roth* [1893], p. 138). Since the whole hymn AV *18.2* contains "Funeral Verses," and deals with the voyage of the soul, that context too would be fitting. (That the souls have to first cross a river "rich with horses" is another matter.)

[21] The shortest abbreviation: the Inca called Gemini "creation time" (Hagar, in *14th International Amerikanisten-Kongress* [1904], p. 599f.). But the very same notion is alluded to, when Castor and Pollux (alpha beta Geminorum) are made responsible for the first fire sticks, by the Aztecs (Sahagún) and, strange to say, by the Tasmanians. (See below, chapter XXIII, "Gilgamesh and Prometheus.")

as being "in mid-air." Many iniquitous characters were banished to this topos; either they were thrown down, or they were sent up —Lilith dwelt there for a while, and King David,[22] also Adonis,[23] even the Tower of Babel itself, and first of all the Wild Hunter (appendix #20). This assembly of figures "in mid-air" helps to give meaning to an otherwise pointless tale, a veritable fossil found in Westphalian folklore: "The Giants called to Hackelberg [= Odin as the Wild Hunter] for help. He raised a storm and removed a mill into the Milky Way, which after this is called the Mill Way."[24] There are other fossils, too, the wildest perhaps being that of the Cherokee who called the Galaxy "Where the dog ran." A very unusual dog it must have been, being in the habit of stealing meal from a corn mill owned by "people in the South" and running with it to the North; the dog dropped meal as he ran and that is the Milky Way.[25] It is difficult here to recognize Isis scattering ears of wheat in her flight from Typhon.[26] And yet, the preference of the very many mythical dogs, foxes, coyotes—and even of the "way-opening" Fenek in West Sudan—for meal and all sorts of grain—more correctly "the eight kinds of grain"—a trait which is hardly learned by eavesdropping on Mother Nature, could have warned the experts to beware of these doggish characters. They are not to be taken at their pseudo-zoological face value.

Thus, everybody and everything has left the course, Wild Hunter, dog and mill—at least its upper half, since through the hole in the lower millstone the whirlpool is seething up and down.

[22] See J. A. Eisenmenger, *Entdecktes Judenthem* (1711), vol. *1*, p. 165; vol. *2*, pp. 417ff.

[23] "Es ton ēera," see F. K. Movers, *Die Phönizier* (1967), vol. *1*, p. 205.

[24] J. Grimm, TM, pp. 1587f.

[25] Mooney, pp. 253, 443.

[26] See R. H. Allen, *Star Names* (1963), p. 481; W. T. Olcott, *Star Lore of All Ages* (1911), p. 393.

CHAPTER XIX

The Fall of Phaethon

Quel del sol, che suiando, fu combusto
Per l'orazion della terra devota
Quando fu Giove arcanamente giusto

DANTE, *Purg.* XXIX.*118*

THE GREAT AND OFFICIAL myth concerning the Galaxy is Phaethon's transgression and the searing of the sky in his mad course. Manilius tells it in his astrological poem:[1]

> ... *this was once the Path*
> *Where Phoebus drove; and that in length of Years*
> *The heated track took Fire and burnt the Stars.*
> *The Colour changed, the Ashes strewed the Way,*
> *And still preserve the marks of the Decay:*
>
> *Besides, Fame tells, by Age Fame reverend grown,*
> *That Phoebus gave his Chariot to his Son,*
> *And whilst the Youngster from the Path declines*
> *Admiring the strange Beauty of the Signs,*
> *Proud of his Charge, He drove the fiery horse,*
> *And would outdo his Father in his Course.*
> *The North grew warm, and the unusual Fire*
> *Dissolv'd its Snow, and made the Bears retire;*
> *Nor was the Earth secure, each Contrey mourn'd*
> *The Common Fate, and in its City's burn'd.*
> *Then from the scatter'd Chariot Lightning came,*

---

[1] *1*.730–49. Anonymous translation (T.C.) London, 1697; reprinted 1953 by National Astrological Library, Washington, D.C., p. 44.

> *And the whole Skies were one continued Flame.*
> *The World took Fire, and in* new kindled Stars
> *The bright remembrance of its Fate it bears.*

The myth of Phaethon has been told broadly and with magnificent fantasy by Ovid (*Met. 1.747–2.400*) and by Nonnos (*Dionysiaka* Book *38*). Gibbon in his old age, commemorating his own adolescence, speaks of his rapt discovery of the beauty of Latin poetry as he read Ovid's description of the tragic venture of Phaethon. The story goes on that Helios, taking his oath by the waters of Styx, promised to fulfill any wish of his rash young son Phaethon, who was visiting him for the first time. The boy had only one desire, to drive the Sun's chariot once, and the most desperate requests of his father could not move him to change his mind. Although knowing well that nothing could prevent the fatal ending of this adventure, Helios did his best to teach Phaethon all the dangers lurking at every step of the way—a welcome occasion for both poets to elaborate the paternal admonitions into some kind of "introduction to astronomy." As the father feared, Phaethon was incapable of managing the horses and came off the proper path; Ovid has it that the boy dropped the reins at the sight of Scorpius. Unbelievable confusion results; no constellation remains in its place, and the Earth is terribly scorched. In despair "she" cries aloud to Jupiter to make him act immediately: "Look how your heavens blaze from pole to pole—if fire consumes them the very universe will fall to dust. In pain, in worry, Atlas almost fails to balance the world's hot axis on his shoulders."[2] And Nonnos states (*38.350ff.*): "There was tumult in the sky *shaking the joints* of the immovable universe; the very *axle bent* which runs through the middle of the revolving heavens. Libyan Atlas could hardly support the self-rolling firmament of stars, as he rested on his knees with bowed back under this greater burden."

Zeus has to intervene and hurls his thunderbolt at the boy. Phaethon falls into the river Eridanus where, according to Apollonios

---

[2] *Met.* 2.294–97: circumspice utrumque:/ fumat uterque polus quos si vitiaverit ignis/atria vestra ruent Atlas en ipse laborat/ vixque suis umeris candentem sustinet axem.

Rhodios, the stench of his half-burned corpse made the Argonauts sick for several days when they came upon it in their travels (4.619–23).

The Phaethon story has often been understood to commemorate some great flashing event in the skies, whether comet or meteor. Everyone rushes by instinct—more accurately, habit—for a so-called natural explanation. But on examination, the case turns out not to be so easy. The narrating of the cataclysm may be fanciful and impressionistic, as if the poets enjoyed an emotional release from the regularity of celestial orbs, but their account also makes technical sense, as anyone would suspect who has read Stegemann's[3] solid inquiry into Nonnos as the heir to Dorotheos of Sidon's tight-knit astrology. As for Ovid, his standing as a scholar is by now unchallenged and, in fact, he hints at rigid cosmological formulae with surprising authority. In his description of the "hidden mountains" emerging from the waves, when the seas shrank into sand (2.260ff.)—they rise as "new islands." How much better does this image of "mountain peaks" and "islands" illustrate the stars of a constellation rising, one after the other (at vernal equinox), than, for instance, the Icelandic wording of the emerging of "a new earth"!

In any case, an independent confirmation emerges in Plato's version of the crisis, as he gives it in *Timaeus* 22CE. The Egyptian priest talking with Solon states that the legend of Phaethon "has the air of a fable; but the truth behind it is a *deviation* [*parallaxis*] of the bodies that revolve in heaven round the earth, and a destruction, *occurring at long intervals*, of things on earth by a great conflagration." This is a clear statement, and one in accordance with Ovid and with Nonnos, as it should be, since it has to do with a *Pythagorean* tradition: Aristotle tells us so.[4]

---

[3] *Astrologie und Universalgeschichte* (1930).

[4] *Meteorologica* 1.8.345A: "The so-called Pythagoreans give two explanations. Some say that the Milky Way is the path taken by one of the stars at the time of the legendary fall of Phaethon; others say that it is the circle in which the sun once moved. And the region is supposed to have been scorched or affected in some other such way as a result of the passage of these bodies." See also H. Diels, *Doxographi*, pp. 364f. = Aetius III.1. (In former times when classical authors were not yet eagerly prefixed with as many "pseudos" as possible, this was Plutarch, *De placitis* 3.1.)

The Pythagoreans were neither idle storytellers, nor were they even mildly interested in unusual sensational "catastrophes" caused by meteors, and the like. Actually, the Egyptian priest said to Solon, concerning the legend of Phaethon, "the story current *also* in your part of the world." Where, then, is the story in Egypt? Since the Egyptian cosmological language was more technical, in the old sense, than that of the Greeks, it will take some time to find out the exact parallel. Anyhow, in Egypt the down-hurled Phaethon would have been termed "the lost eye," or rather one among the "lost eyes." The eye was "lost" in the so-called "mythical source of the Nile," the source of all waters. So it is surprising that Ovid knew (*Met. 2.254ff.*) that because of Phaethon's fall, "Nile ran in terror to the end of the earth to hide its head which now is still unseen."[5] Leaving the Egyptian case for the time being, it is appropriate here to cite two widely separated survivals concerned with the Phaethon theme. They are useful because they come from points far removed from the Greek landscape and consequently cannot be connected with any local catastrophes which are supposed to have made such a tremendous impression on the Greek mind. The Fiote of the African Loango Coast, already mentioned, say: "The Star Way [Galaxy] is the road for a funeral procession of a huge star which, once, shone brighter from the sky than the Sun."[6] Conveniently short, and no technicalities. The Northwest American version is broader. Because of the absence of chariots in pre-Columbian America,[7] the Phaethon figure of the Bella Coola Indians, who had come to visit his father Sun by means of an arrow-chain, wants to carry Sun's torches in his stead. Helios agrees, but he warns his son not to make mischief and burn people. "In the morning," he says, "I light one torch, slowly increasing their number until high-noon. In the afternoon I put them out again little by little." The next morning, "Phaethon," climbing the path of the Sun, not only kindled all the torches he had, he did so much too early, so that the earth became red hot: the woods began to

---

[5] Nilus in extremum fugit perterritus orbem/ occuluitque caput, quod adhuc latet.

[6] E. Pechuel-Loesche, *Volksunde von Loango* (1907), p. 135.

[7] See H. S. Gladwin, *Men out of Asia* (1947), pp. 356–59, for this "feature."

burn, the rocks split, many animals jumped into the waters, but the waters began to boil, too. "Young woman," the mother of the Bella Coola Phaethon, covered men with her coat and succeeded in saving them. But Father Sun hurled his offspring down to earth, telling him: "From now on you shall be the Mink!"[8]

It is necessary to revive some other very ancient ideas lost to our time. That Eridanus was the river Po in northern Italy was a common and simple notion in the Greece of Euripides. In one of his great tragedies (*Hippolytus*), the chorus yearns for a flight away from the world of guilt, to mountains and clouds, to lands far off:

> *Where the waters of Eridanus are clear*
> *And Phaethon's sad sisters by his grave*
> *Weep into the water, and each tear*
> *Gleams, a drop of amber, in the waves.*

Any hearer would have understood that Phaethon's sad sisters were the poplars lining the banks of the river, and that the "drop of amber" was an allusion to the riches of the "amber route" which led from the Baltic Sea to the familiar reaches of the Adriatic. So far so good. But what can be made of Strabo, a still later author (5.215) who called Eridanus "nowhere on earth existing" and thus referred clearly to the constellation Eridanus in heaven, and what does Aratus (360) mean when he talks of "those poor remains of Eridanus" because the river was "burnt up through Phaethon's fall." Is this the very same river, ample and lined with poplars, which runs into the delta of the Po?

Apollonios of Rhodes, in recounting the heroic travels of the Argonauts, carefully preserved the double level of meaning, for the adventures are set in an earthly context, yet they make, geographically speaking, no sense at all. The explorers do sail up the Po, where they are confronted, as was said, with the stench of Phae-

---

[8] W. Krickeberg, *Indianermärchen aus Nordamerika* (1924), pp. 224f., 396. Cf. E. Seler, *Gesammelte Abhandlungen*, vol. 5, p. 19. A mere mink might appear to us, today, as insignificant, like the tapir, or as the "Mouse-Apollo"—we fall for mere "words" and "names" only too easily. *This* particular Mink introduces the tides, steals the fire, fights with the "winds," playing Adapa, Prometheus, Phaethon all at the same time.

thon's remains—but those might be located higher up in a waterfall in the Alps, near the Dammastock, as one distinguished scholar would like to suggest. For the *Argo* moves from the Po into Lake Geneva and the Rhône, goes down it to the sea again and sails out following the same longitude; then, by a considerable feat of portage crosses the Sahara all the way to the coast of West Africa, and reaches Fernando Po. This is at least how those who understand the text as geography read it without blinking. Surely, it is closer to common sense to treat Eridanus as a feature of the skies, where it is already clearly marked together with Argo; and to treat the other features accordingly will give at least a significant story, although it will not dispel the mystery of the Argonauts.

Thus tradition holds that after the dreadful fall of Phaethon, and when order was re-established, Jupiter "catasterized" Phaethon, that is, placed him among the stars, as Auriga (Greek Hēniochos and Erichthonios); and at the same time Eridanus was catasterized. Manilius hinted at this event only with the lines "The world took fire, and in new kindled stars / the bright remembrance of its fate it bears." Nonnos gave a more detailed report (*38*.424–31):[9]

> But father Zeus fixed Phaethon in Olympus, like a charioteer, and bearing that name. As he holds in the radiant Chariot of the heavens with shining arm, he has the shape of a Charioteer starting upon his course, as if even among the stars he longed again for his father's car. The fire-scorched river also came up to the vault of the stars with consent of Zeus, and in the starry circle rolls the meandering stream of burning Eridanus.

Now, in times when myth was still a serious form of thought, objects were not identified in heaven which did not belong there in the first place. The problem which arose later is the one raised by Richard H. Allen, who remarks that "the Milky Way was long known as Eridanus, the Stream of Ocean,"[10] and by the translator of Nonnos, W. H. D. Rouse, who shortly annotated Eridanus as "the Milky Way." It takes some nerve to say of the

[9] See also F. X. Kugler, *Sibyllinischer Sternkampf und Phaethon* (1927), pp. 44, 49.
[10] *Star Names* (1963), p. 474.

Galaxy that it meanders—actually the Greek text has it that it moves like a helix (*helissetai*). But apart from this incongruent image of the "helixing" Milky Way, the myth of Phaethon was meant by the Pythagoreans to tell of the departure of the Sun and planets from their former path, and the enthroning of Eridanus, which together with Auriga was to take over the function of the Milky Way: that is why they were "catasterized" together. Admittedly, one faces a frightening confusion between the rivers in heaven and those on earth, and the names which were given to both kinds of streams, but with patience the threads can be disentangled.

Taking the rivers of our globe first, it was not only the Po that received the name of Eridanus, but the Rhône,[11] and the Nile and the Ganges. Finally in Higgins' *Anacalypsis* there is a quote, without the ancient source but reasonably reliable: "Ganges which also is called Po."[12] Thus it is not surprising that much later, in medieval times, several redactions of the Alexander Romance show different opinions about the river used by the king to travel to paradise in order to win immortality. In a French prose novel of the 14th century Alexander sails the Nile upstream, whereas in a Latin version of the 12th century, he uses the Ganges: as the Indians had told him, the Ganges had its source in paradise.[13] So have, indeed, all great rivers of myth.

In the sky, the number of candidates for election is three. Besides the Milky Way, Eratosthenes' authoritative Catasterisms called the constellation Eridanus Nile or Ocean.[14] But the astrologers Teukros and Valens listed Eridanus among the *paranatellonta* of Aquarius. Paranatellonta are the constellations that "rise at the same time" as a given one, i.e., in this instance, as Aquarius. That is, they called the gush from the jug of Aquarius Eridanus. More awkward still,

[11] For Po and Rhône and the joining of their waters, see A. Dieterich, *Nekyia* (1893), p. 27, quoting Pliny and Pausanias.

[12] (1927 repr.), p. 357: Ganges qui et Padus dicitur. As concerns the general idea of Eridanus being in India, see O. Gruppe, *Griechische Mythologie* (1906), p. 394, referring to Ktesias.

[13] F. Kampers, *Mittelalterliche Sagen vom Paradiese* (1897), pp. 72f.

[14] No. 37 (Robert ed. [1878], pp. 176f.).

this gush from Aquarius' jar was meant to join *our* constellation Eridanus below Piscis Austrinus.[15] Says Manilius (*1*.438ff.):

> Next swims the Southern Fish, which bears a Name
> From the South-Wind, and spreads a feeble Flame.
> To him the Flouds in spacious windings turn
> One fountain flows from cold Aquarius' Urn;
> And meets the other where they joyn their Streams
> One Chanel keep, and mix the starry Beams.

Eratosthenes' Catasterisms bring one more complication into the picture, but it is one which leads, finally, to the decisive insight. Differing from those of Aratus (360f.) and from Ptolemy, it counts Canopus in the constellation Eridanus, instead of Argo, and thus gives the river a different direction.[16] The whole "Gordian knot" of misapprehensions hinges upon the name *Eridanus*, and one can do nothing better than to follow the good example set by Alexander and "pull out the pole pin." Eridanus, lacking a decent Greek etymology, finds a reasonable derivation from *Eridu*, as was proposed by Kugler, Eridu being the seat of Enki-Ea, Sumerian ^{mul}NUN^{ki} = Canopus (alpha Carinae).[17] Eridu marked, and *meant*, the "confluence of the rivers," a topos of highest importance, to which, beginning with Gilgamesh, the great "heroes" go on a pilgrimage trying in vain to gain immortality—including Moses according to the 18th Sura of the Koran. Instead of this unobtainable boon, they gain "the measures," as will be seen. "Eridu" being known as the "confluence of the rivers," Eridanus had to join, by definition so to speak, some "river" somewhere in the South, or it had to flow straightaway into Eridu-Canopus, as the Catasterisms claimed. There have been more drastic "solutions" still. The first is given by Servius (to *Aeneid* 6.659) who pretends Eridanus and

[15] F. Boll, *Sphaera* (1903), pp. 135-38.
[16] See L. Ideler, *Sternnamen* (1809), p. 231; see also E. Maass, *Commentariorum in Aratum Reliquiae* (1898), p. 259.
[17] B. L. van der Waerden, JNES 8 (1949), p. 13; see also P. F. Gössmann, *Planetarium Babylonicum* (1950), 306; J. Schaumberger, *3. Erg.* (1935), pp. 334f.

Phaethon were one and the same.[18] The second, presented by
Michael Scotus,[19] agrees with Servius concerning the identity of
Phaethon and Eridanus, but does much more. He places into the
"sign" Eridanus the "Figura sonantis Canoni"—consisting of seventeen stars—which he calls Canopus and claims that Canopus touches
Argo. And about this enigmatic personage Scotus says that he "hindered the work of the Sun by the tone of his lute, because the
horses listened to it, and enraged Jupiter pierced him with the
lightning."[20]

Eridanus was understood by the astrologers to be the whirlpool
(*zalos*), as has been seen, flowing through the underworld with
its many realms, including those from which one sees the celestial
South Pole. Virgil wrote in the *Georgics* (*1.242f.*): "One pole is
ever high above us, while beneath our feet is seen the other, of black
Styx and the shades infernal." But why was Auriga catasterized at
the same time as Eridanus, and what is the "function" which these
two constellations had to take over from the Milky Way? The
Galaxy was and remains the belt connecting North and South,
above and below. But in the Golden Age, when the vernal equinox
was in Gemini, the autumnal equinox in Sagittarius, the Milky
Way had represented a visible equinoctial colure; a rather blurred
one, to be true, but the celestial North and South were connected
by this uninterrupted broad arch which intersected the ecliptic at
its crossroads with the equator. The three great axes were united,
the galactic avenue embracing the "three worlds" of the gods, the
living and the dead. This "golden" situation was gone, and to Eridanus was bequeathed the galactical function of linking up the
"inhabited world" with the abode of the dead in the (partly) invisible South. Auriga had to take over the northern obligations of the

---

[18] Fabula namque haec est: Eridanus Solis filius fuit. hic a patre inpetrato curru
agitare non potuit, et cum eius errore mundus arderet, fulminatus in Italiae fluvium
cecidit: et tunc a luce ardoris sui Phaethon appellatus est, et pristinum nomen
fluvio dedit: unde mixta haec duo nomina inter Solis filium et fluvium invenimus.

[19] Cf. appendix #10, Vainamoinen's Kantele.

[20] See Boll, pp. 273-75, 540-42: Alii dicunt quodcum impediret opus solis sono
canoni, quia equi attendebant dulcedini sonorum, iratus Jupiter eum percussit
fulmine.

Galaxy, connecting the inhabited world with the region of the gods as well as possible. There was no longer a visible continuous bond fettering together immortals, living and dead: Kronos alone had lived among men in glorious peace.

And here there is a proposition to be made. In order to evaluate it, one has to consider the fact that alpha Aurigae is Capella, the Goat. This remarkable figure was the nurse of infant Zeus in the Dictaean Cave, and out of her skin Hephaistos was later to make the Aegis: Amaltheia. Capella-Amaltheia's Horn was the Horn of Plenty for the immortals, and the source of Nectar and Ambrosia. Mortals called it "second table," dessert so to speak.[21] But there are two shreds of Orphic tradition which seem to be revealing, both handed down to us by Proclus. The first says that Demeter *separated* the food of the gods, splitting it up, as it were, into a liquid and a solid "part," that is, into Ambrosia and Nectar.[22] The second declares that Rhea became Demeter after she had borne Zeus.[23] And *Eleusis*, for us a mere "place name," was understood by the Greeks as "Advent"—the New Testament uses the word for the Advent of Christ. Demeter, formerly Rhea, wife of Kronos, when she "arrived," split up the two kinds of divine food having its source in alpha Aurigae. In other words, it is possible that these traditions about Demeter refer to the decisive shifting of the equinoctial colure to alpha Aurigae.

But one should also look at some other traditions. Turning to India, which is often helpful in its abundance, it was the Ganges that stood for the Galaxy, almost as a matter of course,[24] but the *Mahabharata* and the *Puranas* tell at least how the link was conceived: Ganga was born of the Milky Way. Says the *Vishnu Purana*:[25]

[21] See Athenaeus, *Deipnosophistai* 643a; also 783c, 542a.

[22] *Orphicorum Fragmenta*, ed. O. Kern (1963), frg. 189, p. 216 (Proclus in Cratylus 404b, p. 92, 14 Pasqu.); cf. G. Dumézil, *Le Festin d'Immortalité* (1924), p. 104. See also Roscher, in Roscher s.v. Ambrosia: sitos kai methy, sithos kai oinos, etc.

[23] *Orphicorum Fragmenta*, frg. 145, p. 188.

[24] The same goes for the Jaxartes and Ardvī Surā Anāhitā of Iranian tradition; see H. S. Nyberg, *Die Religionen des Alten Iran* (1966), pp. 260f.

[25] 2.8 (Wilson trans., p. 188).

Having the source in the nail of the great toe of Vishnu's left foot, Dhruva (Polaris) receives her, and sustains her day and night devoutly on his head; and thence the seven Rishis practise the exercise of austerity in her waters, wreathing their braided locks with her waves. The orb of the moon, encompassed by her accumulated current, derives augmented lustre from this contact. Falling from on high, as she issues from the moon, she alights on the summit of Meru (the World Mountain in the North), and thence flows to the four quarters of the earth, for its purification . . . The place whence the river proceeds, for the purification of the three worlds, is the third division of the celestial regions, the seat of Vishnu.

It was, in fact, a colossal event to have "the stream Air-Ganges fall down from Heaven," and its violence was only restrained by Shiva's receiving it in the curls of his hair. One might add that he bore it there "for more than 100 years, to prevent it from falling too suddenly upon the mountain." The Indian imagination is freewheeling, and cares little for time sequence, but it is clear that the flow is perpetual. Were it not for Shiva's hair acting as a catchment, the earth would have been flooded by the Waters Above. They come, as was just quoted, from the third region of the sky, the "path of Vishnu" between Ursa Major and the Pole Star. Wilson stated in 1840: "The situation of the sources of the Ganges in heaven identifies it with the Milky Way."[26]

But if the flow is perpetual, it still had a point of "beginning" and this is found in the *Bhagavata Purana* (Wilson, p. 138, n. 11): "The river flowed over the great toe of Vishnu's left foot, which had previously, as he lifted it up, made a fissure in the shell of the mundane egg, and thus gave entrance to the heavenly stream." How can the Milky Way pour its waters over Polaris? And how can it

[26] The Chinese report as given by Gustave Schlegel (*L'Uranographie Chinoise* [1875; repr. 1967], p. 20) is shorter but it points to the same fanciful conception. "La fleuve céleste se divise en deux bras près du pôle Nord et va de là jusqu'au pôle Sud. Un de ses bras passe par l'astérisme Nan-teou (lambda Sagittarii), et l'autre par l'astérisme Toung-tsing (Gemeaux). Le fleuve est l'eau céleste, coulant à travers les cieux et se précipitant sous la terre."

Nan-teou is the "Southern Bushel": mu lambda phi sigma tau zeta Sagittarii; the Northern Bushel = the Big Dipper.

Although we agree with Phyllis Ackerman's view (in *Forgotten Religions* [1950], p. 6): "The Nile, however, (like many, if not all sacred rivers originally—compare the Ganges) is the earthly continuation of the Milky Way," we maintain that the mere recognition does not help to restore sense and meaning to the myth.

flow to the four quarters of the earth? Indian diagrams remained fanciful, in the same way as Western medieval ones. It takes some time for one who looks at the great tympanon at Vézelay to realize that here is a space-time diagram, as it were, of world history centered on the figure of Christ. The effect is all the greater for the transpositions. It was not wholly absurd, either, for archaic cosmology to have double locations, one, for instance, on the ecliptic and one circumpolar. If Tezcatlipoca drilled fire at the pole to "kindle new stars," if the Chinese Saturn had his seat there too, so could Vishnu's toe have bilocation: one "above" in the third region, the other in beta Orionis-Rigel (the Arabian word for "foot"), the "source" of Eridanus. (And might not Rigel-the-source stand also for Oervandil's Toe, catasterized by Thor?) For Rigel marked the way to Hades in the tradition of the Maori of New Zealand as well as in the Book of Hermes Trismegistos.

Fanciful, assuredly, but neither the real Milky Way nor the terrestrial Ganges offered any basis for the imagery of a river flowing to the four quarters of the earth "for the purification of the three worlds." One cannot get away from the "implex" and it is now necessary to consider the tale of a new skeleton map, alias skambha: the equinoctial colure had shifted to a position where it ran through stars of Auriga and through Rigel. *Skambha*, as we have said, was the World Tree consisting mostly of celestial coordinates, a kind of wildly imaginative armillary sphere. It all had to shift when one coordinate shifted.

There are stylistic means other than "catasterizations," that is, being promoted to heaven among the constellations, to describe changed circumstances in the sky. Thus, a Babylonian cuneiform tablet states: "The Goat-Star is also called the witch-star; the divine function of Tiamat it holds in its hands." The Goat-Star (^{mul}UZA $=$ enzu), apart from representing Venus, "rises together with Scorpius" and has been identified with Vega.[27] If one can rely on this identification, it seems to describe the situation as seen from across the sky: the shifting from Sagittarius to Scorpius, and Vega

[27] Gössmann, 145; van der Waerden, JNES *8*, p. 20.

taking over the northern part of the "function" of the Galaxy. That Tiamat is the Milky Way, and no "Great Mother" in the Freudian sense, any more than Ganga, Anahita and others, seems by now obvious. And the same is true of Egyptian Nūt,[28] but the story has different terms there: Mother Nūt is changed into a cow and ordered to "carry Rā." (It is, by the way, a "new" Rā: the older Rā made it quite clear that he wanted to retire for good, going somewhere "where nobody could reach" him) (appendix #21).

[28] The Arabian name of the Galaxy is sufficiently tale-telling: "Mother of the Sky" (*um as-sama*), and in northern Ethiopia it is called "Em-hola," i.e., "Mother of the Bend [Mutter der Kruemmung]." See E. Littmann, "Sternensagen und Astrologisches aus Nordabessinien," ARW *11* (1908), p. 307; Ideler, p. 78.

CHAPTER XX

The Depths of the Sea

Hast thou entered into the springs of the sea?
Or hast thou walked in the search of the depth?

Job XXXVIII.*16*

IT WILL HELP NOW to take a quick comparative look at the different "dialects" of mythical language as applied to "Phaethon" in Greece and India. The Pythagoreans make Phaethon fall into Eridanus, burning part of its water, and glowing still at the time when the Argonauts passed by. Ovid stated that since that fall the Nile hides its sources. *Rigveda* 9.73.3 says that the Great Varuna has hidden the ocean. The *Mahabharata* tells in its own style why the "heavenly Ganga" had to be brought down.[1] At the end of the Golden Age (Krita Yuga) a class of Asura who had fought against the "gods" hid themselves in the ocean where the gods could not reach them, and planned to overthrow the government. So the gods implored Agastya (Canopus, alpha Carinae = Eridu) for help. The great Rishi did as he was bidden, drank up the water of the ocean, and thus laid bare the enemies, who were then slain by the gods. But now, there was no ocean anymore! Implored by the gods to fill the sea again, the Holy One replied: "That water in sooth hath been digested by me. Some other expedient, therefore, must be thought of by you, if ye desire to make endeavour to fill the ocean." It was this sad state of things which made it necessary to bring the Galaxy "down." This is reminiscent of the detail in the Jewish tradition about Eben Shetiyyah, that the waters sank down so

---

[1] Mbh. 3.104-105 (Roy trans., vol. *2*, pt. 2, pp. 230f.); see also H. J. Jacobi, ERE, vol. *1*, p. 181A; S. Sörenson, *Mahabharata Index* (1963), p. 18A.

deeply that David had to recite the "fifteen songs of ascension" to make them rise again.

Now Agastya, the great Rishi, had a "sordid" origin similar to that of Erichthonios (Auriga), who was born of Gaia, "the Earth," from the seed of Hephaistos who had dropped it while he was looking at Athena.[2] In the case of the Rishi:

He originated from the seed of Mitra and Varuna, which they dropped into a water-jar on seeing the heavenly Urvashi. From this double parentage he is called Maitrāvaruni, and from his being born from a jar he got the name Khumbasambhava."[3] [Khumba is the name of Aquarius in India and Indonesia, allegedly late Greek influence.]

On the very same time and occasion there also was "born" as son of Mitra and Varuna—only the seed fell on the ground not in the jar—the Rishi Vasishtha. This is unmistakably zeta Ursae Majoris, and the lining up of Canopus with zeta, more often with Alcor, the tiny star near zeta (Tom Thumb, in Babylonia the "fox"-star) has remained a rather constant feature, in Arabic Suhayl and as-Suha. This is the "birth" of the valid representatives of both the poles, the sons of Mitra and Varuna and also of their successors. To follow up the long and laborious way leading from Rigvedic Mitrāvaruna (dual) to the latest days of the Roman Empire where we still find a gloss saying "mithra funis, *quo navis media vincitur*" —"mithra is the rope, by which the middle of the ship is bound," would overstep the frame of this essay by far. Robert Eisler[4] relying upon his vast material, connected this fetter of "rope," mithra,

[2] Besides Greece and India, the motif of the dropped seed occurs in Caucasian myths, particularly those which deal with the hero Soszryko. The "Earth" is replaced by a stone, Hephaistos by a shepherd, and Athena by the "beautiful Satana," who watches carefully the pregnant stone and who, when the time comes, calls in the blacksmith who serves as midwife to the "stone-born" hero whose body is blue shining steel from head to foot, except the knees (or the hips) which are damaged by the pliers of the smith. The same Soszryko seduces a hostile giant to measure the depth of the sea in the same manner as Michael or Elias causes the devil to dive, making the sea freeze in the meantime.

[3] RV 7.33.13-14; *Brihad-Devata* 5.152ff.; Sörenson, p. 18B. Let us mention that the Egyptian Canopus is himself a jar-god; actually, he is represented by a Greek *hydria* (see RE s.v. Kanopos).

[4] *Weltenmantel und Himmelszelt* (1910), pp. 175f.

right away with the "ship's belt" from the tenth book of Plato's *Republic*.

Of the inseparable dual Mitrāvaruna, Varuna is still of greater relevance, particularly because it is he who "surveyed the first creation" (RV *8*.41.10), he who hid the Ocean—Ovid had it that the sources of the Nile were hidden—and he who is himself called "the hidden Ocean" (RV *8*.41.8). Varuna states about himself: "I fastened the sky to the seat of the Rita" (RV *4*.42.2). And at that "seat of Rita" we find *Svarnara*, said to be "the name of the celestial spring . . . which Soma selected as his dwelling.[5] This is no other "thing" than *Hvarna* (Babylonian *melammu*) which the "bad uncle" Afrasiyab attempted to steal by diving to the bottom of Lake Vurukasha, although Hvarna belonged to Kai Khusrau (see above, pp. 40, 201). Thus in whichever dialect the phenomenon is spelled out, the fallen ruler of the Golden Age is held to dwell nearest to the celestial South Pole, particularly in Canopus which marks the steering oar of Argo, Canopus at the "confluence of the rivers." This is true whether Varuna fastened the sky to the seat of the Rita (and his own seat), whether Enki-Ea-Enmesharra, dwelling in Eridu, held all the norms and measures (Rita, Sumerian *me:* Akkadian: *parsu*)—Thorkild Jacobsen called him very appropriately the "Lord modus operandi"—or whether Kronos-Saturn kept giving "all the measures of the whole creation" to Zeus while he himself slept in Ogygia-the-primeval.

And there is little doubt, in fact none, that Phaethon (in the strange transformation scenes of successive ages) came to be understood as Saturn. There is the testimony of Erastosthenes' Catasterisms,[6] according to which the planet Saturn was Phaethon who

---

[5] See H. Lüders, *Varuna*, vol. 2: *Varuna und das Rita* (1959), pp. 396–401 (RV *4*.21.3; *8*.6.39; *8*.65.2f.; *9*.70.6). Soma is addressed as "lord of the poles," and to Agni is given the epithet *svarnaram* thrice (RV *2*.2.1; *6*.15.4; *8*.19.1; cf. Lüders, p. 400). But we did hear before about "Agni, like the felly the spokes, so you surround all the gods," and Soma and Agni supplement each other, as will come out eventually, but *not* in this essay; the proportions Mitra: Varuna, Agni: Soma, Ambrosia: Nectar are not as easily computed as wishful thinking might expect.

[6] No. 43 (Robert ed., pp. 194f.). E.g., Hyginus II 42, dealing with the planets, beginning with Jupiter: "Secunda stella dicitur Solis, quam alii Saturni dixerunt; hanc Eratosthenes a Solis filio Phaetonta adpellatam dicit, de quo complures dixerunt, ut patris inscienter curru vectus incenderit terras; quo facto ab Iove fulmine percussus in Eridanum deciderit et a Sole inter sidera sit perlatus."

fell from the chariot into Eridanus, and Stephanus of Byzantium[7] calls Phaethon a Titan. There is, moreover, the Orphic wording of the case: "After Kronos had emasculated Ouranos, Zeus threw his father [Kronos] from the chariot and 'entartarosed' him" right away, if we translate the word literally.[8] Essential key words are easily mistaken for petty details, as in this case the "chariot," from which Kronos/Phaethon was thrown into "Tartaros." The vehicle in question is the two-wheeled race car, Greek *harma*, Latin *currus*, Babylonian *narkabtu*. It is the chariot of Auriga in Babylonia, surviving in the "Sphaera barbarica" of astrologers,[9] whereas in our Sphere the Charioteer is bereft of any vehicle. And, indeed, no other than Erichthonios (a Greek name for Auriga, besides Hēniochos) is claimed to have invented the two-wheeled race car drawn by four horses (Erat. Catast. no. 13, pp. 98–101) which has to be distinguished carefully from the even more important four-wheeled truck, the Big Dipper, that is, Greek *hamaxa*, Latin *plaustrum*, Sumerian[mul] MAR.GID.DA = Charles' Wain.

Slightly perplexing traditions have come down in cuneiform texts, but they clearly allude to the same "event." So, for instance, "The Elamitic chariot, without seat, carries the corpse of Enmesharra. The horses which are harnessed to it are the death-demon of Zu. The king who stands in the chariot is the hero-king, the Lord Ninurta." Leaving aside the two last sentences which are, in reality, not so pitch dark as they look at first glance, the translator, Erich Ebeling,[10] leaves no doubt that the "Elamitic chariot" is identical with the constellation "Chariot of Enmesharra," which the authorities on Babylonian astronomy have identified with beta and zeta Tauri.[11] This Enmesharra now has a "telling" name: En.*ME*.SARRA is "Lord of all the *me*," that is, he is Lord of "norms and measures," also called "Lord of the World Order,"

[7] s.v. *Eretria* (Eretrios, "Son of Phaethon, and this was one of the Titans"). See M. Mayer, *Giganten und Titanen* (1887), pp. 70, 124.

[8] *Hieronymi et Hellanici theogonia* (Athenagoras), see Kern frg. 58, p. 138; cf. also R. Eisler, *Weltenmantel und Himmelszelt* (1910), p. 338.

[9] Cf. Boll, *Sphaera*, pp. 108ff. (Teukros and Valens).

[10] *Tod und Leben nach den Vorstellungen der Babylonier* (1931), pp. 29, 33f.

[11] Gössmann, p. 89; Schaumberger, *3. Erg.*, p. 327; E. F. Weidner, in RLA *3*, p. 77.

"Lord of the Universe = Ea" and, this is important, "the weighty one in the underworld" and "the sovereign of the underworld."[12]

The "underworld" is misleading, though; the word is *Arallu*. The experts generally—not the Assyriologists alone—prefer to talk of *names*, in plural, given to the *one* "underworld," instead of trying to find out the precise whereabouts of the several provinces of that huge country, and to establish which name might properly fit every quarter. As if one did not know of the plurality of "hells" and "heavens." Here, however, it is not necessary to bring order into the quarters of the Mesopotamian Hades, and for the time being, it suffices that the Lord of the World Order, Enmesharra, is Enki-Ea, because it is known anyhow that he dwells "at the seat of Rita": Eridu-Canopus. And since "Enmesharra's chariot" is the vehicle of Auriga, beta zeta Tauri, there can be little doubt that the tradition of Phaethon's fall was already a Sumerian myth (appendix #22). And as in Greece, where the drastic version of the Orphics, of Hesiod and others are found side by side with those of Plutarch and Proclus, according to which Kronos gives with paternal grace "all the measures of the whole creation" to his son Zeus,[13] so, too, we have in Mesopotamia cruel-sounding variations besides "reasonable" ones. For example, when Marduk builds his "world" and receives fifty new names, his father Ea gives him his very own name, stating (EE 7.141f.): "His name shall be Ea. All my combined rites he shall administer; all my instructions he shall carry out." And as concerns Ea under the name of Enmesharra, Edzard states: "An incantation of Neo-Assyrian times, using an epithet of Enmesharra 'Who transferred scepter and sovereignty to Anu and Enlil' possibly hints to the voluntary abdication of the god."[14]

One of the questions begging answers is, *which* measures are meant, and how does Saturn accomplish his assignment "to give

---

[12] D. O. Edzard, "Die Mythologie der Sumerer und Akkader," Wb. Myth., vol. *1*, p. 62; P. Michatz, *Die Götterliste der Serie Anu ilu A-nu-um* (Phil. Diss.; 1909), p. 12; K. Tallqvist, *Sumerische Namen der Totenwelt* (1934), p. 62, and *Akkadische Götterepitheta* (1938), pp. 304, 437.

[13] See also Lucian who makes Kronos say: "No, there was no fighting, nor does Zeus rule his empire by force; I handed it to him and abdicated quite voluntarily."

[14] Edzard, p. 62.

them continuously" to Jupiter? And, even if it is accepted that his "seat" is Canopus, how can he possibly give the measures from there? Without pretending to understand the scheme well for the time being, there are some explanations which seem to be the most plausible ones.

Above (p. 136), attention was called to the significance of the revolution of that Trigon which is built up by "Great Conjunctions" of Saturn and Jupiter, and was still understood by Kepler (see figure). Now, whoever tries to imagine the degree of difficulty which faced the oldest "mythographers" will realize how welcome it must have been to find periods which fitted into each other at least approximately. This Trigon of Great Conjunctions presented itself as the instrument by which one could "narrow down" the almost imperceptible tempo of the Precession. To move through the whole zodiac, one of the angles of the Trigon needs approximately $3 \times 794\frac{1}{3} = 2383$ years. That comes tolerably near to one double-hour of the greatest "day" of the Precession of 25,900 years (appendix #23). A new zodiacal sign was termed to "rule" starting from the day of a great conjunction at the place of the "passage." The marginal point of Greek time-reckoning was the date of the first Olympic Games: they had been founded in memory of the wrestling of Kronos and Zeus, Pausanias said. The celestial constellation, however, ruling the different traditional dates of the first Olympic Games does not justify this claim; in other words, it is not known yet which particular great conjunction it was in the memory of which the Games were supposed to have been introduced. Our own era, the Age of Pisces, started with a great conjunction in Pisces, in the year 6 B.C.

By means of this Trigon, Saturn does give *panta ta metra* continuously to his "son" Zeus, and this same Trigon appears to be called "genus" in the Orphic fragment already quoted (155 Kern), where Zeus addresses Kronos with the words, "Set in motion our genus, excellent Demon." And Proclus alluded to it in his statement (*ibid.*), "And Kronos seems to have with him the highest causes of junctions and separations." And still according to Macrobius he

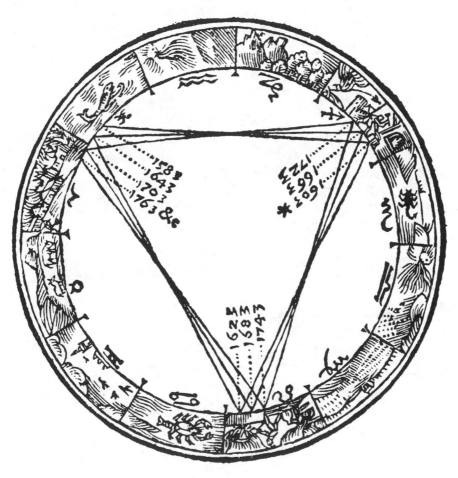

A detailed illustration of the motions of the Trigon of Great Conjunctions from
1583–1763.

was the "originator" of time (*Sat. 1.22.8*: "Saturnus ipse qui auctor est temporum.").[15]

So much for Saturn the unalterable planet gliding along his orb. Saturn as the fallen ruler of the Golden Age and retired to Eridu is a much harder proposition. Although there is also evidence to the contrary, there are many indications that the South Pole—Canopus—was taken for static, exempted from the Precession.[16] And this

[15] See R. Klibansky, E. Panofsky, and R. Saxl, *Saturn and Melancholy* (1964), pp. 154f. Cf. pp. 333f., with quotations from the Latin translation of Abū Ma'shar, where Saturn "significat . . . quantitates sive mensuras rerum," and where "eius est . . . rerum dimensio et pondus."

[16] We have neither time nor space to deal sufficiently with the relevant and copious information on the "joyful" South Pole (see L. Ideler, *Sternnamen*, pp. 265f.), the "Kotb Suhayl" of the Arabians, called thus after Canopus, which is recognized in Fezzan as "l'étoile primordiale Sahel, identifié au premier ciel contenant les constellations à venir" (V. Pâques, *L'arbre cosmique* [1964], p. 36)—the primordial star, "presented under the form of an egg that contained all the things that were to be born" (Pâques, p. 47). To begin discussing the static South Pole, one might well start with the "Seven Sleepers of Ephesus," who were thought to be on board the *Argo*—even if this is explicitly stated only in very late Turkish tradition (16th century)—particularly from Louis Massignon's article, "Notes sur les Nuages de Magellan et leur utilisation par les pilotes arabes dans l'Océan Indien: sous le signe des VII Dormants" (*Revue des Études Islamiques* [1961], pp. 1–18 = part VII of Massignon's series of articles on the Seven Sleepers in Islamic and Christian tradition; part I appeared in the same review in 1955, part VIII in 1963), and in the very substantial review article by T. Monod, "Le ciel austral et l'orientation (autour d'un article de Louis Massignon)" (*Bulletin de l'Institut Français d'Afrique Noire* [1963], vol. 25, ser. B, pp. 415–26). In both articles one finds, besides the surprising notion of the happy South, noteworthy information about human migrations directed toward the South in several continents. Massignon derived the "lucky" significance of the Kotb Suhayl and the Magellanic Clouds from historical events; i.e., from the expectations of exiled and deprived peoples escaping from the perpetual wars and raids in the northern countries: "Nomades ou marins, ces primitifs expatriés n'eurent pour guides, dans leur migrations et leur regards désespérés, que les 'étoiles nouvelles du ciel austral'" (1961, p. 12). Monod (p. 422), however, pointed to the crucial key word as given by Ragnar Numelin (*Les Migrations Humaines* [1939], p. 270n.), who remarked: "Il est possible que beaucoup de ces mystérieuses pérégrinations se proposaient comme but de trouver 'l'étoile immobile' dont parle la tradition. Le culte de l'Étoile Polaire peut avoir provoqué de tels voyages," annihilating thus with the second sentence the treasure which he had detected in the first. But Massignon and Monod also missed the decisive factor, namely, that *the South Pole of the ecliptic* is marked by the Great Cloud, and that Canopus is rather near to this south ecliptic pole, whereas the immovable center in the North of the universe is not distinguished by any star, as has been said previously.

would mean—at least it might mean, because it fits so well into those notions of "time and the rivers"—that expired periods return "home" into timelessness, that they flow into eternity whence they came. Access to the Confluence of the Rivers, Mouth and Source of aeons and eras, the true seat of immortality, has always been denied to any aspect of "time, the moving likeness of eternity." For eternity excludes motion. But from this desired motionless home, source and mouth of times, the world-ruler has to procure the normed measures valid for his age; they are always based on time, as has been said. Again it is the same whether it was Marduk who first "crossed the heavens and surveyed the regions. He squared Apsu's quarter, the abode of Nudimud [= Ea]. As the lord measured the dimensions of Apsu," and then erected his palace as the "likeness" of Apsu, or whether it was Sun the Chinese Monkey who fetched his irresistible weapon from the "navel of the deep"— an enormous iron pillar by means of which, once upon a time, Yü the Great had plumbed out the utmost depth of the sea. In any case, whether the description is sublime or charmingly nonsensical, it is literally the "fundamental" task of the Ruler to "dive" to the topos where times begin and end, to get hold of a new "first day." As the Chinese say, in order to rule over space one has to be master of time.

The reader may suspect by now that Hamlet has been forever forgotten. The way has been long and circuitous, but the connection is still there. Even in so late and damaged a tradition as that of Saxo Grammaticus, every motif once made sense in high and far-off times. If it is difficult to recognize the central significance of the "oar" of Odysseus,[17] how much more difficult is it to spot the

---

For the fun of it, a note of Monod's should be quoted here (p. 421): "Quand Voltaire nous dit que Zadig 'dirigeait sa route sur les étoiles' et que 'la constellation d'Orion, et le brillant astre de Sirius le guidaient vers le pôle de Canope', nous retrouvons dans cette dernière expression un témoignage du rôle joué par Canopus dans l'orientation astronomique. Il n'y a pas lieu, bien entendu, de vouloir la corriger en 'port de Canope'; cf. Voltaire, *Romans et contes*, éd. Garnier 1960, note 49, p. 621." Where shall we ever find security from the "improvements" of philologists?

[17] Sooner or later, one more object will have to be admitted to the assembly of imperial measuring oars, or *gubernacula:* the enigmatical Egyptian *hpt*, the so-

"steering oar of Argo" = Canopus-Eridu, in the childish riddle of Amlethus? And yet, the "measuring of the depth of the sea" is there all the time; infant-Kullervo dared to do it with a ladle, coming to the startling result of "three ladles and a little bit more." And there is an even less suitable measure to be had, a veritable stylus. Jacob Grimm gives the story: "The medieval Dutch poem of Brandaen . . . contains a very remarkable feature: Brandaen met on the sea a man of thumb size, floating upon a leaf, holding in his right hand a small bowl, in the left hand a stylus; the stylus he kept dipping into the sea and letting water drip from it into the bowl; when the bowl was full, he emptied it out, and began filling it again. It was imposed on him, he said, to measure the sea until Judgment-day."[18] This particular kind of "instrument" seems to reveal the surveyor in charge in this special case. Mercury was the celestial scribe and guardian of the files and records, "and he was the inventor of many arts, such as arithmetic and calculation and geometry and astronomy and draughts and dice, but his great discovery was the use of letters," as Plato has it (*Phaedrus* 274).

It remains to be seen whether or not all the measuring planets can be recognized by their particular methods of doing the measuring. It is known how Saturn does it, and Jupiter. Jupiter "throws," and Saturn "falls." But, as was said before, Saturn giving the measures as resident in Canopus is hard to imagine. Maybe all the available keys to this door have not been tried? Observing so many characters occupied with measuring the depth of the sea, one stumbles over the strange name given to Canopus by the Arabs: they call it

---

called "ship's device" (Schiffsgerät) of obscure literal meaning, which the Pharaoh brought running to a deity in the ritual of the "oar-race." There was also a "jar-race" and a "bird-race," the Pharaoh carrying a water jar or a bird, respectively. In several Pyramid Texts the soul of the dead ruler takes this ship's device and brings it to another celestial department, while the actual rowing of the boat is done by the stars (Pyr. 2173A, D; see also 284A, 873D, 1346B). See Aeg. Wb., vol. 3, pp. 67–71; A. Gardiner, *Egyptian Grammar* (1957), p. 581; M. Riemschneider, *Augengott und Heilige Hochzeit* (1953), pp. 255f. For the different imperial races, see the (unsatisfying) investigation by H. Kees, *Der Opfertanz des Aegyptischen Königs* (1912), pp. 74–90, the "oar-race."

[18] *Deutsche Mythologie* (1953), pp. 420/373. The English translation (TM, p. 451) makes it "pointer" instead of "stylus"; Grimm has "Griffel." Cf. K. Simrock, *Handbuch der Deutschen Mythologie* (1869), § 125, p. 415.

"the weight," and the Tables of Alphonsus of Castile spell it "Suhel ponderosus," the heavy-weighing Canopus.[19] This "weight" is the plumb at the end of the plumb line, by means of which this depth was measured. So far so good. But where does Saturn come in? He can be understood as the "living" plumb line.[20] This would be hard to believe if the story of this surveying were not told by the plumb line itself, Phaethon. Only when he told it, he had another name, as belongs to the manners and customs of celestial characters: Hephaistos.[21]

In the first book of the *Iliad* (*1.589ff.*), Hephaistos tries to appease his mother Hera who is very angry with her husband Zeus, and says to her:

> "It is hard to fight against the Olympian. There was a time once before now I was minded to help you and he caught me by the foot and threw me from the magic threshold, and all day long I dropped helpless, and about sunset I landed in Lemnos, and there was not much life left in me."

Hephaistos mentions the event once more, when Thetis asks him to forge the shield for her son Achilles (*18.395ff.*):

> "She saved me when I suffered much at the time of my great fall through the will of my own brazen-faced mother,[22] who wanted to hide me, for being lame. Then my soul would have taken much

[19] "Suhail al wazn." The epithet "wazn" has been given also to other stars of the southern sky. For ample discussions of this name, see Ideler, pp. 249–52, 263; Allen, pp. 68f.; J. N. Lockyer, *The Dawn of Astronomy* (1964), p. 294; W. T. Olcott, *Star Lore of All Ages* (1911), p. 133.

[20] The strange "beacon" in Aeschylus' *Agamemnon*, which announced the Fall of Troy, must have been something of this kind; the context excludes absolutely any possible devices of the signal corps.

[21] To avoid misunderstanding, we do not wish to insist upon the absolute identity of the fall of Phaethon and the account of the fall as told by Hephaistos in the first book of the *Iliad*. We suspect that the verbal image "Jupiter-hurls-down-Saturn" describes the shaping of the Trigon of great conjunctions, not, however, of the Trigon generally but of that *new* Trigon whose first angle was established by a conjunction of the Big Two at the beginning of a new world-age. On the other hand, this picturesque formula might cover the shifting of the Trigon of conjunctions from one Triplicity to the next (cf. appendix #23); these highly technical problems cannot be solved yet.

[22] That is not what Homer says, it is *kunōpis*, dog-eyed; Hera seems to have been near Sirius at that time.

"The shepherd is shown on the left sighting first the pole star, and on the right observing the transit through the meridian of the stars forming the easily recognized W of Cassiopeia."

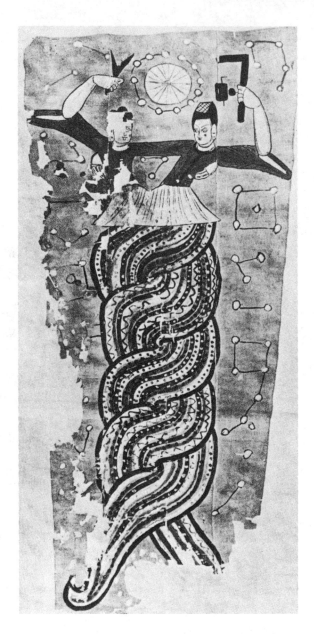

The Chinese picture illustrates in true archaic spirit (which means that only hints are given, and the spectator has to work out for himself the significance of the details) the surveying of the universe. The two characters surrounded by constellations are Fu Hsi and Nu Kua, i.e., the craftsman god and his paredra, who measure the "squareness of the earth" and the "roundness of heaven" with their implements, the square with the plumb bob hanging from it, and the compass. The intertwined serpent-like bodies of the deities indicate clearly enough, although in a peculiar "projection," circular orbits intersecting each other at regular intervals.

suffering had not Eurynome and Thetis caught me and held me. Eurynome, daughter of Ocean, whose stream bends back in a circle. With them I worked nine years as a smith."

Indentured as a smith again, like Kullervo.

Krates of Pergamon[23] explains this feature in the sense that Zeus aspired to the measurement of the whole world (*anametrēsin tou pantos*). He succeeded in determining the measures of the cosmos by "two torches moving with the same speed": Hephaistos and the Sun. Zeus hurled the former down from the threshold to earth at the same moment when the latter was starting from point east on his way to the west. Both reached their goal at the same time: the Sun was setting when Hephaistos struck Lemnos.

Krates felt convinced that Homer spoke of a sphere, and since he himself was most interested in the coordinate system of the sphere he did not find it strange to interpret in his own sense the shield of Agamemnon (*Iliad 11.*32f.) and of Achilles (*18.*468ff.).[24] He also conceived Odysseus' sailing from Circe's island to Hades as a voyage from the Tropic of Capricorn to the South Pole. The idea is not so strange as it might seem. Zeus, establishing the equinoctial colure by hurling down the fictitious "Phaethon," introduced a new *skambha*—one remembers Plato about this: "It has the air of a fable . . ."[25] But there is also Cornford's idea of the

---

[23] It is to the credit of Hans Joachim Mette and his work, *Sphairopoiia, Untersuchungen zur Kosmologie des Krates von Pergamon* (1936), that we find collected every relevant testimonial and fragment concerned with Krates and his topics.

[24] See Mette, pp. 30–42, and his introduction.

[25] We cannot discuss here the Homeric wording of the topos from which Zeus threw down Hephaistos: "magic threshold" means nothing, anyhow (apo bēlou thespesioio); there were ancient scholars who claimed that Krates connected this "bēlos" with the Chaldean "Bel"/Baal = Marduk. We leave it at Auriga's chariot, Babylonian *narkabtu*, the more so, as Marduk, too, used it when tipping over Tiamat. The "Babylonian Genesis" does not tell that Marduk hurled people around, but there is a cuneiform text (VAT 9947) called by Ebeling (*Tod und Leben*, 37f.) "a kind of a calendar of festivals," where it says: "the 17th is called (day) of moving in, when Bel has vanquished his enemies. The 18th is called (day) of lamentation, at which one throws from the roof Kingu and his 40 sons." Kingu had the epithet "Enmesharra," i.e., "Lord of norms and measures"; he was the husband of Tiamat—as Geb was husband of Nūt—who gave him the "tablets of fate," which Marduk was going to take away from him after his victory, and 40 is the

vision of Er,[26] according to which Plato's "souls actually see in their vision not the universe itself, but a model, a primitive orrery in a form roughly resembling a spindle . . ."

It is sad to observe, and certainly odd, how little scholars trust their own eyes and words—as in the case of Jane Harrison who remarked on the Titans: "They are constantly driven down below the earth to nethermost Tartaros and always reemerging. The very violence and persistence with which they are sent below shows that they belong up above. They rebound like divine india-rubber balls."[27] It is rather evident that these divine india-rubber balls were not really sent below: what was overthrown were the expired ages together with the names of their respective rulers.

But now the galactical stage is empty and it is almost time to watch the working of the next *skambha* grinding out the "destiny" for the first postdiluvian generation. But before facing the hero of the oldest, the most difficult, and by all means the oddest of epics, there is an interval. We seize the occasion to insert a chapter on methods, presented by means of a well-known episode.

---

number of Enki-Ea (see below, p. 288). The rest is easy to calculate. We are hampered by our inappropriate ideas about "names," and by the misleading labels settled upon celestial characters by the translators who make Tiamat, Kingu and their clan into "monsters."

[26] *The Republic of Plato*, p. 350.

[27] *Themis*, p. 453f. Cf. for a similar sort of mistrusting one's own evidence, M. Mayer, *Giganten und Titanen*, p. 97.

Chapter XXI

The Great Pan Is Dead

Everyone has once read, for it comes up many times in literature, of that pilot in the reign of Tiberius, who, as he was sailing along in the Aegean on a quiet evening, heard a loud voice announcing that "Great Pan was dead." This engaging myth was interpreted in two contradictory ways. On the one hand, it announced the end of paganism: Pan with his pipes, the demon of still sun-drenched noon, the pagan god of glade and pasture and the rural idyll, had yielded to the supernatural. On the other hand the myth has been understood as telling of the death of Christ in the 19th year of Tiberius: the Son of God who was everything from Alpha to Omega was identified with *Pan* = "All."[1]

Here is the story, as told by a character in Plutarch's dialogue "On why oracles came to fail" (419 B–E):

> The father of Aemilianus the orator, to whom some of you have listened, was Epitherses, who lived in our town and was my teacher in grammar. He said that once upon a time in making a voyage to Italy he embarked on a ship carrying freight and many passengers.

[1] O. Weinreich ("Zum Tode des Grossen Pan," ARW *13* [1910] pp. 467–73) has collected the evidence for such strange notions, first found in 1549 (Guillaume Bigot), then three years later in Rabelais' *Pantagruel*, and ridiculed in later times, e.g., by Fontenelle, in the beginning of the 18th century: "Ce grand Pan qui meurt sous Tibere, aussi bien que Jésus-Christ, est le Maistre des Demons, dont l'Empire est ruiné par cette mort d'un Dieu si salutaire à l'Univers; ou si cette explication ne vous plaist pas, car enfin on peut sans impieté donner des sens contraires à une mesme chose, quoy qu'elle regarde la Religion; ce grand Pan est Jesus-Christ luy-mesme, dont la mort cause une douleur et une consternation générale parmy les Demons, qui ne peuvent plus exercer leur tirannie sur les hommes. C'est ainsi qu'on a trouvé moyen de donner à ce grand Pan deux faces bien differentes" (Weinreich, pp. 472–73).

It was already evening when, near the Echinades Islands, the wind dropped and the ship drifted near Paxi. Almost everybody was awake, and a good many had not finished their after-dinner wine.

Suddenly from the island of Paxi was heard the voice of someone loudly calling Thamus, so that all were amazed. Thamus was an Egyptian pilot, not known by name even to many on board. Twice he was called and made no reply, but the third time he answered; and the caller, raising his voice, said, "When you come opposite to Palodes, announce that Great Pan is dead." On hearing this, all, said Epitherses, were astounded and reasoned among themselves whether it were better to carry out the order or to refuse to meddle and let the matter go. Under the circumstances Thamus made up his mind that if there should be a breeze, he would sail past and keep quiet, but with no wind and a smooth sea about the place he would announce what he had heard. So, when he came opposite Palodes, and there was neither wind nor wave, Thamus from the stern, looking toward the land, said the words as he heard them: "Great Pan is dead." Even before he had finished there was a great cry of lamentation, not of one person, but of many, mingled with exclamations of amazement. As many persons were on the vessel, the story was soon spread abroad in Rome, and Thamus was sent for by Tiberius Caesar. Tiberius became so convinced of the truth of the story that he caused an inquiry and investigation to be made about Pan; and the scholars, who were numerous at his court, conjectured that he was the son born of Hermes and Penelope.

Plutarch has not been accepted, and a "simple" explanation was suggested. As the ship drifted along shore by a coastal village, the passengers were struck by the ritual outcry and lamentations made over the death of Tammuz-Adonis, the so-called grain god, as was common in the Middle East in high summer. Other confused shouts were understood by the pilot Thamus as directed to him.[2] Out of that, gullible fantasy embroidered the tale, adding details for credibility as usual. This sounded good enough. The story had been normalized, that is, disposed of as insignificant.

One is still allowed to wonder why such a fuss was made at the time about exclamations which must have been familiar to contemporaries, and why, unless Plutarch be a liar, that most learned of mythologists, the Emperor Tiberius himself, thought the matter worth following up.

[2] See F. Liebrecht, *Des Gervasius von Tilbury Otia Imperialia* (1856) pp. 179–80; J. G. Frazer, *The Dying God (Golden Bough 3)*, pp. 7f.

Therefore, with all due respect for the scholars involved, it is worth trying a different tack. One can assume that it was not all background noise, as we say today, but that there was an actual message filtered through: "The Great Pan is dead," *Pan ho megas tethnēke*, and that it was Thamus who had to announce it.

It was enough of a message for Tiberius' committee of experts (*philologoi*) to decide that it referred to Pan, the son of Penelope and of Hermes, number 3 in Cicero's list given in *De natura deorum* 3.56.[3] Penelope, whoever she really was, must have had quite a life after the events narrated in the *Odyssey*.[4] Mythology seems to have been a careful science in those circles.

If it is decided to credit the message, one is led to consider a number of similar stories, some of them collected by Jacob Grimm, but the bulk by Mannhardt.[5] They are strictly on the level of folktale, which at least preserved their innocence from literate interference. There is a whole set of stories from the Tyrol concerning the "Fanggen," a kind of "Little People" (or giants), dryads or tree spirits whose existence is bound to trees, so that the felling of such a tree would annihilate a Fangga. They were once willing to live with peasants in the form of servant maids and would bring blessings to the home,[6] but would also vanish unaccountably. A favorite story is that of the master of the house coming home and telling the family of a strange message that he has heard from a voice, such as "Yoke-bearer, yoke-bearer, tell the *Ruchrinden* [Rough-bark] that Giki-Gäki is dead on the Hurgerhorn," or "Yoke-bearer,

[3] Tertius Jove tertio natus et Maia, ex quo et Penelopa Pana natum ferunt. Cf. also *Herodotus 2.145*.

[4] As concerns the version according to which Pan was the son of Penelope and all the suitors, Preller remarks (*Griechische Mythologie* [1964], vol. *1*, p. 745): "the repulsive myth."

[5] J. Grimm, TM, pp. 453n., 1413f.; cf. pp. 989, 1011–12 ("The Devil's dead, and anyone can get to heaven unhindered"); W. Mannhardt, *Wald- und Feldkulte*, vol. *1* (1875), pp. 89–93; vol. *2* (1877), pp. 148ff.

[6] Generally, however, they are claimed to show rather revolting habits, such as eating children or disposing of them in another peculiar manner, as, for instance, pulverizing them into snuff. Thus, of one Fangga it is said: "Wenn sie kleine Buben zu fassen bekam, so schnupfte sie dieselben, wie Schnupftabak in ihre Nase, oder rieb sie an alten dürren Bäumen, die von stechenden Aesten starrten, bis sie zu Staub geraspelt waren." It seems to be a very deep-seated desire of "higher powers" to change divine or human beings into powder and dust.

yoke-bearer, tell the *Stutzkatze* [also Stutzamutza, i.e., Docked Cat] Hochrinde [High-bark] is dead." At which point the housemaid breaks into a loud lament and runs away forever.

Or it might be that while the family was sitting at dinner, a voice called three times through the window: "Salome, come!" and the maid vanished. This story has a sequel: some years later a butcher was coming home at midnight from Saalfelden through a gorge, when a voice called to him from the rocky wall: "Butcher, when you come to such and such a place [*zur langen Unkener Wand*], call into the crack in the rock: 'Salome is dead.' " Before dawn the man had come to that point, and he shouted his message three times into the crack. And at once there came from the depths of the mountain much howling and lamentations, so that the man ran home in fear. Sometimes the message delivered is followed by the "flyting" of whole tribes of Little People: it was their "king" whose death had been announced.[7] It is remarkable that in most of the cases registered, the master was addressed as "Yoke-bearer." No one knows why. But the wild woodmaid invariably vanished.

Felix Liebrecht speaks of the ways of certain ghostly were-wolves, the "Lubins," that haunted medieval Normandy. These timid ghosts hunted in a pack, but to little point, for instead of turning on the intruder, they would disperse at the slightest noise, howling: "Robert est mort, Robert est mort."[8] This meaningless yarn gains perspective once its trail is followed back to the "Wolf-Mountain" in Arcadia and the Lycaean "Wolf-games"—the parent-festival of the Lupercalia in Rome—held on this Mountain Lykaios. Pan is said to have been born here,[9] and here he had a sanctuary. Here also Zeus tilted a "table"—whence the place had the name Trapezous—because Lykaon had served him a dish of human meat,

[7] "No is Pippe Kong dod" (Schleswig); otherwise "König Knoblauch" (King Garlic), "King Urban"; "Hipelpipel is dead" (Lausitz); "Mutter Pumpe is tot" (Hessen). Cf. Grimm, p. 453; K. Simrock, *Handbuch der Deutschen Mythologie* (1869), § 125, pp. 416f.; F. Liebrecht, *Zur Volkskunde* (1879), p. 257*n.*, who gives additional references. See also P. Herrmann, *Deutsche Mythologie* (1898), pp. 89f.

[8] Liebrecht, *Zur Volkskunde*, p. 257*n.*

[9] Pindar frg. 100 (68); Rhea had borne Zeus there also (Paus. 8.38.2f.), and on top of the mountain was a *temenos* of Zeus, where nothing and nobody cast a shadow.

consisting of his own son. Zeus turned Lykaon into a werewolf, and in tilting the "table" caused the Flood of Deukalion, the "table," of course, being the earth-plane through the ecliptic. This is the significant event of the tale, and the whole is so long no sensible person would try to summarize it.

Next, there is the case of Robert, known as Robert le Diable, allegedly a historical character who was supposed to have spells as a werewolf and then to do penance by "lying in the guise of a dog under the ladder." And thereby hangs the puzzle of the dynasty of the Scaligeri in Verona (we all remember Prince Escalus in *Romeo and Juliet*) whose powerful founder was Can Grande della Scala, "Great Dog of the Ladder," who became a host to Dante wandering in exile, and a patron of the *Divine Comedy*. His successors, Mastino, Cansignorio, had dog names too.[10] Now, for the purposes of this essay, this is the end of this line of approach, except for two hints for the future. First, Pythagoras called the planets "the dogs of Persephone." Second, there is only one huge ladder, the Galaxy, and only one canine character lying under this ladder, Sirius. But at this point we are only ringing bells at random.

What matters here is the tenacious survival of motifs in simple surroundings. Moving one step down in folklore, there is a story spread all over northern Europe (Mannhardt *1*, 93) of which this is the English version (the end is from a German variant). A clowder of cats have met in an abandoned broken-down house, where a man is watching them unobserved. A cat jumps on the wall and cries: "Tell Dildrum that Doldrum is dead." The man goes home and tells his wife. The house cat jumps up and yowls: "Then *I* am king of cats!" and vanishes up the chimney.

This is how the "body" of tradition survives the death of its "soul," fractured, with all ideas gone, preserved like flies in amber. Greek gods have become cats and housemaids among illiterate folk; the Powers pass, but the information remains. By checking on the repeats, one has the message of the Voice in the canonic form: "Wanderer, go tell Dildrum that the Great Doldrum is dead." The

[10] See O. Höfler, "Cangrande von Verona und das Hundesymbol der Langobarden," in *Festschrift Fehrle* (1940), pp. 107–37.

bearer of the message may be an unknown pilot, a passerby, an animal, a watcher. The substance is that a Power has passed away, and that the succession is open. The cosmos has in its own way registered some key event.

For another example of hardly credible survival, there are also the findings of Leopold Schmidt on "Pelops and the Hazel Witch,"[11] a collection of tales from the Alpine valleys of Southern Tyrol. It is again about housemaids among peasants. The story goes that a farm servant accidentally watches the dinner of some witches, in which a housemaid is boiled and consumed by her fellow witches. A rib is thrown at the young man, and when after the meal the witches rebuild and revive the girl, this rib is missing and has to be replaced by a hazel branch. At the very moment that the farmhand tells his master that his housemaid is a hazel witch, the housemaid dies. This is no witch hazel trick— it is simply a rehearsing of the archaic tale of Pelops, son of Tantalos, the Titan, who had been boiled and served for dinner by his evil father at the table of the gods. The gods, it is said, kept away from the food that looked suspicious, all except for Demeter, who, lost in her grief for the death of Persephone, absently ate a shoulder blade believing it to be mutton. The gods brought the child back to life. But a shoulder blade was missing and it was replaced by ivory. Pelops went on to become a famous hero, from whom the Peloponnesos was named, and he won the foot race at Olympia from King Oinomaos, thus inaugurating the Olympian games. The two are portrayed before the race on the metope of the temple of Zeus in Olympia. Oinomaos stands there looking stuffy, Pelops relaxed, and above the two the great figure of Apollo with arms outstretched, as if to consecrate the event. But Olympia became holy because it was the site where Zeus overcame his father Kronos[12] and threw him down out of the royal chariot. Near Olympia you can see the

[11] "Pelops und die Haselhexe," *Laos 1* (1951), pp. 67–78.

[12] Paus. 5.7.10. It is not from mere "religious" motifs that "in the hippodrome the pillar which marked the starting point had beside it an altar of the *Heavenly Twins*" (Pind. *Olympian Odes 3.36*; Paus. *5.15*); cf. F. M. Cornford in Harrison (*Themis* [1962], p. 228); see also above, pp. 206f. n. 5, for the Circus Maximus in Rome.

Kronion hill, which still bears the imprint of the celestial posterior. Exeunt the official characters. Only the great Olympic Games remained an "international" event which took place every four years and became the Greek way of counting time. What has all that to do with a little fairy housemaid in the Austrian Alps, thousands of years later? Nothing at first glance, and yet, if one dug deeper into the story of this shoulder blade, there would be a good case history to be made.[13] Tradition goes on tenaciously, even through ages of submerged knowledge. At least, by now, some distance has been made well away from the fertility rites of Frazer and others, which accounted for things too patly. This is an important gain.

Returning to Plutarch's text the dialogue's chatty style gives an impression of casualness, but in these matters Plutarch usually knew more than he cared to discuss. There was a pilot, a *kybernetes*, giving an announcement from the stern deck (*prumne*) of his ship. These details seem not to be casual. For there is one stern and one pilot which cannot be overlooked in mythology. The stern is that of the constellation Argo, a ship which consists of a stern and little else. It is understood to be the Ship of the Dead with Osiris on board (he is the *strategos* of the ship, according to Plutarch's *Isis and Osiris* 359 EF), and the Pilot star in the stern is Canopus itself, the site of the great Babylonian god Ea (Sumerian Enki), its name in Sumerian being $^{mul}NUN^{ki}$, and Enki is the father of Tammuz, which might lead back to the trail.

But the striking thing is that Mesopotamian Canopus bears the

[13] There is not only "moskhou omon chryseion," the golden shoulder of the ox in the hands of Mithras (Egyptian Maskheti, the Bull's Thigh, Ursa Major), and *Humeri*, an antiquated Latin name of Orion, as we know from Varro; the highest god, Amma, of the West Sudanese Dogon (or the *Clarias senegalensis*, the shadfish, an avatar of the Dogon's "Moniteur Faro," whose emblem is the very same as that of ityphallic Min, the Egyptian Pan) carries in his *humeri* the first "eight grains," and these 8 sorts of grain (stereotypically including beans) play their cosmogonic role from the Dogon to China (cf. for another striking similarity of West Sudan and China, the chapter on the "shamanistic" drums, but there are many more). There is also the tale from modern Greece (see J. G. von Hahn, *Griechische und Albanische Märchen* [1918], vol. 1, pp. 181–84) of the "Son of the shoulder-blade," one of those "Strong Boys" who, after adventures in spirit land, grinds his mother to porridge on a hand mill. How these and other traditions are connected with the shoulder blade oracle, if they are connected at all, cannot be made out yet.

name "Yoke-star of the Sea"[14]—the "Yoke-star of the Sky" being Draco. Here then there is a death fate, a pilot, and a yoke-bearer in an unsuspected but suggestive complex. Dealing with such pro-found experts of archaic myth as Plato and Plutarch, one is not likely to overlook the "Egyptian king Thamus" in *Phaedrus* (274C–275B, see below, chapter XXIII), who drives it home to Thot-Hermes, who was very proud of just having invented writing, that this new art was an extremely questionable gain. It must have been a mighty "king" who dared to criticize Mercury's merits. But then, the chapter on the Galaxy and the fall of Phaethon will have shown that geographical terms are not to be taken at their face value, least of all "Egypt," a synonym of the ambiguous Nile.

To find something more about the substance of the message we shall move many centuries back, to a text certainly ancient, but of undetermined date. It is the so-called "Nabataean Agriculture" which has little to do with farming but much to do with agrarian rites. The author, Ibn Wa'shijja, claimed to have derived it from an almost primordial Chaldean source.[15] Modern critics have decided it was a fabrication of uncertain origin, a so-called falsification. Whatever else it may be, original it is not. Such things are built out of traditional material. Maimonides judged it worth quoting at length, Chwolson and Liebrecht analyzed it, comparing it with An-Nadim's report on the Tammuz festival of the Harranians, held in the month of July and called el-Bûqat, the "weeping women."[16] Here first is a passage studied by Liebrecht:[17]

[14] See P. F. Gössmann, *Planetarium Babylonicum* (1950), 281; J. Schaumberger, in Kugler's *3. Ergänzungsheft* (1935), p. 325, and n. 2 (one version: the "yoke of Ea"); P. Jensen, *Die Kosmologie der Babylonier* (1890), pp. 16ff., 25; F. Boll-C. Bezold, *Farbige Sterne* (1916), p. 121.

[15] Actually, he (and others) claimed that the book was written by three (or even more) authors, namely Ssagrît, Janbûshâd, and Qutâmâ. The first living in the seventh thousand of the 7000 years of Saturn—which he ruled together with the Moon—the second at the end of the same millennium, the third appeared after 4000 years of the 7000-year cycle of the Sun had passed; so that between the beginning and the end of the book 18,000 solar years have passed (according to Maqrîzî). See D. Chwolson, *Die Ssabier und der Ssabismus* (1856), vol. *1*, pp. 705f. (cf. p. 822 for the special alphabet used by Janbûshâd). So we are up to another "Tris-megistos," three *times* great, not just "thrice." Time is involved. Hermes is repeated three times historically.

[16] Chwolson, vol. *2*, pp. 27f., 207, 209.

[17] *Zur Volkskunde*, pp. 251f.

It is said that once the Sakaîn (angels) and the images of the gods lamented over Janbûshâd, just as all Sakaîn had lamented over Tammuz. The tale goes that the images of the gods gathered from all corners of the earth in the temple of the Sun, around the great golden image, which hung between heaven and earth. The great image of the Sun was in the middle of the temple, surrounded by the images of the Sun from everywhere, and also by the images of the Moon, then those of Mars, then those of Mercury, of Jupiter, of Venus, and finally of Saturn.[18]

Chwolson's part of the text goes on:

This idol (that hung between earth and heaven) fell down at this point and began to lament Tammuz and to recount his story of sorrows. Then all the idols wept and lamented through the night; but on the rising of the morning star, they flew off and returned to their own temples in all corners of the world.

Such is the story which, Liebrecht says, was rehearsed in the temples after prayers, with more weeping and lamentations. This is then the archaic setting. It concerns planetary gods, the great cult of Harran. Two of them stand out, almost *ex aequo:* Tammuz and Janbûshâd. Now this latter is no other than Firdausi's Jamshyd.[19] It has been seen already (p. 146) that Jamshyd is in Avestic Yima xsaēta, the name from which came Latin Saturnus. There is no question then, this is about Saturn/Kronos, the God of the Beginning, Yima (Indian Yama), the lord of the Golden Age. A lament over the passing of Kronos would have been in order even in Greece,[20] since he had been dethroned and succeeded by Zeus.

---

[18] Let us note that the planets are not given in the astronomical order of their periods, but in the order given by the heptagram, which describes the days of the week.

[19] See Liebrecht, p. 251*n:* "The Babylonian Izdubar [= Gilgamesh] is called by Ibn Wa'shijja's Book on the Nabataean Agriculture 'Janla-Shad' (Janbûshâd), i.e. Jamshid . . . Thus Rawlinson in *Athenaeum* December 7, 1872."

[20] Cf. the report by Plutarch (*Isis and Osiris* 363E) on Egypt: "There is also a religious lament sung over Cronus. The lament is for him that is born in the regions of the left, and suffers dissolution in the regions on the right; for the Egyptians believe that the eastern regions are the face of the world, the northern the right, and the southern to the left. The Nile, therefore, which runs from the south and is swallowed up by the sea in the north, is naturally said to have its birth on the left and its dissolution on the right." Kronos having been the ruler of "galactical times" (Geb "inside" Nūt), this makes more sense than meets the eye. See also chapter XIII, "Of Time and the Rivers."

But who was Tammuz? The grain god dying with the season, the rural Adonis, would hardly fit into such exalted company. Now it is clear he was astronomical first of all. So much has been written about his fertility rites that it took time to locate the real date, given by Cumont.[21] The lament over Adonis-Tammuz did not fall simply in "late summer": it took place in the night between July 19 and 20, the exact date which marked the opening of the Egyptian year, and remained to determine the Julian calendar. For 3000 years it had marked the heliacal rising of Sirius.

Tammuz was extremely durable, for he is found in Sumer as Dumuzi, already the object of the midsummer lamentations. It was seen that he was worshiped as the son of Enki, who was the Sumerian Kronos. The cult went on in Harran as late as the 13th century, long after Mohammedanism had engulfed the Ssabian population. Notwithstanding the severe displeasure of the Caliph of Baghdad, it went through sporadic but intense revivals in an area that spread from Armenia to Quzistan.[22] As mentioned, the celebration was called el-Bûqat, "the weeping women." And the lament was mainly over the god who was cruelly killed by being ground between millstones, just like John Barleycorn in the rhyme we quoted earlier:[23]

They roasted o'er a scorching fire
The marrow of his bones
But a miller used him worst of all
For he ground him between two stones.

What kind of grinding could it have been? Surely, the lament referred in popular consciousness to the death of a corn god, called also Adonis (the Lord), slain by a wild boar, but the celestial aspect

[21] "Adonis et Sirius," *Extrait des Mélanges Glotz*, vol. 1 (1932), pp. 257–64. But see for the different dates of the Adonia, F. K. Movers, *Die Phönizier* (1841), vol. 1, pp. 195–218, esp. p. 205.

[22] See Liebrecht, *Gervasius von Tilbury*, pp. 180–82; *Zur Volkskunde*, pp. 253ff.; W. Robertson Smith, *The Religion of the Semites* (1957), p. 412 (lamentations over "the king of the Djinns," and over "Uncûd, Son of the grape cluster").

[23] It was Felix Liebrecht who first felt reminded of John Barleycorn.

is predominant compared with the agrarian one, and more ancient, too; the more so as that "wild boar" was Mars.[24]

This leaves a knotted story to untangle. It is hampered considerably by too many "identifications" taken for granted by the scholars who with magnificent zeal have extirpated the dimension of time in the whole mythology. Actually, it is not known yet who Tammuz is.[25] He looks almost like a title, just as "Horus" was a title. There is doubt of his "identity"—as taken in the current sense—with Adonis, and with Osiris,[26] Attis, Balder,[27] and others. The "Nabataean Agriculture" leaves no doubt that there were lamentations over Tammuz and Janbûshâd/Jamshyd. The Egyptians lamented on account of Kronos and Maneros[28] (*Herodotus* 2.79). Tammuz, after all, is not the only star who came to fall in the course of the Precession. (And was not King Frodhi a repetition of Freyr, Kai Khusrau a repetition of Jamshyd, as Apis was the repetition of Ptah [the Egyptian Saturn-Hephaistos], and Mnevis that of Ra?)

This is a long way from Great Pan, and it is not clear yet who or what was supposed to have passed away in the time of Tiberius, that is, which "Pan." Creuzer[29] claimed right away that he was Sirius —and any suggestion from Creutzer still carries great weight—the first star of heaven and the kingpin of archaic astronomy. And

[24] See Nonnos *41*.208ff. on Aphrodite: "Being a prophet, she knew, that in the shape of a wild boar, Ares with jagged tusk and spitting deadly poison was destined to weave fate for Adonis in jealous madness." Cf. for the other sources, Movers, vol. *1*, pp. 222ff.

[25] To give tiniest minima only: Tammuz = Saturn (Jeremias in Roscher s.v. Sterne, col. 1443); Tammuz = Mars (W. G. Baudissin, *Adonis und Esmun* [1911], p. 117, quoting the Chronicle of Barhebraeus). For the unheard-of number of names given to "Tammuz" in Mesopotamia, see M. Witzel, *Tammuz-Liturgien* (1935). For his name "Dragon of the Sky" (Ušungal-an-na) = Sin (the Moon) see K. Tallqvist, *Akkadische Götterepitheta* (1938), p. 482; see also p. 464, where Tammuz = "Mutterschafbild" ("mothersheep-image").

[26] It is worth noticing that the death of Osiris, in his turn, was announced by "the Pans and Satyrs who lived in the region around Chemmis (=*Pan*opolis), and so, even to this day, the sudden confusion and consternation of a crowd is called a panic" (Plutarch: *Isis and Osiris*, c. 14, 356D).

[27] All the gods of the North came together, in best "Nabataean" fashion, to weep over Balder's death.

[28] We leave aside, though, the cases Linos, Maneros, Memnon, Bormos, etc. See Movers, vol. *1*, p. 244.

[29] *Symbolik und Mythologie der Alten Völker* (1842), vol. 4, pp. 65ff.

Aristotle says (*Rhet. 2. 24*, 1401 a 15) that, wishing to circumscribe a "dog," one was permitted to use "Dog-star" (Sirius) or Pan, because Pindar states him to be the "shape-shifting dog of the Great Goddess" (*O makar, honte megalas theou kyna pantodapon kaleousin Olympioi*).[30] But this is far enough for now. The amazing significance of Sirius as leader of the planets, as the eighth planet,[31] so to speak, and of Pan, the dance-master (*choreutēs*) as well as the real *kosmokrator*, ruling over the "three worlds,"[32] would take a whole volume. The important point is that the extraordinary role of Sirius is not the product of the fancy of silly pontiffs, but an astronomical fact. During the whole 3000-year history of Egypt Sirius rose every fourth year on July 20 of the Julian calendar. In other words, Sirius was not influenced by the Precession, which must have led to the conviction that Sirius was more than just one fixed star among others. And so when Sirius fell, Great Pan was dead.

Now, Creuzer had no monopoly on deriving from Egypt the ideas connected with Pan, nor has the derivation been invented independently here. W. H. Roscher undertook this task in his article on "The Legend of the Death of the Great Pan,"[33] being convinced

[30] See also Plato's *Cratylus* 408B: ton Pana tou Hermou einai hyon diphyē echei to eikos.

[31] Creuzer takes Pan-Sirius for Eshmun/Shmun, "the eighth," great god of Chemmis.

[32] Cf. the Orphic Hymn to Pan (no. 11; see also Hymn 34.25): Pana kalō krateron, nomion, kosmoio to sympan/ ouranon ēde thalassan ide chthona pambasileian/ kai pyr athanaton . . . Echous phile . . . pantophyēs, genetōr pantōn, polyōnyme daimon/ kosmokratōr . . . As concerns his love for Echo, Macrobius (*Sat. 1.22.7*) explains it as harmony of the spheres: quod significat harmoniam caeli, quae soli amica est, quasi sphaerarum omnium de quibus nascitur moderatori, nec tamen potest nostris umquam sensibus deprehendi. But then, Macrobius was the first among the "sun-struck" mythologists, harmlessly claiming Saturn and Jupiter and everybody else, including Pan, to be the Sun. It is not the echo itself which is the harmony of the spheres but the syrinx—Pan makes it out of the reeds into which his beloved Echo had changed—and the seven reeds of Pan's pipe are indeed the seven planets, the shortest representing the Moon, the longest Saturn. (It is worth consideration that in China the echo was understood as the acoustical pendant to the shadow, so that under the pillar or tree, in the very center of the world, the kien-mu, there is no echo and no shadow.)

[33] "Die Legende vom Tode des Grossen Pan," in *Fleckeisens Jahrbücher für klassische Philologie* (1892), pp. 465–77. Referring to the "Panic" element in Mannhardt's stories about the Fanggen, Roscher declares it "an accidentally similar motif."

that the myth could not be understood by means of Greek ideas and opinions, the less so, as Herodotus (2.145) informs us of the following:

> In Greece, the youngest of the gods are thought to be Heracles, Dionysos, and Pan; but in Egypt Pan is very ancient, and once one of the "eight gods" who existed before the rest;[34] Heracles is one of the "twelve" who appeared later, and Dionysos one of the third order who were descended from the twelve. I have already mentioned the length of time which by the Egyptian reckoning elapsed between the coming of Heracles and the reign of Amasis; Pan is said to be still more ancient, and even Dionysos, the youngest of the three, appeared, they say, 15000 years before Amasis. They claim to be quite certain of these dates, for they have always kept a careful written record of the passage of time. But from the birth of Dionysos, the son of Semele, daughter of Cadmus, to the present day is a period about 1600 years only; from Heracles, the son of Alcmene, about 900 years; from Pan, the son of Penelope—he is supposed by the Greeks to be the son of Penelope and Hermes—not more than about 800 years, a shorter time than has elapsed since the Trojan war.[35]

These details are given, without meddling with them, in order to draw attention to the modest numbers; whoever takes these elapsed years for historical ones,[36] presupposes a special Egyptian (and Babylonian, Indian, etc.) frame of mind, a human nature, in fact, which is fundamentally different from ours, forgetting that we are all members of the very same species, Homo sapiens.

[34] Archaiotatos kai tōn oktōn tōn prōtōn legoumenōn theōn.
[35] Cf. A. Wiedemann, *Herodots zweites Buch* (1890), pp. 515–18.
[36] See J. Marsham, *Canon chronicus Aegypticus, Ebraicus, Graecus* (1672), p. 9: "Immensa Aegyptiorum chronologia astronomica est, neque res gestas sed motus coelestes designat!" See also Ideler (*Beobachtungen*, 1806), p. 93. Apart from the sensible 17th century, at the beginning of the 19th century still, the progressive delusion was remarkably underdeveloped.

CHAPTER XXII

The Adventure and the Quest

THE EPIC OF GILGAMESH in its first recorded version goes back to Sumerian times.[1] It has been rehearsed with variants by Hurrians and Hittites, by Babylonians and Assyrians. Even in the best recensions there are large gaps, many tablets are damaged beyond repair, and to aggravate these detrimental conditions one must add the efforts of a goodly number of specialists which have not helped to clarify matters.

The story has been told many times over and it appears to be fairly secure in its main lines, a patrimony of world literature. Misleading as this appearance is—the way through those texts being incredibly slippery—it is best to leave it at face value for the present, giving only a brief outline of the accepted scheme in Heidel's version. Then it will be possible to examine certain difficult points which may eventually upset the scheme altogether.

Gilgamesh is claimed to have been one of the earliest kings of Uruk (or Erech). The circumstances of his fabled birth make him two-thirds god and one-third man, which makes him—in the sexagesimal system of Mesopotamia—two-thirds of 60 (= Anu) = 40, the number which characterized Enki-Ea, whence the latter's denomination of "Shanabi (= $\frac{2}{3}$, i.e., of 60), and Nimin (Sumerian = 40)."[2] Be that as it may, it is told that he lives in splendor and dissoluteness, and makes a nuisance of himself until the gods bring relief to his people by rearing a human being, either twin or counter-

---

[1] See, for example, S. N. Kramer, "The Epic of Gilgamesh and Its Sumerian Sources," JAOS *64* (1944), p. 11: "the poem was current in substantially the form in which we know it, as early as the first half of the second millennium B.C."

[2] E. Weidner, RLA, vol. *2*, p. 379.

part,[3] who can stand up to him. It is Enkidu, the man of the Wilds, a kind of wolf-child as simple as the beasts he plays with, a happy son of nature, hairy all over, grown to enormous strength. A harlot is sent out to seduce him, and through her he learns love and the ways of man, and is lured into the city (appendix #24).

His first encounter with Gilgamesh is a fierce battle which rocks the community house and seems to damage the doorpost (appendix #25) until the king manages to subdue Enkidu and decides he is worthy of becoming his friend and playmate.

Together they plan an expedition to the great Forest, to overcome the terrible ogre Huwawa or Humbaba,[4] whom the god Enlil, the so-called "god of storm" or "god of air," had appointed as its guardian. Indeed, "Enlil has appointed him as a sevenfold terror to mortals . . . his roaring is (like that of) a flood-storm, his mouth is fire, his breath is death!"[5]

Even if it is taken for granted that fights with dragons or ogres were a popular subject once upon a time, some dry data on this "monster" would do no harm. He "is invariably called a god in the texts"[6] and appears to correspond to the Elamitic god Humba or Humban, who shares the title "the prevalent, the strong" with the planets Mercury and Jupiter and with Procyon (alpha Canis Minoris). He occurs, moreover, in a star list, carrying the determinative [mul] (Babylonian [kakkab]) announcing stars, as [mul]Humba (appendix #26). The identification with Procyon may eventually turn out to be the decisive clue which will reconcile the Sumerian version with the many others.

Ancient texts do not become more lucid if every strange-looking aspect is silently omitted, and so it is well to mention that Humbaba

---

[3] Actually, the goddess Aruru makes him "in the likeness of Anu," literally "a *zikru* of Anu she conceived in her heart." But Enkidu is also said to look like Gilgamesh "to a hair." See A. L. Oppenheim, "Mesopotamian Mythology," *Orientalia* 17 (1948), pp. 24, 28.

[4] Huwawa in the Old Babylonian and Hittite versions, Humbaba in the Assyrian version.

[5] Tabl. 3.136f., 109–11, Heidel trans., p. 35.

[6] Langdon, *Semitic Mythology* (1931), p. 253. See also F. Hommel, *Ethnologie und Geographie des Alten Orients* (1926), pp. 35, 42, claiming *hum* to mean "creator," and talking of Humbaba (= Hum-is-the-father) as of the "guardian of the cedar of paradise."

is some kind of a "god of intestines." More than that, his head or face is built of intestines, and Langdon (MAR 5, 254) draws attention to the fact that "the face of this monster . . . is designed by a single winding line, except eyes." Böhl, moreover, in his inquiry on the Babylonian origin of the labyrinth,[7] pointed out the Babylonian notion of the entrails as a labyrinthic "fortress of intestines."

This much about the "person" Humbaba, who is, evidently, no primitive monster at all, the less so, as his unattractive face strikingly resembles the features of Tlaloc, the so-called "rain-god" of the Aztecs, whose face is built up of two serpents. Precipitate identifications lead only to mischief,[8] and the "Case Humbaba" is not even partly solved, despite many efforts.

The only established features of the story seem to be that the heroes reached the forest of cedars, which is said to extend for "ten thousand double-hours" (say, 70,000 miles), and that they cut off the head of Humbaba after having felled, apparently, the largest of the cedars entrusted to Humbaba's guard by Enlil, but the feat is not accomplished without the powerful help of Shamash-Helios "who sends a great storm to blind the monster and put him at their mercy."

Returned to Uruk, Gilgamesh washes his hair and garbs himself in festive attire. As he puts on his tiara, Ishtar, the goddess of love (in Sumerian, Inanna), is entranced with his looks and asks him to marry her. Gilgamesh rejects her, reminding her in scornful words of what happened to her previous mates, including the hapless Tammuz, later known as Adonis.

It is not unusual for a hero to refuse the love, and the unheard-of presents, offered by a goddess. In every such case only two celestial personalities are possible candidates for this role: the planet Venus, and Sirius, alias Sothis, who has some of the reputation of a harlot.

[7] F. M. Böhl, "Zum babylonischen Ursprung des Labyrinths," in Festschrift Deimel (1935), pp. 6–23.

[8] Hommel, Ethnologie und Geographie, p. 35, dealing with an Elamitic star list, makes "Amman-ka-sibar (derived from Chumban-uk-sinarra . . . i.e., Chumban, king of the bolt? . . .) = Ninib-Mars." We would hazard a premature guess that apart from Procyon, Mercury would be the safest bet, the second candidate being Jupiter; but the latter would never make a convincing lord of entrails, nor would any other outer planet: their orbits do not allow for such notions—and Venus is much too regular for this role.

This terra-cotta mask shows the unlovely face of Humbaba/Huwawa, the guardian of the cedar felled by Gilgamesh and Enkidu. The title of "God of the fortress of intestines" is also given to him, and some scholars conclude from this title, as well as from the pictorial evidence, that Humbaba was the inhabitant and lord of the labyrinth, a predecessor of Minotaurus.

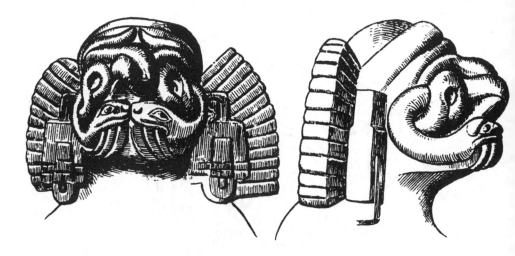

As revolting as the face of Humbaba are the features of Tlaloc, the so-called "rain-god" of Mexico. They are revealing, however: constructed out of two serpents, Tlaloc's head represents, as it were, the caduceus of Hermes/Mercury.

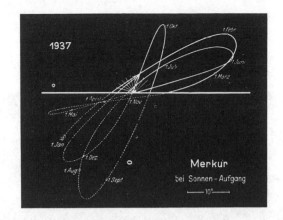

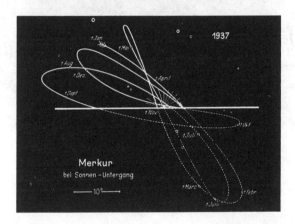

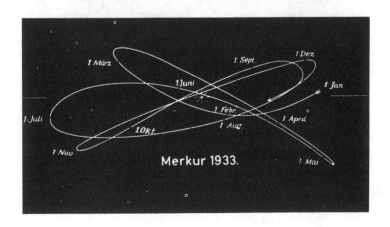

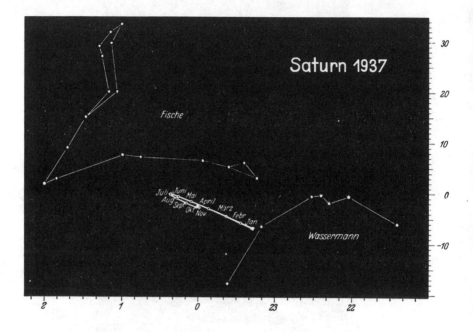

The reason the caduceus, the face of Tlaloc, and the notion of a "god of intestines" point exclusively to Mercury, one of the inferior planets (together with Venus) which are nearer to the sun than our planet Earth, becomes evident from figures 3–5, which illustrate Mercury's movements; it becomes even more evident if the representations of Mercury's racing feats within one solar year are compared to those of one of the superior planets within the same time (figure 6). It takes Saturn thirty years to accomplish his revolution around the sun, returning to the same fixed star; Mercury comes around in eighty-seven days (sidereal revolution; 115 days synodical).

The Egyptian goddess Serqet, or Selket.

A green jasper scarab of Greco-Phoenician origin (6th–5th century B.C.) shows the goddess with "the rear of a four-legged, winged scorpion"; the archaeologists recognize "the fore part of Isis," but the Scorpion lady is clearly in evidence.

In the Maya Codex Tro-Cortesianus we meet again the "old goddess with the scorpion's tail," although with a very different graphic convention. In Nicaragua and Honduras, "Mother Scorpion, who dwells at the end of the Milky Way," is described as many-breasted.

There is the story of Ugaritic Aqht, who shows mocking haughtiness to Anat;[9] of Picus who flatly turns down the offer of Circe and who is subsequently turned into the woodpecker by the angry goddess; there is Arjuna—a "portion of Indra"—who rejected the heavenly Urvashi, whom he regarded as the "parent of my race, and object of reverence to me . . . and it behoveth thee to protect me as a son."[10]

There is also Tafa'i of Tahiti (Maori: Tawhaki) who went with his five brothers courting an underworld princess. As a test, the suitors "were told to pull up by the roots an *ava* tree which was possessed by a demon, and which had caused the death of all who had attempted to disturb it." Three of the brothers were devoured by the demon; Tafa'i revived them, and then gladly renounced the hand of the princess.[11] (Ava = Kawa, and stands for the "next-best-substitute" for Amrita, the drink of immortality which is the property of the gods; mythologically Polynesian Kawa resembles almost exactly the Soma of Vedic literature; even the role of the "Kawa-filter" is an ancient Indian reminiscence; and, as befits the pseudo-drink-of-immortality, it is stolen, by Maui, or by Kaulu, exactly as happens in India, and in the *Edda*, and elsewhere).

Meanwhile Ishtar, scorned, goes up to heaven in a rage, and extracts from Anu the promise that he will send down the Bull of Heaven to avenge her.[12] The Bull descends, awesome to behold.

---

[9] See C. H. Gordon, *Ugaritic Literature* (1949), pp. 84–103. This "Legend of Aqht" is the more relevant, in that the goddess wants nothing but "the *Bow*," made by the Deus Faber and in Aqht's possession, and promises everything including immortality, if the youth will hand over ^mulBAN, that being its fateful name. Cf. above, pp. 215f., for this bow.

[10] Mbh. 3.45–46. Urvashi, the goddess, "trembling with rage" condemned the hero to pass his time "among females unregarded, and as a dancer, and destitute of manhood and scorned as a eunuch." She raged the more, as she had, in anticipation, before actually visiting Arjuna, "mentally sported with him on a wide and excellent bed laid over with celestial sheets." Arjuna had to suffer the curse of Urvashi in the thirteenth year of the exile of the Pandava, but he regained his power on the expiration of that year.

[11] T. Henry, *Ancient Tahiti* (1928), pp. 561ff.

[12] Ugaritic Anat, after having been rebuked by Aqht, goes to her father too, asking for revenge, and she goes "toward 'Il, at the course of the Two Rivers / (at the midst of the streams) of the Two Deeps" (Gordon, p. 91). Ginsberg translation (ANET, p. 152): "Towards El of the Source of the Floods (in the midst of the headwaters) of the Two Oceans."

With his first snort he downs a hundred warriors. But the two heroes tackle him. Enkidu takes hold of him by the tail, so that Gilgamesh as *espada* can come in between the horns for the kill. The artisans of the town admire the size of those horns: "thirty pounds was their content of lapis lazuli." (Lapis lazuli is the color sacred to Styx, as we have seen. In Mexico it is turquoise.)

Ishtar appears on the walls of Uruk and curses the two heroes who have shamed her, but Enkidu tears out the right thigh of the Bull of Heaven and flings it in her face, amid brutal taunts (appendix #27). It seems to be part of established procedure in those circles. Susanowo did the same to the sun-goddess Amaterasu, and so did Odin the Wild Hunter to the man who stymied him.

A scene of popular triumph and rejoicings follows. But the gods have decided that Enkidu must die, and he is warned by a somber dream after he falls sick.[13]

The composition of the epic has been hitherto uncouth and repetitious and, although it remains repetitious, it becomes poetry here. The despair and terror of Gilgamesh at watching the death of his friend is a more searing scene than Prince Gautama's "discovery" of mortality.[14]

"Hearken unto me, O elders, [and give ear] unto me!
It is for Enk[idu], my friend, that I weep,
Crying bitterly like unto a wailing woman
[My friend], my [younger broth]er (?),[15] who chased
the wild ass of the open country [and] the panther of the steppe.
Who seized and [killed] the bull of heaven;

[13] Gen. xlix.5–7 is frequently brought into the play here—the "twins" Simeon and Levi mutilating the bull—but we leave aside this whole chapter xlix bristling with allusions to lost knowledge.

[14] The quotation marks that enclose the word "discovery" are a measure of precaution, advisable in our times ruled by Euhemerism; the most edifying among the relevant model cases we found in Diakanoff's review article on Böhl's translation of GE (see below): "F. M. Th. de Liagre Boehl shares the opinion of A. Schott that the problem of human mortality was originally raised in the reign of Shulgi" (= Third Ur Period, between 2400 and 2350 B.C., according to T. Jacobsen: *The Sumerian King List* [1939], Table II). This "originally" is enough to show what happens to Orientalists once evolutionist platitudes have taken hold of them.

[15] See appendix #28.

Who overthrew Humbaba, that [dwelt] in the [cedar] forest—!
Now what sleep is this that has taken hold of [thee]?
Thou hast become dark and canst not hear [me]."
But he does not lift [his eyes].
He touched his heart, but it did not beat.
Then he veiled [his] friend like a bride [. . .]
He lifted his voice like a lion
Like a lioness robbed of [her] whelps . . .

"When I die, shall I not be like unto Enkidu?
Sorrow has entered my heart
I am afraid of death and roam over the desert . . .
[Him the fate of mankind has overtaken]
Six days and seven nights I wept over him
Until the worm fell on his face.
How can I be silent? How can I be quiet?
My friend, whom I loved, has turned to clay."[16]

Gilgamesh has no metaphysical temperament like the Lord Buddha. He sets out on his great voyage to find Utnapishtim the Distant, who dwells at "the mouth of the rivers" and who can possibly tell him how to attain immortality. He arrives at the pass of the mountain of Mashu ("Twins"), "whose peaks reach as high as the banks of heaven—whose breast reaches down to the underworld—the scorpion people keep watch at its gate—those whose radiance is terrifying and whose look is death—whose frightful splendor[17] overwhelms mountains—who at the rising and setting of the sun keep watch over the sun."[18]

---

[16] Tabl. *8*, col. 2; Tabl. *9*, col. 1, 3–5 (Heidel, pp. 62–64); Tabl. *10*, col. 2, 5–7, 11–12 (Heidel, p. 73).

[17] The word which Heidel translates as "frightful splendor" and Speiser (ANET) as "halo" is *melammu*, the Babylonian equivalent of Iranian *hvarna*, the so-called "glory" for the sake of which the bad uncle Afrasiyab dived in vain, because it belonged to Kai Khusrau.

[18] That the Mashu mountain(s) does so "every day," as translated by Heidel, Speiser, and others, is obviously wrong. Even if we stipulate, for the sake of peace, the idea of a terrestrial mountain, the sun is not in the habit of rising on the same spot every day, and it needs no profound astronomical knowledge to become aware of this fact.

The hero is seized "with fear and dismay," but as he pleads with them, the scorpion-men recognize his partly divine nature. They warn him that he is going to travel through a darkness no one has traveled, but open the gate for him.

"Along the road of the sun [he went?]—dense is the dark[ness and there is no light]" (Tabl. 9, col. 4, 46). The successive stretches of 1, then 2, then 3 and so on to 12 double-hours he travels in darkness. At last it is light, and he finds himself in a garden of precious stones, carnelian and lapis lazuli, where he meets Siduri, the divine barmaid, "who dwells by the edge of the sea."

Under the eyes of severe philologists, slaves to exact "truth," one dare not make light of this supposedly "geographical" item with its faint surrealistic tang. Here is a perfectly divine barmaid by the edge of the sea, called by many names in many languages. Her bar should be as long as the famed one in Shanghai, for she has along her shelves not only wine and beer, but more outlandish and antiquated drinks from many cultures, drinks such as honeymead, soma, sura (a kind of brandy), kawa, pulque, peyote-cocktail, decoctions of ginseng. In short, from everywhere she has the ritual intoxicating beverages which comfort the dreary souls who are denied the drink of immort 'ity. One might call these drinks *Lethe*, after all (appendix #29).

Earnest translators have seriously concluded that the "sea" at the edge of which the barmaid dwells must be the Mediterranean, but there have also been votes for the Armenian mountains. Yet the hero's itinerary suggests the celestial landscape instead, and the scorpion-men should be sought around Scorpius. The more so as lambda ypsilon Scorpii are counted among the Babylonian *mashu*-constellations, and these twins, lambda ypsilon, play an important role also in the so-called Babylonian Creation Epic, as weapons of Marduk.

In any case, Siduri, who must be closely related to Aegir and Ran of the *Edda* with their strange "Bierstube,"—as well as to the nun Gertrude, in whose public house the souls spent the first night after death (see above, p. 208, n. 9)—takes pity on Gilgamesh in his ragged condition, listens to his tale of woe but advises him to return home and make the best of his life. Even Shamash comes to him and tells him: "The life which thou seekest thou wilt not find." But Gilgamesh goes on being afraid of eternal sleep: "Let mine eyes see the sun, that I may be sated with light."[19] And he insists on being shown the way to Utnapishtim. Siduri warns: "Gilgamesh, there never has been a crossing; and whoever from the days of old has come thus far has not been able to cross the sea, [but] who besides Shamash crosses [it]? Difficult is the place of crossing . . . And deep are the waters of death, which bar its approaches." And she warns him that, at the waters of death, "there is Urshanabi, the boatman of Utnapishtim. Him let thy face behold."[20]

Siduri-Sabitu sits "on the throne of the sea" (*kussu tamtim*), and W. F. Albright,[21] picking up a notion of P. Jensen, thoroughly compares Siduri and Kalypso, whose island Ogygia is called by Homer "the Navel of the Sea" (*omphalos thalassēs*). Moreover, Albright points to "the similar figure of Ishara tamtim," Ishara of the sea, the latter being the goddess of Scorpius,[22] corresponding to the Egyptian Scorpius-goddess Selket, and to "Mother Scorpion . . . dwelling at the end of the Milky Way, where she receives the souls of the dead; and from her, represented as a mother with many breasts, at which children take suck, come the souls of the new-born." This last-mentioned "Mother Scorpion" is a legitimate citizen of ancient Nicaragua and Honduras,[23] an offshoot of the Mayas' "Old Goddess with a scorpion's tail."[24]

At this point there is still another recurrence of the disheartening breakdown in communication between scholars. The Orientalists, taking the story of Gilgamesh as a "normal epic," search for traces

---

[19] Old Babyl. Version, Tabl. *10*, col. 1, 8, 13 (Heidel, p. 69).

[20] Assyrian Version, Tabl. *10*, col. 2, 21–28 (Heidel, p. 74). *Shanabi* meaning *40*, Ur-shanabi means something like "he of 40"; Hommel rendered it "priest of 40."

[21] W. F. Albright, "The Goddess of Life and Wisdom," AJSL *36* (1919–20), pp. 258–94.

[22] See appendix #30. The name of the goddess is pronounced Ish-khara.

[23] H. B. Alexander, *Latin American Mythology* (1920), p. 185.

[24] The many-breasted Mother Scorpion of Central America goes well with the farmer's calendar of ancient Rome which attributes Scorpius to Diana (see F. Boll, *Sphaera* [1903], p. 473; W. Gundel, RE s.v. Scorpius, p. 602). It remains still dark, however, what caused Athanasius Kircher to localize the many-breasted Diana of the Ephesians into Aquarius, calling, moreover, this celestial department "Regnum Canubicum."

of him in the physical landscape of the Near East, ignoring the work of equally learned scholars, experts in the heavens, whose well-prepared and organized tool kits have long been available to help in the search. One wonders whether the Orientalists, intent on reconstructing texts, have ever even heard of Boll and Gundel and men like them. Perhaps not, because they pass them by without a word. In any case, it is appropriate here to mention once again two valuable tool kits assembled by Franz Boll,[25] who presents the whole tradition on the constellations "Hades," "Acherusian lake," "ferryman," with many more details than are needed now, as they have survived in astrological tradition. These topoi are found together around the southern crossroads of Galaxy and ecliptic, between Scorpius and Sagittarius. Boll points out that, instead of the Scorpion people,[26] Virgil (*Aeneid 6*.286) and Dante posted centaurs at the entrance of the underworld, representing Sagittarius. And so back to the quest; Gilgamesh faces Urshanabi, expecting to be ferried over the waters of death. The boatman demurs: the "stone images" have been broken by Gilgamesh (appendix #31). But at length he instructs the pilgrim to cut down 120 poles, each sixty cubits (thirty yards) in length. With these he must punt the boat along, so that his hands may not touch the waters of death. At last they reach the far shore; there is Utnapishtim the Distant. The hero is puzzled: "I look upon thee, thine appearance is not different, thou art like unto me. My heart had pictured thee as one perfect for the doing of battle; [but] thou liest (idly) on (thy) side, (or) on thy back; [Tell me], how didst thou enter into the company of the gods and obtain life (everlasting)?"[27]

---

[25] *Sphaera*, pp. 19f., 28, 48, 173, 246–51; *Aus der Offenbarung Johannis* (1914), pp. 71ff., 143. See also W. Gundel, *Neue Texte de Hermes Trismegistos* (1936), esp. pp. 235ff. (on p. 207 he votes for Centaurus as guardian of the netherworld instead of Sagittarius).

[26] The coronation mantle of Emperor Heinrich II shows woven in the statement: *Scorpio dum oritur, mortalitas ginnitur* (= *gignitur*). E. Maass, *Commentariorum in Aratum Reliquiae* (1898), p. 602; R. Eisler, *Weltenmantel und Himmelszelt* (1910), p. 13; Boll, *Aus der Offenbarung*, p. 72. We might also point to Ovid's description of the fall of Phaethon, according to which the son of Helios lost his nerve, and let go the reins when Scorpius drew near, and to the death of Osiris on Athyr 17, the month when the Sun went through Scorpius (Plutarch, *De Is. Os.*, c. 13, 356c).

[27] Tabl. *11*, 3–7 (Heidel, p. 80).

Utnapishtim is spry enough to tell in great detail the story of the Deluge. He tells how Enki-Ea has warned him of Enlil's decision to wipe out mankind, and instructed him to build the Ark, without telling others of the impending danger. "Thus shalt thou say to them: [I will . . . go] down to the *apsu* and dwell with Ea, my [lor]d." He describes with great care the building and caulking of the ship, six decks, one *iku* (acre) the floor space, as much for each side, so that it was a perfect cube, exactly as Ea had ordered him to do. This measure "1-iku" is the name of the Pegasus-square, and the name of the temple of Marduk in Babylon, as is known from the New Year's Ritual at Babylon, where it is said: "Iku-star, Esagil, image of heaven and earth."[28] Shamash had let Utnapishtim know when to enter the ship and close the door. Then the cataracts of heaven open, "Irragal [= Nergal] pulls out the masts [appendix #32]; Ninurta comes along (and) causes the dikes to give way; The Anunnaki[29] raised (their) torches, lighting up the land with their brightness; The raging of Adad reached unto heaven (and) turned into darkness all that was light.[30] . . . (Even) the gods were terror-stricken at the deluge. They fled (and) ascended to the heaven of Anu; The gods cowered like dogs (and) crouched in

---

[28] Trans. A. Sachs, ANET, p. 332, ll. 275ff. Concerning the Rectangle of Pegasus, see B. L. van der Waerden, "The Thirty-Six Stars," JNES *8* (1949), pp. 13–15; C. Bezold, A. Kopff, and F. Boll, *Zenit- und Aequatorialgestirne am babylonischen Fixsternhimmel* (1913), p. 11.

[29] These divine beings of the "underworld" (their equivalent "above": the Igigi) were also written A-nun-na-nun[ki] (Deimel, PB, pp. 57f.), i.e., they belong to NUN[ki] = Eridu (Canopus), the seat of Enki-Ea. The Sumerian name Anunna is interpreted as "(Gods who are) the seed of the 'Prince,'" according to A. Falkenstein ("Die Anunna in der sumerischen Uberlieferung," in *Festschrift Landsberger* [1965], pp. 128ff.). See also D. O. Edzard, "Die Mythologie der Sumerer und Akkader," in *Wörterbuch der Mythologie*, vol. *1*, p. 42: "Die 'fürstlichen' Samens [sind]," the "Prince" (NUN) being Enki-Ea of Eridu. Concerning NUN = "Prince," defined by T. Jacobsen as "one of authority based on respect only, settling disputes without recourse to force," Falkenstein, politely, mentions: "Ganz abweichend K. Oberhuber: Der numinose Begriff ME im Sumerischen, S.6f." The title of this opus (Innsbruck 1963, Innsbrucker Beiträge zur Kulturwissenschaft. Sonderheft 17) expresses sufficiently the hair-raising propositions that it contains, concerning ME, NUN, and other termini.

[30] Speiser, ANET, p. 94, n. 207, remarks: "The term šuharratu . . . does not mean 'rage,' but 'stark stillness, bewilderment, consternation,'" and he translates *11*.105–06: "Consternation over Adad reaches to the heavens, Who turned to blackness all that had been light."

distress (?).[31] Ishtar cried out like a woman in travail; the lovely-voiced lady of the g[ods] lamented . . . 'How could I command (such) evil in the assembly of the gods! How could I command war to destroy my people, For it is I who bring forth (these) my people . . .' The Anunnaki-gods wept with her; The gods sat bowed (and) weeping."

The end of the story is almost exactly that of Noah's landing on the mountain, except that Noah sends out a raven and twice the dove, whereas Utnapishtim let fly dove, swallow, raven.[32]

When Enlil was still wroth because one family did escape, Enki-Ea, "who alone understands every matter" (*11.176*), took him to task: "How, o how couldst thou without reflection bring on (this) deluge?" He added severely that Enlil could have punished only the sinful, and spared the innocent. The remark is one that pious exegetes of the Bible are still left to ponder. Then Enlil went up to the Ark and apologized. He granted Utnapishtim and his wife "to be like unto us gods. In the distance, at the mouth of the rivers, Utnapishtim shall dwell" (*11.194–95*).

"But now as for thee," the old man concludes his tale (*11.197ff.*), turning to Gilgamesh, "who will assemble the gods unto thee, that thou mayest find the life which thou seekest? Come, do not sleep for six days and seven nights."

We gather a gentle hint there from the Ancient of Days (Sumerian: Ziusudra, with Berossos: Xisuthros), also called Atrahasis, "the exceedingly wise." It amounts to this: "Young man, you have come to the land where time has come to a stop, and the immortality granted to us consists in remaining conscious and partaking of truth while not being wholly awake. Now you try." But Gilgamesh cannot. "As he sits (there) on his hams, sleep like a rainstorm blows upon him" (*11.200ff.*).

---

[31] Speiser: "The gods cowered like dogs, crouched against the outer wall" (*11.115*).

[32] One is usually inclined to take such motifs as that of the sending out of birds —not to mention the particular species—for minor matters, but A. B. Rooth can teach us a remarkable lesson by means of her thorough inquiry: *The Raven and the Carcass: An Investigation of a Motif in the Deluge Myth in Europe, Asia, and North America* (1962).

One can imagine how Atrahasis-Utnapishtim would explain some essentials during Gilgamesh's sleep. The Exceedingly Wise would point to "his like," to Kronos sleeping in his golden cave in Ogygia, as described by Plutarch,[33] and yet continuously giving "all the measures of the whole creation" to his beloved son Zeus, as described by Proclus.[34] The Exceedingly Wise would refer freely to characters faraway in time[35] and in geographical space, as only *he* is entitled to do—to Kiho-tumu, for instance, creator god of the Tuamotu islands, Kiho-tumu "the-All-Source" who sleeps, face downward, in "Great-Havaiki-the-Unattainable," and yet takes action when the "administration" oversteps the "laws" and measures given by him. In the most amiable words Utnapishtim would admonish the children of our century to perceive the divine mummies of Ptah and Osiris—Osiris the "*strategos*" of the Ship *Argo*—and to start to think about the mummies of gods, generally, about the idea of the Seven Sleepers of Ephesus on board the *Argo*, about the data given in the *Liber Hermetis Trismegisti* concerning the relevant degrees (in Taurus, dealing with latitudes south of the Ship) belonging to Saturn, meaning "continua vero delectatio, diminutio substantiae, remissio malorum." Atrahasis would tell of the Chinese "Ancient Immortal of the Celestial South Pole," of the numerous sleeping Emperors in Mountain Caves (appendix #33)—and the hours would pass like seconds, but one knows that Utnapishtim, half-dreaming, half-teaching, had all the time an eye on the sleeping "hero." He says to his wife: "Is this the strong man who wants life

---

[33] *De facie in orbe lunae* 941A.

[34] See frg. no. 55, *Orphicorum Fragmenta*, ed. O. Kern (1963).

[35] The oldest and most exact traits have a perplexing talent of surviving, and of turning up at unexpected places. Says R. S. Loomis (*Arthurian Literature in the Middle Ages* [1959] pp. 70–71): "We have a unique version of Arthur's survival alluded to by Godfrey of Viterbo, secretary to Frederick Barbarossa, about 1190. Merlin prophesies that though the king will perish from his wounds, he will not perish wholly but will be preserved *in the depths of the sea* and will reign for ever as before." *How* could the secretary to Frederick Barbarossa—an emperor who was himself bound to that place where expired ages and their rulers sleep—get hold of the "right" version? (We should be glad to learn, moreover, where the archaeologist Pierre Plantard [quoted by Gérard de Sède: *Les Templiers sont parmi nous* (1962), p. 280] got hold of the information on "Canopus, l'oeuil sublime de l'architecte, qui s'ouvre tous les 70 ans pour contempler l'Univers.")

everlasting?" And then, he wakens the man, on the seventh day, and the startled Gilgamesh reacts thus: "Hardly did sleep spread over me, when quickly thou didst touch and rouse me."

Les jeux sont faits. Gilgamesh is given a change of raiment and told to go home; Urshanabi, the boatman, is told to escort him, and there is, evidently, no return again for him to *pī nārāti*, to the mouth of the rivers. But at the last minute, Utnapishtim's wife says to her husband: "Gilgamesh has come hither, he has become weary, he has exerted himself, What wilt thou give (him wherewith) he may return to his land?" Utnapishtim takes compassion and addresses the hero:

> "Gilgamesh, I will reveal (unto thee) a hidden thing . . . There is a plant like a thorn . . . Like a rose (?) its thorn(s) will pr[ick thy hands]. If thy hands will obtain that plant, [thou wilt find new life]" (*11.264–70*).

"New life" sounds misleading, and Speiser remarks: "Note that the process is one of rejuvenation, not immortality."[36]

To get hold of this plant, growing apparently in a tunnel leading to the Apsu which the hero has to open, Gilgamesh dives deep, weighting himself with stones. But then as he travels home with the boatman, he stops to take a bath in a well, a serpent (literally, earth-lion) comes up from the water, snatches the plant and, returning into the water, sloughs its skin. The last hope is gone—at least so it looks in the translations.

Since this is not a manual on the Epic of Gilgamesh, this whole affair of the plant, the diving, the fateful bath in the well, must stand as it is, even though every word in it is no better than a man-trap (appendix #34), to come to the point which is of particular relevance here.

Leaving the boat on the shore, Gilgamesh and the boatman walked for another 50 double-hours on the way home.

> When they arrived in Uruk, the enclosure, Gilgamesh said to him, to Urshanabi, the boatman: "Urshanabi, climb upon the wall of Uruk (and walk about); inspect the foundation terrace and examine the brickwork, (to see) if its brickwork is not of burnt bricks, And (if) the seven wise men did not lay its foundation!" (*11.301–305*).

[36] Speiser, p. 96, n. 227.

The Mesopotamian cylinder seal shows in the upper part the "God Boat"; in the lower part people are building a ziggurat, the proposition being that the boat is bringing the *me* from Eridu-Canopus, the measures of creation.

The "God Boat" surrounded by the crescent moon, three single stars, and constellations; recognizable are the Scorpion, the Plow (=^{mul}APIN, Triangulum), and perhaps the Lion following directly behind the boat; the jug above the Lion might indicate Aquarius.

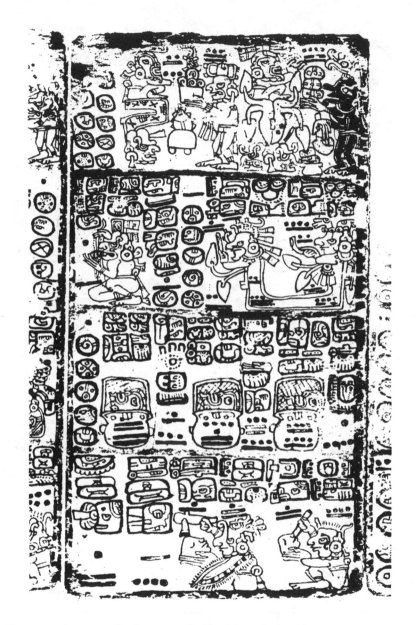

The same character clearly occurs in the Maya Codex Tro-Cortesianus, second row.

The "God Boat" in the appropriate surroundings on the Arabian celestial globe made by Tabarī (in the year 684 of the Hijra), after the catalogue of stars of 'Abd ar Rahman as Sufi. The name of Suhayl (Canopus) is written below the oar of this personified boat, i.e., of Argo, closest to the celestial South Pole.

But before the Epic started (Tabl. *1*, 19), it was said that "the Seven Wise Ones" had laid the foundation of ramparted Uruk. So the ring has been closed.

But what does it mean? Why is Urshanabi, of all people, asked to survey Uruk, enclosed—according to the rule—by seven walls? And what have the Seven Sages to do with the foundation of Gilgamesh's city?

To take the latter question first: the Seven Sages are the stars of the Big Dipper, the Indian Saptarshi, the Seven Rishis.[37] The solstitial colure, called the "Line of the Seven Rishis," happened to run through one after the other of these stars during several millennia (starting with ēta, around 4000 B.C.): and to establish this colure is "internationally" termed "to suspend the sky"—the Babylonians called the Big Dipper "bond of heaven," "mother bond of heaven," the Greeks spelled it "Omphaloessa."

Next, why is it the business of the boatman from the "confluence of the rivers" (that is what *pī nārāti* is) to check the measures of Uruk? It is established that the boatman's name was "servant (or priest) of 40 or of ⅔,"[38] and that makes him a "piece," or whatever one prefers to designate it, of Enki-Ea, called Shanabi = ⅔ (of 60 = 40). Enki's residence is Eridu, at the confluence of the rivers, at ^{mul}NUN^{ki} = Canopus (alpha Carinae), the seat of the *me*, the norms and measures. From there these *me* have to be procured. Urshanabi, however, seems to be bound with close family ties to Enki-Ea, in fact to be his son-in-law, husband of Nanshe.[39]

Numerous texts and inscriptions show that Enki-Ea, Lord of the Apsu, was responsible for the ground plan of "temples," whether celestial or terrestrial ones. The one who actually drew up the plans, with the "holy stylus" of Eridu, was Nanshe, Enki's

---

[37] And exactly as the Indian texts have a lot to say about the Seven Rishis with their sister (and wife) *Arundati*, so the Mesopotamian ones talk about the "Sebettu with their sister *Narundi*" (see H. Zimmern, "Die sieben Weisen Babyloniens," ZA *35* [1923], p. 153; Edzard, vol. *1*, p. 55; H. and J. Lewy, "The Origin of the Week," HUCA *17* [1942–43], p. 44). Arundati = Alcor, the tiny star near zeta Ursae Majoris.

[38] See also Langdon, p. 213.

[39] T. Jacobsen, "Parerga Sumerologica," JNES *2* (1943), pp. 117f. See also Edzard, p. 109.

daughter. And to her, the wife of Urshanabi the boatman, the "holy stern of the ship was consecrated."[40]

Considering that Argo is a stern only, that Eridu-Canopus marks the steering oar of Argo, it is fair to conclude that Gilgamesh, bringing with him Urshanabi in person, had procured "the *me* from Eridu." This is how it is styled in the Sumerian "dialect";[41] in the international mythical language the terminus technicus reads "to measure the depth of the sea." (Odysseus, more advanced and accordingly considerably more modest than Gilgamesh, did not even take over a veritable oar from Teiresias. He only procured the latter's advice, according to which he was, later, to take an oar and carry it inland until he found people who had never heard of or seen a ship) (appendix #35).

Now that Gilgamesh "surveys" the world is stated explicitly in a text. (That this truth is uttered involuntarily by the translator who meant to express "that he saw everything" makes it the more enjoyable.) The invocation, quoted by Lambert, says:

Gilgamesh, supreme king, judge of the Anunnaki,
Deliberative prince, the . . . of the peoples,
Who surveys the regions of the world, bailiff of the underworld,
                 lord of the (peoples) beneath,
You are a judge and have vision like a god.
You stand in the underworld and give the final verdict.
Your judgement is not altered, nor is your utterance neglected.
You question, you inquire, you give judgement, you watch and you
                 put things right.
Shamash has entrusted to you verdicts and decisions.
In your presence kings, regents and princes bow down.
You watch the omens about them and give the decision.[42]

---

[40] Gudea Cylinder A XIV, in A. Falkenstein and W. von Soden, *Sumerische und Akkadische Hymnen* (1953), p. 152; see also F. Hommel, *Die Schwur-Göttin Esch-Ghanna und ihr Kreis* (1912), p. 57.

[41] See, for example, S. Kramer's *Sumerian Mythology* (1944), pp. 64–88; and his *Enmerkar and the Lord of Aratta* (1952), p. 11. We feel strongly inclined to accuse the much discussed "God Boat" (Dieu Bateau) of many seal cylinders of "bringing the *me* from Eridu," particularly when the seals show a ground plan, or a stage tower in the making. See P. Amiet, *La Glyptique Mésopotamienne Archaïque* (1961), pp. 177–86, plates 106–109; H. Frankfort, *Cylinder Seals* (1939), pp. 67–70, plates xiv, xv, xix.

[42] In *Gilgamesh et sa légende* (1960), p. 40. Cf. E. Ebeling, *Tod und Leben nach den Vorstellungen der Babylonier* (1931), p. 127: "der die Welträume überschaut."

That neither this nor other clear hints make the slightest impression upon once-and-for-all Euhemerists goes without saying. Any unprejudiced student would at least ponder for some minutes about that opened water-tunnel or the well in which the "hero" was taking a bath, once he comes to learn about a text, also mentioned by Lambert (p. 43), dealing with the digging of wells, where "an instruction is given for the utterance of the words '*the well of Gilgamesh*' . . . , as the well is being dug. Since, when water is reached, it must be libated to Shamash, the Anunnaki, and any known spirit, the well is thought of as a connexion with the underworld" (appendix #36).

It seems obvious that sooner or later the data on Gilgamesh —incompatible as they sound for the time being—have to be assembled on a common denominator. But this is not likely to be accomplished unless the specialists renounce several of their firm preconceptions and make up their minds to a thorough re-examination of the whole case.

For the time being it is worth paying attention to information such as that given by Strabo (*16.1.5*) on the tomb of Bel (*ho tou Bēlou taphos*) in Etemenanki, the Tower of Babylon, and to mind the baffling Mesopotamian texts dealing with gods cutting off each other's necks and tearing out each other's eyes. It well might be rewarding to look at the tombs of Anu and of Marduk,[43] to consider the fundamental role of the Abaton in Philae, tomb of Osiris,[44] and of divine sepulchers generally. The basic difficulty which has to be

---

[43] Ebeling, pp. 25f., 39; see also G. Meier, "Ein Kommentar zu einer Selbstprädikation des Marduk aus Assur," ZA *47* (1942), pp. 241–48. H. Zimmern, "Zum babylonischen Neufahrsfest," BVSGW *58* (1906), pp. 127–36. S. A. Pallis, *The Babylonian Akitu Festival* (1926), pp. 105–108, 200–43.

[44] Apart from the Shabaka Inscription, the end of which is of the utmost relevance, the highest Egyptian oath was taken by "Osiris who lies in Philae," as we know from Diodorus; the Greek gods took their most solemn oaths by the waters of Styx. We remember Virgil's information on Styx who sees the celestial South Pole, and of the followers of Zeus who, before attacking Kronos, took their oath by Ara. "Oath-stars" are to be found, rather regularly, among the southern circumpolar constellations. As concerns swearing by Gilgamesh, see Ebeling, p. 127. Compare also Pallis (*Akitu Festival*, p. 238) who compares the "Mysteries of Osiris" in Abydos with the Babylonian New Year Festival built around the "dead" Marduk (who sits during the ceremonies "in the midst of Tiamat").

overcome is our ignorance of the concrete meaning of the technical term "tomb," whether one has to do with the Omphalos of Delphi, grave of Python,[45] with the "burial mound of dancing Myrina" (*Iliad* 2.814), with the burial mound of Lugh Lamhfada's foster mother, around which the Games of Taillte were performed, or with many others.

What is haunting is the suspicion that "Uruk" stands for a "new" realm of the dead, and that Gilgamesh is the one who was destined to "open the way" to this abode and to become its king, and the judge of the dead, like Osiris, and also Yama, of whom the *Rigveda* states (*10.14.1–2*): (1) "Him, who followed the course of the great rivers, and who discovered the way for many, the Son of Vivasvat, the gatherer of peoples—King Yama we honor with sacrifice. (2) Yama is the one who first discovered the way; this trodden path is not to be taken away from us; on that way that our forefathers travelled when they left us, on that way the later born follow each his trail."[46]

That neither Yama's nor Gilgamesh's "way" was, originally, meant to last forever and ever, goes without saying. Again and again the *me* must be brought from Eridu, the Depth of the Sea must be measured respectively, and again and again the sky has to be "suspended" by means of the "Line of the Seven Rishis"—the huge precessional clock does not stop. What has been stopped, instead, is the understanding among the heirs of the mythical language who, out of ignorance, failed to adapt this idiom to "preceded" situations. Without thinking, they changed a movie into a set of stills, projected a complex motion into conventional posters, and de-

---

[45] *Omphalos* belongs among the words which are easily said and hard to "imagine." Yet, during the Middle Ages, Jerusalem, with the Holy Sepulcher, was understood as the Omphalos of the earth and, moreover, the tomb of Adam localized under the Cross in Golgatha, "in the middle of the earth." (See, for example, *Vita Adae et Evae*, in F. Kampers, *Mittelalterliche Sagen vom Paradiese und vom Holze des Kreuzes Christi* (1897), pp. 23, 106f.; W. H. Roscher, *Omphalos* (1913), pp. 24–28.

[46] Cf. *Atharva Veda* 18.1.50 (Whitney trans.): "Yama first found for us a track, that is not a pasture to be borne away; where our former Fathers went forth, there (go) those born (of them), along their own roads."

stroyed, by this measure, all the sense of a carefully considered system.[47]

This might be dismissed as a minor tragedy, but it is just one of those "progressive measures" which violently interrupt the continuity of tradition. There must have been several such eruptive and reckless "corrections of style"—otherwise it would be utterly incomprehensible that all our most ancient texts consist of "Scholia" interpreting one or the other "antediluvian" "Book with seven seals." In the case of that neglected tragedy just mentioned,[48] a tragedy coming from absentmindedness, the final blow was dealt to the tradition that had established "us," mankind, as a unity. And if we did not have Plato's *Timaeus*, it would be a hopeless task altogether to understand the reason which made it obligatory in those "archaic" times to watch the immense cosmic clock most carefully. Plato himself, to be sure, started on the way of all intellect—moving from thought to literature, from literature into philology, before flowing into nothing; but let us make it clear, this official "trend" is not going to detract us from our own unconditional respect.

This essay could spend many chapters on the *Timaeus*, that "topos" from which come and to which return all "rivers" of cosmological thought, and several more chapters on *Phaedrus* and

[47] In our most unheeding times, nobody will even notice when in the not too remote future Leo will be drowned in the sea when he arrives at the autumnal equinox: the constellation of Leo, undisputed "king" of the hot plains, was coined at a time when His Majesty of the Zodiac ruled the summer solstice, highest and hottest "point" of the sun's orbit; and who will care for pitiable Aquarius having no more water to shed from his jars, once he has arrived at the vernal equinox—but, after all, who has considered poor Pisces, lying "high and dry" since the times of Christ, the opener of the Age of Pisces? His title "Fish," i.e., Greek *Ichthys*, is officially explained as being the first letters of "*Iēsous Chreistos Theou Yios Sotēr*"—Jesus Christ God's Son Savior.

[48] Without going into details, we think it possible that it was this very change from "constellations" to "signs" and, more generally, the enthronement of that astronomical language which alone is recognized as "scientific" by contemporary historians, i.e., the terminology of "positional astronomy," which interrupted Homeric tradition; the Greeks quoted Homer all day long, they interpreted him, they broke their heads about the significance of details: his terminology had died long ago.

Politikos, on the *Epinomis* (entartarosed by the label "Pseudo-"), but we make it short. We leave aside the very "creation" which Timaeus styles like the manufacturing of a planetarium—which is exactly what makes this creation difficult for non-mathematicians. But it can be done without here. What counts is this: When the Timaean Demiurge had constructed the "frame," *skambha,* ruled by equator and ecliptic—called by Plato "the Same" and "the Different"—which represent an X (spell it Khi, write it X) and when he had regulated the orbits of the planets according to harmonic proportions, he made "souls." In manufacturing them, he used the same ingredients that he used when he had made the Soul of the Universe, the ingredients however, being "not so pure as before." The Demiurge made "souls in equal number with the stars (*psychas isarithmous tois astrois*), and distributed them, each soul to its several star."

> There mounting them as it were in chariots, he showed them the nature of the universe and declared to them the laws of Destiny (nomous tous heimarmenous). There would be appointed a first incarnation one and the same for all, that none might suffer disadvantage at his hands; and they were to be sown into the instruments of time; *each one into that which was meet for it,* and to be born as the most god-fearing of living creatures; and human nature being twofold, the better sort was that which should thereafter be called "man."

> And he who should live well for his due span of time should journey back to the habitation of his consort star and there live a happy and congenial life; but failing of this, he should shift at his second birth into a woman; and if in this condition he still did not cease from wickedness, then according to the character of his depravation, he should constantly be changed into some beast of a nature resembling the formation of that character, and should have no rest from the travail of these changes, until letting the revolution of the Same and Uniform within himself draw into its train all that turmoil of fire and water and air and earth that had later grown about it, he should control its irrational turbulence by discourse of reason and return once more to the form of his first and best condition.

> When he had delivered to them all these ordinances, to the end that he might be guiltless of the future wickedness of any of them, he sowed them, some in the Earth, some in the Moon, some in the other instruments of time (*Timaeus* 41E–42D).

There is no need to engage in the futile task of arguing the fairness of the Demiurge and his statement that all souls had the same chances in their first incarnation. That God must needs be innocent, and that man is guilty, anyway, is not a subject worth arguing with Plato. In fact, this is the hypothesis upon which the whole great edifice of Christian religion, and of our jurisdiction, rests.

In any event, the faultless Demiurge sowed the souls, equal in number to the fixed stars, in the "instruments of time" (i.e., the planets), among which Timaeus counts the earth; he sowed, actually, "each one into that which was meet for it."

What does that mean? Timaeus alludes here to an old system of connecting the fixed and the wandering members of the starry community—and not only the zodiacal "houses" and "exaltations" of the planets are meant, but fixed stars in general. One knows this approach from astrological cuneiform tablets which contain a considerable number of statements on fixed representatives of a planet, and vice versa. But there is not enough to explain the rules of this sophisticated scheme. To say it with Ernst Weidner: "In any case we have to do with a very complicated system. Only a renewed collection and revision of the whole material will perhaps allow us to solve the still existing riddles."[49] Ptolemy records the planetary character of fixed stars in his *Tetrabiblos* (*1*.9 "Of the Power of the Fixed Stars"), and so did all ancient and medieval astrologers. And, one might add, so did Indian and Mexican astrologers. (See above, p. 157, about the privilege enjoyed by Mars and the Pleiades of representing each other in Babylonia and India.)

The souls were, then, taken away from their fixed stars and moved to the corresponding planetary representatives, all according to rules and regulations. The Demiurge retired—turning into the character known under the title "Deus otiosus"—and the Time Machine was switched on.

Cornford, in his translation and commentary on the *Timaeus*, states (p. 146): "In the machinery of the myth it is natural to sup-

[49] RLA *3*, pp. 81f. Cf. Bezold in Boll's *Antike Beobachtung farbiger Sterne* (1916), pp. 102–25 (table, p. 138); A. Jeremias, HAOG (1929), pp. 200ff.

pose that the first generation of souls is sown on Earth, the rest await their turn, unembodied, in the planets."

With all the respect due to Cornford, this is hardly going to work, and no "natural" suppositions can be admitted. The Demiurge of the *Timaeus* is too much of a systematist to allow for this solution. On the contrary, it stands to reason—if one carefully observes the manner in which the Craftsman God gradually and systematically attenuates his original mixture of Existence, Sameness and Difference as described in *Timaeus* 35—that some new principle, some new "dimension," has to be introduced right here.

Eternity abides in unity highest and farthest "outside." Within, Time, its everlasting likeness, moves according to number, doing so by means of the daily turning of the fixed sphere in the sense of "the Same," the celestial equator, and by means of the instruments of time, the planets, moving in the opposite direction along "the Different," i.e., the ecliptic. Taken together, they represent the "eight motions." With the next step, from the planets to the living creatures, the motion according to number is ruled out. The fundamentally different quality of "motion" by generation must replace it (much to Plato's regret).

The planets, albeit "different" from the eternity abiding in unity as well as from the regular motion of the sphere of constellations, remain at least "themselves" and seven in number. The soul of man is not only reincarnated again and again, but it is subdivided further and further, since mankind multiplies, as does the grain to which man is so frequently compared. This simile—misinterpreted time and again by the fertility addicts—ought to be taken very seriously, and literally. The Demiurge did not create the individual souls of every man to be born in all future, he created the first ancestors of peoples, dynasties, etc., the "seed of mankind" that multiplies and is ground to mealy dust in the Mill of Time. The idea of "Fixed Star Souls" from which mortal life started, and to which exceptionally virtuous "souls once released" may return anytime, whereas the average "flour" from the mill has to wait patiently for the "last day" when it hopes to do the same—this idea is not only a vital part of the archaic system of the world, it explains to a certain degree the almost obsessive interest in the celestial goings-on that ruled former millennia.

Although still, in our time, most children are admonished to behave decently, otherwise they may not have a chance to enter heaven, the Christians have abolished the Timaean scheme. They condemned as heresy the opinion of Origen according to which after the Last Judgment, the revived souls would have an ethereal and spherical body (*aitherion te kai sphairoeides*). This fundamental concept has been given voice in many tongues throughout the "Belt of High Civilization." Sometimes the imagery is unmistakable, sometimes it is ambiguous enough to mislead modern interpreters completely, as when the starry "seed" of population groups comes our way under the title of a "totem." But among the unmistakable kind is a rabbinical tradition which says that in Adam were contained the 600,000 souls of Israel like so many threads twisted together in the wick of a candle; the more so, as it is also said: "The Son of David [the Messiah] will not come before all souls that have been on the body of the first man, will come to an end."[50] Unmistakable, also, is the myth of the Skidi Pawnee of the Great Plains dealing with "the last day": "The command for the ending of all things will be given by the North Star, and the South Star will carry out the commands. Our people were made by the stars. When the time comes for all things to end our people will turn into small stars and will fly to the South Star where they belong."[51]

As mentioned in the chapter on India (p. 77), the *Mahabharata* reports how the Pandavas toiled up the snowy mountain and were lost, and how Yudhishthira was finally removed bodily to heaven. Although they were planetary "heroes," the wording of how they came to their end is revealing with respect to mere human beings. The said heroes are called "portions" of the gods, and when the third world comes to its end and the Kali Yuga begins—it could not begin "as long as the sole of Krishna's holy foot touched the earth"—these "portions" are reunited with the gods of whom they are a part. Krishna returns into Vishnu, Yudhishthira into Dharma,

[50] J. A. Eisenmenger, *Entdecktes Judenthum* (1711), vol. 2, p. 16 (Emek hamelech).
[51] H. B. Alexander, *North American Mythology* (1910), p. 117.

Arjuna is absorbed by Indra, Bala Rama by the Shesha-Serpent, and so forth.

These examples will do. What they demonstrate is this: the *Timaeus* and, in fact, most Platonic myths, act like a floodlight that throws bright beams upon the whole of "high mythology." Plato did not *invent* his myths, he used them in the *right* context—now and then mockingly—without divulging their precise meaning: whoever was entitled to the knowledge of the proper terminology would understand them. He did not care much for the "flour" after all.

Living in our days, where nothing is hidden from the press and where every difficult science is "made easy," we are not in the best condition to imagine the strict secrecy that surrounded archaic science. The condition is so bad, indeed, that the very fact is often regarded as a silly legend. It is not. The need for treating science as reserved knowledge is gravely stated by Copernicus himself in his immortal work, the *Revolutions of Celestial Orbs*. An adherent of the Pythagorean conception since his student days in Italy, he acknowledges the inspiration he owes to the great names of the School, like Philolaos and Hicetas, that he had learned from the classics, and who, he says, had given him the courage to oppose the philosophical notions current in his own time. "I care nothing," he writes in his dedication to the Pope, "for those, even Church doctors, who repeat current prejudices. Mathematics is meant for mathematicians . . ." It is the authority of these ancient masters which gave him the independence of judgment to discover the central position of the sun in the center of the planetary system. A shy and retiring scholar, he appeals to that great tradition, which even in the time of Galileo was called "the Pythagorean persuasion," to advance what was commonly considered a revolutionary and subversive theory. But if he did not bring himself to publish until his last years, it was not from fear of persecution, but from an ingrown aversion toward having the subject bandied around among the public. In the first book, he quotes from a "correspondence" among ancient adepts which is probably an ancient pastiche, but shows their way of thinking: "It would be well to remember the Master's precepts, and to communicate the gifts of philosophy to those who have never even dreamt of a purification of the soul. As to those who try to impart these doctrines in the wrong order and without preparation, they are like people who would pour pure water into a muddy cistern; they can only stir up the mud and lose the water."

Creating the language of the philosophy of the future, Plato still spoke the ancient tongue, representing, as it were, a living "Rosetta stone." And accordingly—strange as it may sound to the specialists on Classical Antiquity—long experience has demonstrated this methodological rule of thumb: every scheme which occurs in myths from Iceland via China to pre-Columbian America, to which we have Platonic allusions, is "tottering with age," and can be accepted for genuine currency. It comes from that "Protopythagorean" mint somewhere in the Fertile Crescent that, once, coined the technical language and delivered it to the Pythagoreans (among many other customers, as goes without saying). Strange, admittedly, but it works. It has worked before the time when we decided to choose Plato as Supreme Judge of Appeals in doubtful cases of comparative mythology, for example, when H. Baumann[52] recognized the myth of Plato's Symposium (told, there, by Aristophanes) as the skeleton key to the doors of the thousand and one myths dealing with bisexual gods, bisexual souls, etc.

Plato knew—and there is reason to assume that Eudoxus did, too —that the language of myth is, in principle, as ruthlessly generalizing as up-to-date "tech talk." The manner in which Plato uses it, the phenomena which he prefers to express in the mythical idiom, reveal his thorough understanding. There is no other technique, apparently, than myth, which succeeds in telling structure (again, remember Kipling, and how he tackled the problem by telling of the "ship that found herself"—see above, p. 49). The "trick" is: you begin by describing the reverse of what is known as reality, claiming that "once upon a time" things were thus and so, and worked out in a very strange manner, but then it happened that . . . What counts is nothing but the outcome, the result of the happen-

[52] *Das Doppelte Geschlecht. Ethnologische Studien zur Bisexualitaet in Mythos und Ritus* (1955).

ings told. Generally, it is overlooked that this manner of styling is a technical device only, and the mythographers of old are accused of having "believed" that in former times everything stood on its head (see above, p. 292, n. 14, about the deplorable Mesopotamians who were unaware of their own mortality before the Gilgamesh Epic was written).

Since it is an actual language, the idiom of myth brings with it the emergence of poetry. Every classical philologist has to admit, for instance, that Hyginus and his like report the mythical plots rather faithfully in 3–10 lines of "correct" idiom which sounds hardly more interesting than average abstracts, whereas this instrumental language, when used by Aeschylus, remains soul-shaking even to this day. But, however vast the difference of poetical rank among the mythographers, the terminology as such had been coined long before poets, whose names are familiar to us, entered the stage. To say "terminology," however, sounds too dry and inadequate, for out of this mint have come clear-cut types—surviving for example, in the games of our children, in our chess figures and our playing cards—together with the adventures destined to them. And this spoken imagery has survived the rise and fall of empires and was tuned to new cultures and to new surroundings.

The main merit of this language has turned out to be its built-in ambiguity. Myth can be used as a vehicle for handing down solid knowledge independently from the degree of insight of the people who do the actual telling of stories, fables, etc. In ancient times, moreover, it allowed the members of the archaic "brain trust" to "talk shop" unaffected by the presence of laymen: the danger of giving something away was practically nil.

And now, coming back for a while to "The Adventure and the Quest," one should emphasize that it is, of course, satisfactory to have cuneiform tablets and that it is reassuring that the experts know how to read different languages of the Ancient Near East; but Gilgamesh and his search for immortality was not unknown in times before the deciphering of cuneiform writing. This is the result of that particular merit of mythical terminology that it is handed down independently from the knowledge of the storyteller. (The obvious drawback of this technique is that the ambiguity persists; our contemporary experts are as quietly excluded from the dialogue as were the laymen of old.) Thus, even if one supposes that Plato was among the last who really understood the technical language, "the stories" remained alive, often enough in the true old wording. Accordingly, one can watch how the hero of the "Romaunt of Alexander," in his own right an undisputed historical personality, slipped on Gilgamesh's equipment, while at the same time slipping off whole chapters of sober history.

Alexander had to measure the depth of the sea, he was carried by eagles up into the sky, and he traveled to the most unbelievable "seas" in search of the water of immortality. Expecting it to be in Paradise he sailed up the Nile, or the Ganges—but why repeat the chapter on Eridanus? A true replica of Gilgamesh, Alexander sailed to the magic place whence all waters come and to which they return. And if it was, allegedly, a serpent ("earth-lion") who deprived Gilgamesh of the rejuvenating plant, the Alexander of the fable was defrauded unwittingly by a fish—just a salted one taken along as travel supply. But he was consciously betrayed by the cook Andreas (according to Pseudo-Kallisthenes), who had noticed the fish coming to life when it fell into the brook and who drank of the water himself without telling Alexander of this discovery. The king, in his righteous indignation, had him thrown into the sea with a millstone around his neck, thus effectually preventing the cook from enjoying his immortality.

The range of significant variations of the many Alexander stories precludes anything more than superficial remarks about them, but they are relevant to Gilgamesh who has, perforce, been abandoned in a darkness which is in large part artificial. It is possible to outline some questions that may stir the problem of Gilgamesh out of its stagnation. There is also one detail that points in the direction of a proposition already put forward concerning Gilgamesh (p. 304).

Alexander, says the fable, interrogates the Oracle of Sarapis in Egypt just as Gilgamesh interviewed Utnapishtim. Sarapis answers evasively as concerns the span of life allotted to Alexander, but he points to the foundation of Alexandria and announces that the king will last on in this city "dead and not dead," Alexandria being his sepulcher. In another version Sarapis states: "But after your death, you will be placed among the gods, and receive divine worship and offerings by many, when you have died and, yet, not died. For your tomb will be the very town which you are founding."[53]

The grotesque monster Huwawa appears to be the pointer. Whatever approach is chosen, Huwawa's connection with Procyon, Jupiter and/or Mercury[54] should be taken into consideration, the more so, as the Hurrian fragments seem to know the poem under the title of "Epic of Huwawa."[55] And along with this consideration, the proper attention will have to be paid to the Babylonian name of Cancer, namely Nangar(u), "the Carpenter." This is essential, because in the twelfth tablet of the Gilgamesh Epic, preserved only in Sumerian language, Gilgamesh complains bitterly of having lost his "pukku and mikku," instead of having left them "in the house

[53] Franz Kampers, *Alexander der Grosse* (1901), p. 93f.

The derivation of the name Sarapis from Enki-Ea's name šar apsî, as proposed by C. F. Lehmann-Haupt (see also A. Jeremias in Roscher s.v. Oannes, 3.590) makes sense; the more so as it does not exclude the connection of Sarapis with Apis, since Apis has the title "the repetition of Ptah." Accidentally, a rather revealing shred of evidence fell into our hands, contained in Budge's translation of the Ethiopic Alexander Romance (London 1896, p. 9): "When Nectanebus, king of Egypt and father of Alexander, had escaped to Macedonia, "the men of Egypt asked their god to tell them what had befallen their king." That is what the Ethiopian text says, and Budge adds: "In Meusel's text the god who is being asked is called 'Hephaestus the head of the race of the gods,' and in Mueller's he is said to dwell in the Serapeum."

The common denominator of Ea-šar apsî, Ptah-Hephaistos, "he who is south of his wall," "lord of the triakontaeteris," is and remains the planet Saturn. Admittedly, we knew this before, but we wish to stress the point that those despised "late" traditions of the Romance represent useful "preservation tins"; i.e., if the Romance replaces Utnapishtim of the confluence of the rivers with Sarapis we can trust that there was a valid equation written down somewhere and known to the several redactors—all of them closely related to some "Wagner" and hostile toward any potential "Faust."

[54] It is remarkable that the Tuamotuan "Hiro is said to be Procyon" (M. W. Makemson, *The Morning Star Rises* [1941], p. 270). Hiro (Maori: Whiro), the master-thief, is an unmistakable Mercurian character.

[55] H. Otten, in *Gilgamesh et sa légende* (1958), p. 140.

of the carpenter,"[56] where they would have been safe, apparently.[57]

Whoever reads the Epic in many translations is not likely to overlook the indications of a "fence," or/and a "doorpost," or frame of a door at such an improbable place as Huwawa's great cedar "forest." Why not also try to look out for the "enclosing of Gog and Magog" accomplished by Alexander and told still in the 18th Sura of the Koran, the same Sura which deals with the coming to life of Moses' travel-supply-fish at the "confluence of the rivers?" This "enclosure" is a great theme of medieval folklore, kept fearsomely alive by the sudden appearance of Mongol invasions. The story ran that Alexander had built iron gates over the mountain passes, that the monstrous brood of the Huns, spawning over the limitless plains of the East, had been kept in awe by trumpets sounding from the pass betokening a seemingly immortal conqueror, the "two-horned" hero who watched over the passes. But the trumpets had suddenly fallen silent, and a dwarf from the horde risked his way to the pass, and found the gate deserted. The trumpets were nothing but aeolian harps, stilled by a tribe of owls which had nested in them.[58]

The ancient story of Gog and Magog, revived from the Arabs, plays such a decisive role in the Romance of Alexander that we might rely upon the antiquity of the scheme: actually it ought to occur in our Epic. Considering that Gilgamesh appears to open a new passage, the former one has to be closed. This also was done in the case of Odysseus. Once he arrived in Ithaca with the stipulations for a new treaty with Poseidon, the poor Phaeacians were done for.

---

[56] A careful investigator has to be aware of the numerous traps along his way as, for example, the naughty custom of exchanging Scorpius and Cancer (Cicero for instance calls both constellations *nepa*) which seems to be on account of the similarity between the scorpion and the landcrab (*Geocarcinus ruricula*).

[57] "Pukku and mikku" (see below, p. 441) are lost "at the crying of a little girl" (C. J. Gadd, "Epic of Gilgamesh," RA *30*, p. 132): this sounds slightly improbable. It is laughter, if anything, that wrecks the old, and introduces the new age of the world. Maui lost his immortality because his companions laughed when he passed the "house of death" of the Great-Night-Hina.

[58] Cf. the thorough investigation by A. R. Anderson, *Alexander's Gate, Gog and Magog, and the Inclosed Nations* (1932). An early version of the story comes from a much-traveled Franciscan, Ricoldo da Montecroce.

There was to be no Scheria anymore. This station being closed up, growing mountains were to block off the beautiful island of Nausikaa which was, henceforth, "off limits" for travelers. There are some striking parallels available in Central Polynesia: when the younger Maui stole the fire from "old Maui" (Mauike, Mahuike, etc.) in the underworld, the passage which he had used was shut from that day on. This is particularly remarkable because "it was by the way of Tiki's hole that Maui descended into the home of Mauike in search of fire." Tiki (Ti'i) was the "first man," and "Tiki's hole had been the route by which souls were supposed to pass down to (H)Avaiki." Thus, the souls had to find another way "after this hole had been closed,"[59] that is, after young Maui had accomplished the theft of fire.

The notion of fire, in various forms, has been one of the recurring themes of this essay. Gilgamesh, like Prometheus, is intimately associated with it. The principle of fire, and the means of producing or acquiring it are best approached through them.

[59] W. W. Gill, *Myths and Songs from the South Pacific* (1876), p. 57; cf. R. W. Williamson, *Religious and Cosmic Beliefs of Central Polynesia* (1924), vol. 2, p. 252 (Austral Islands, Samoa).

CHAPTER XXIII

Gilgamesh and Prometheus

> "... quand les esprits bienheureux
> Dans la Voie de Laict auront fait
> nouveaux feux ..."
>
> *Agrippa d'Aubigné*

Fire is, indeed, a key word, deserving a special inquiry. For the time being, however, it is not essential to understand everything about the different norms and measures, rules and regulations which have to be procured by gods or heroes who are destined to open "new ways." One can ignore here the true nature and identity of the various "treasures," whether they are called "oar" or "ferry man," or "hvarna-melammu," or "golden fleece," or "fire." This is not to say that all these terms are different names for the same thing, but that they identify several parts of the frame.[1]

It will be useful to recapitulate the ideas of the frame, as it has been traced through the Greek precedents. It started out, innocently enough, with the frame of a ship (see above, pp. 230f.), as the Greeks did, and finally ended up with the bewildering "world tree" called the *skambha*, which even Plato might have found intractable. In the end, it is nothing more than the structure of world colures, even if it rustles with many centuries of Hindu verbiage.

---

[1] Even a superficial study of the Chinese novel *Feng Shen Yen I* (i.e., Popular Account of the Promotion to Divinity) which, under the disguise of "historiography" dealing with the end of the Shang Dynasty and the beginning of the Chou, presents us with a fantastic description of a major crisis between world-ages, will reveal to the attentive reader the amount of "new deities"—responsible for old cosmic functions—who have to be appointed at a new Zero, beginning with 365 gods, 28 new lunar mansions, etc.

Another point to bear in mind is the cosmological relevance of "way-openers" and "path-finders" like Gilgamesh. They are the ones who bring the manifold measures from that mysterious center, called Canopus or Eridu, or "the seat of Rita." One can illustrate the general scheme by means of two adventures.

The Argonauts, with the Golden Fleece on board, had to pass the Symplegades, the clashing rocks. Once a ship with its crew came through unharmed[2]—so the "blessed ones" (*makaroi*) had decided long ago—the Symplegades would stay fixed, and be clashing rocks no longer.[3] After that "accepting the novel laws of the fixed earth," they should "offer an easy passage to all ships, once they had learnt defeat."[4] This is only *one* station on the long "opening travel" of the Argonauts transporting the Golden Fleece (of a ram), undertaken in all probability to introduce the Age of Aries,[5] but it demonstrates best the relevant point, namely, "the novel laws."

Another instance—in fact, a crucial one—of an Opening of the Way comes to us from the Catlo'Itq in British Columbia.[6] We would call it a pocket encyclopedia of myth:

A man had a daughter who possessed a wonderful bow and arrow, with which she was able to bring down everything she wanted. But she was lazy and was constantly sleeping. At this her father was angry and said: "Do not be always sleeping, but take thy bow and shoot at the navel of the ocean, so that we may get fire."

[2] The Symplegades cut off, however, the ornament of the ship's stern (aphla-stoio akra korymba), where the "soul" of the ship was understood to dwell. We do not know yet the precise meaning of this trait. Cf. H. Diels, "Das Aphlaston der antiken Schiffe," in *Zeitschrift des Vereins für Volkskunde* (1915), pp. 61–80. It should be emphasized that, contrary to a widespread opinion, the *planktai* and the *symplegades* are not identical.

[3] Apollonios Rhodios, *Argonautica* 2.592–606; Pindar, *Pyth.* 4.210: "but that voyage of the demigods made them stand still in death."

[4] Claudianus 26.8–11.

[5] See the First Vatican Mythographer (c. 24, ed. Bode, vol. *1*, p. 9) stating about "Pelias vel Peleus" that he sent Jason to Colchis, "ut inde detulisset pellem auream, in qua Juppiter in caelum ascendit," i.e., to fetch the Golden Fleece, in which Jupiter climbs the sky. See also A. B. Cook, "The European Sky-God," *Folk-Lore 15* (1904), pp. 271f., for comparable material.

[6] F. Boas, *Indianische Sagen von der Nord-Pacifischen Küste Amerikas* (1895), pp. 8of. Cf. Frazer, *Myths from the Origin of Fire* (1930), pp. 164f.; also L. Frobenius, *The Childhood of Man* (1960), pp. 395f.

The navel of the ocean was a vast whirlpool in which sticks for making fire by friction were drifting about. At that time men were still without fire. Now the maiden seized her bow, shot into the navel of the ocean, and the material for fire-rubbing sprang ashore.

Then the old man was glad. He kindled a large fire; and as he wanted to keep it to himself, he built a house with a door which snapped up and down like jaws and killed everybody that wanted to get in. But the people knew that he was in possession of the fire, and the stag determined to steal it for them. He took resinous wood, split it and stuck the splinters in his hair. Then he lashed two boats together, covered them with planks, danced and sang on them, and so he came to the old man's house. He sang: "O, I go and will fetch the fire." The old man's daughter heard him singing, and said to her father: "O, let the stranger come into the house; he sings and dances so beautifully."

The stag landed and drew near the door, singing and dancing, and at the same time sprang to the door and made as if he wanted to enter the house. Then the door snapped to, without however touching him. But while it was again opening, he sprang quickly into the house. Here he seated himself at the fire, as if he wanted to dry himself, and continued singing. At the same time he let his head bend forward over the fire, so that he became quite sooty, and at last the splinters in his hair took fire. Then he sprang out, ran off and brought the fire to the people.

Such is the story of Prometheus in Catlo'ltq. It is more than that. For the stag has stood for a long time for Kronos. In the Hindu tradition he is Yama who has been met before as Yama Agastya, and who, "following the course of the great rivers, discovered the way for many." This stag is spread far and wide in the archaic world, with the same connotations. And he is the archaic Prometheus-Kronos, "you who consume all and increase it again by the unlimited order of the Aion, wily-minded, you of crooked counsel, venerable Prometheus." In Greek, *semnē Prometheu*. It leaves no doubts. The Orphic invocation to Kronos, quoted in the very beginning on p. 12, defines him as "venerable" and couples him with the name of Kronos the Titan, and we did not go on to quote the awful name of Prometheus so as not to confuse the issue. To avoid confusing matters gratuitously, the name Prometheus has so far been used sparingly. It summons up a formidable implex. The scholiast of Sophocles who gave the reference, quoting Polemon and Lysimachides who are now lost sources, explains: "Prometheus was the first and the older who held in his right hand the scepter, but Hephaistos later and second."[7]

These are the underground regions of Greek mythology, still barely noticed by the school of Frazer and Harrison in their search for prehistoric cults and symbols in the classical world. Yet here ancient Greek myth suddenly emerges in full light among Indian tribes in America, miraculously preserved. The very unnaturalness of the narrative shows how steps were telescoped or omitted through the ages. In one moment the Whirlpool emerges as the bearer of the fire-sticks of Pramantha and Tezcatlipoca. But why should they be in the whirl? Myth has its own shorthand logic to relate those floating fire-sticks to the cosmic whirl. And that logic goes on tying together the basic themes, the bow and the arrow of celestial kingship, the bow and arrow aimed at (or ending in) Sirius, *stella maris* (compare appendix #2 on Orendel).

The singing and dancing of the stag is intricately involved with a proto-Pythagorean theme. And the theme appears full-fledged in still another tale from the Northwest. The Son of Woodpecker, before shooting his bow, intoned a song, and as soon as he had found the right note, the flying arrows stuck in each other's necks until they built the bridge of arrows to heaven; Sir James Frazer himself identified this theme with that of the scaling of Olympus in the Gigantomachy. But there is more. Although it is not stated explicitly that the "clashing doors" (the precessing equinoxes) of the old owner of fire ceased to clash, surely the stag opened a new passage by passing the door at the predestined right moment in his quest for the "fire."

There was little room for invention and variation in this solemn play with the great themes, although imagination did retain some freedom. Thus one might feel tempted to see pure imagination in the feckless laziness of the Old Man's Daughter. And yet, was it imagination, if one discovers in her the prototype of Ishtar, of whom it was said (see above, p. 215) that she "stirs up the apsu before Ea"? Lady-archers being a rare species, it is worth considera-

[7] Schol. Soph. O. C. 56 (Mayer, *Giganten und Titanen*, p. 95).

tion that the great Babylonian astronomical text, the so-called "Series ^mulAPIN" (= Series Plough-Star, the Plough-Star being Triangulum), states: "the Bow-star is the Ishtar of Elam, daughter of Enlil." There has been mention of the constellation of the Bow, built by stars of Argo and Canis Major, Sirius serving as "Arrow-Star" (see above, p. 216 and figure on p. 290). It is no less significant that the Egyptian divine archeress, Satit, aims her arrow at Sirius, as can be seen on the round Zodiac of Dendera.

When one discovers a brief tale that miraculously encapsulates great myths in a few words, one is led to the suspicion that such tales are fragments of long and intricate recitals meant to hold their audience for hours; that, actually they represent something like "Apollodorus" or "Hyginus" who passed on the essential information in brief abstracts. But behind them stood a fully shaped and powerful literary tradition along with the Greek poets to give the ideas flesh and blood, whereas with an illiterate neolithic people such as the Catlo'ltq only the bare skeleton, even "Hygini Fabulae," appears to have survived, unless we assume the informants withheld from the ethnologists the richer versions. (A colleague once told us about a Tibetan minstrel who, bidden to recite the saga of Bogda Gesser Khan, asked whether he should do the large version or the small one: the large would have taken weeks to recite properly.)

It was stated earlier and should be re-stated here that *"fire" was thought of as a great circle reaching from one celestial pole to the other*, and also that the fire sticks belong to the *skambha* (*Atharva Veda 10.8.20*), as an essential part of the frame. Among the things which helped us to recognize *"fire" as the equinoctial colure*, only one fact needs mention here, that the Aztecs took Castor and Pollux (alpha beta Geminorum) for the first fire sticks, from which mankind learned how to drill fire. This is known from Sahagún.[8] Considering that the equinoctial colure of the Golden Age ran through Gemini (and Sagittarius), the fire sticks in Gemini offer a correct

---

[8] *Florentine Codex* (trans. Anderson and Dibble), vol. 7, p. 60. See also R. Simeon, *Dictionnaire de la Langue Nahuatl* (1885) s.v. "mamalhuaztli: Les Gemeaux, constellation," who does not mention, though, that Sahagún identified mamalhuaztli with "astijellos," fire sticks. Also, the Tasmanians felt indebted to Castor and Pollux for the first fire (see J. G. Frazer, *Myths of the Origin of Fire* [1930], pp. 3f.).

rhyme to a verse in a Mongolian nuptial prayer which says: "Fire was born, when Heaven and Earth separated";[9] in other words, before the falling apart of ecliptic and equator, there was no "fire," the first being kindled in the Golden Age of the Twins.

There is no certainty yet whether or not there are fixed rules, according to which one fire has to be fetched from the North, and the other from the South; both methods are employed. The Finns, for example, insist on the fire's "cradle on the navel of the sky," whence it rushes through seven or nine skies into the sea, to the bottom if it, in fact.[10] And Tezcatlipoca is claimed to be sitting at the celestial North Pole also, when drilling fire in the year 2-Reed, after the flood. Whereas it is said of the so-called fire-god of Mesopotamia:

> Gibil, the exalted hero whom Ea adorned with terrible brilliance [= melammu], who grew up in the pure apsu, who in Eridu, the place of (determining) fates, is unfailingly prepared, whose pure light reaches heaven—his bright tongue flashes like lightning; Gibil's light flares up like the day.[11]

Gibil is also called, briefly, "hero, child of the Apsu." If the "fire," adorned with "terrible brilliance"—melammu/hvarna—is prepared in Eridu, one should be permitted to conclude that it has to be procured from there, just as the Rigvedic Agni-Matarishvan, one among the Agnis, "fires," had to be sought at the "confluence of the rivers" (appendix #38).

But whether the "fire" comes from "above" or from "below," the divine or semidivine (or two-thirds divine as Gilgamesh) beings who bring it from either topos could all be named after their common function, as is done in Mexico, where Quetzalcouatl is also called "Ce acatl" = 1-Reed,[12] and Tezcatlipoca "Omacatl (Ome

---

[9] U. Holmberg, *Die religiösen Vorstellungen der altaischen Völker* (1938), p. 99.

[10] K. Krohn, *Magische Ursprungsrunen der Finnen* (1924), p. 115.

[11] W. F. Albright, "The Mouth of the Rivers," AJSL 35 (1919), p. 165; see also K. Tallqvist, *Akkadische Götterepitheta* (1939), p. 313.

[12] Acatl/Reed represents, indeed, the arrow-stick, the drill stick of the fire drill and the "symbol of juridical power." See E. Seler, *Gesammelte Abhandlungen* (1960–61), vol. 2, pp. 996, 1102; vol. 4, p. 224.

acatl) = 2-Reed. In the same way we might call the corresponding heroes of the Old World "1-Narthex," "2-Narthex," and so forth, after the "reed," in which the stolen fire was brought by the most famous Titan, Prometheus, a "portion" of Saturn.

Without taking part in the heated discussion on the interpretation of the very name Gilgamesh—ᵈGIS.GIN.MEZ/MAS, and other forms—one can mention that GIS means "wood, tree," and MEZ/MAS a particular kind of wood,[13] and that there are reasons for understanding our hero as a true Prometheus.

Here it is worth turning briefly to a text recently translated anew and edited by P. Gössmann, the tablets of the *Era-Epos*. This is a grim poem, whose appalling fierceness emerges in almost every word, dedicated as it is to the god of Death, Era (also spelled Irra), a part of Nergal. The subject matter is wholly mythological, handling the end of a world in terms which would hardly disgrace the *Edda,* and dealing again with the Flood to end all floods in the gloomy spirit of Genesis. But here something shines out unmistakably that the commentators on Genesis have missed. They have missed it so completely that even in our day some well-intentioned Fundamentalists applied for permission to search for the remains of the Ark on Mount Ararat. They were impatiently denied access by the Soviet authorities, who suspected espionage with a CIA cover name. No one, they figured, could be that simpleminded. The simplemindedness obviously extends to the researchers of the Sumerological Institutes, who went looking for Eridu in the Persian Gulf, and for the dwelling of the divine barmaid Siduri on the shores of the Mediterranean. But it is evident that the events of the Flood in the Era Epic, however vivid their language, apply unmistakably to events in the austral heavens and to nothing else.

It becomes evident that all the adventures of Gilgamesh, even if ever so earthily described, have no conceivable counterpart on earth. They are astronomically conceived from A to Z—even as the

---

[13] See R. Labat, *Manuel d'Epigraphie Akkadienne* (4th ed., 1963), nos. 296, 314; also F. Delitzch, *Assyrisches Handwörterbuch* (1896), p. 420 s.v. miskannu; Tallqvist, s.v. Gilgamesh. Albright calls Gilgamesh "torch-fecundating hero" (JAOS *40*, p. 318).

fury of Era does not apply to some meteorological "Lord Storm" but to events which are imagined to take place among constellations. The authors of Sumer and Babylon describe their hair-raising catastrophes of the Flood without a thought of earthly events. Their imagination and calculations as well as their thought belong wholly among the stars. Their capacity for transposition seems to have been utterly lost to us earthlings, of the earth earthy, who think only of "primitive" images and primitive experiences, which could account for the narrative so intensely and humanly projected. Perhaps they are mutants from our type. In any case they seem beyond the comprehension of mature contemporary intellects, who have adjusted comfortably to the mental standards of Desmond Morris' Naked Ape of their own devising.

These phantoms being now laid to rest, one finds oneself dealing with utterly unknown ancestors, whose biblical rages and passions have to be read in an entirely new context. To be sure, the planets are still neighbors: Mars, who is Era and Nergal, is only a few light-minutes away, Marduk-Jupiter about eight minutes, Saturn an hour. But they are all equally lost in cosmic space, their optical evidence, like that of ghosts, equally unseizable, equally potent or impotent in terms of present physical standards, equally and dreadfully present according to *those* other standards.

Era is sternly reprimanded by Jupiter/Marduk for having sent his weapons forth to destroy what remained after the Flood (Ea once spoke in the same vein to Enlil after the earlier Flood) but Marduk saved seven wise ones (*ummâni*) by causing them to descend to the Apsu or Abyss, and to the precious *mes*-trees by changing their places. "Because of this work, O hero, which thou didst command to be done, where is the *mes*-tree, flesh of the gods, adornment of kings?" "The *mesu*-tree," says Marduk, "had its roots in the wide sea, in the depth of Arallu, and its top attained to high heaven." He asks Era where are the lapis lazuli, the gods of the arts, and the seven wise ones of the Apsu. He might well ask where is Gilgamesh himself, that deceptively human hero, now transformed into a beacon of light from a *mes*-tree of other-worldly dimensions. Such is the fate of heroes, as they have been followed from Amlodhi onwards, whether they come as a spark hiding in a narthex like Prometheus, or fire from the wood splinters in Stag's hairs, or become a beam from Canopus-Eridu. Lost in the depths of the Southern Ocean, they were capable of giving the Depths of the Sea to our forefathers, and now are able to have the directional systems of our missiles lock on them for interplanetary flight—they remain points, circles, geometries of light to guide mankind past and future on its way.

And so under the present circumstances it is necessary to leave Era's somber prophecy unfulfilled, relating as it does to a coming world age:

> *Open the way, I will take the road,*
> *The days are ended, the fixed time has past.*

But with it comes the clearest statement ever uttered by men or gods concerning the Precession. Says Marduk:

> *When I stood up from my seat and let the flood break in,*
> *then the judgement of Earth and Heaven went out of joint . . .*

> *The gods, which trembled, the stars of heaven—*
> *their position changed, and I did not bring them back.*

The Lost Treasure

by GIORGIO DE SANTILLANA

> ... while each art and science has often been developed as far as possible, and then again perished, these opinions, with others, have been preserved until the present like relics (*leipsana*) of the ancient treasure.
>
> ARISTOTLE, *Metaphysics*
> Bk. *Lambda 1074b*

I

As WE* WERE MOVING toward the conclusion of this essay, some chance or accident or kind intention brought to our eyes, after many years, the work of an author who was our guide when we tried for a first understanding of the early consciousness of man. It was Cassirer's opus on mythical thought. And with all the respect one owes the great historian of Renaissance philosophy, we were astounded. We went through the persuasive and limpid prose, tracing the gradual growth of the concept from wild and uncouth beginnings to the height of Kantian awareness. We gazed again at the stately cortege of great scholars and researchers, Humboldt, Max Muller, Usener and Wissowa, Frazer and Cumont and so many others, the imposing phalanx in which philology, ethnology, history of religion, archaeology, and not least philosophy, display their

* Throughout the text the pronoun *we* has been used as little as possible because it is so difficult to know what it means from one usage to the next. For the next several pages, *we* necessarily will appear often and will refer solely to us, the authors.

well-knit progress in good order, to be finally sifted and cleared up by the modern historian of culture. And then, as we reflected further that here was the material that was going to provide advanced survey courses in the immense universities of the future, to build the gleaming machinery of electronic-printed and audiovisual General Humanities for the Masses, we were suddenly overcome with the haunting memory of that unwearied, dedicated, and ridiculous pair, Bouvard and Pécuchet. The merciless irony of Flaubert was surely not called for in the case of Cassirer, but the same genius who created Madame Bovary was suddenly showing us again the shape of certain things to come. A noble enterprise was due to fail, worse, was slated already for the coming *Dictionnaire des Idées Reçues*. What Flaubert's pathetic little self-taught characters had in common with the sovereign cultural historian was clear: it was intellectual pride, judging from the height of Progress, which telescopes the countless centuries of the archaic past into artless primitive prattle, to be understood by analogy with the surviving "primitives" around us. Too much of that primitiveness lies in the eye of the beholder. It took the uncanny penetration of trained observers like Griaule to uncover suddenly the universe of thought which remained hidden to generations of modern Africanists.

The great merit of Ernst Cassirer lies in his tracing the existence of "symbolic forms" from the past in the midst of historic culture. Who but he should have been able to discern the lineaments of archaic myth? Yet he remained blinded by condescension. Evolution, a brilliant biological idea of our own past, construed into a universal banality, held him in thrall. He could not follow up his insight because of the fatal confusion which has established itself between biological time, the time of evolution, and the time of mankind. The time of man, in which he has lived the life of the mind, goes back into the tens of millennia, but it is not the same as biological time. Again and again, in our text, we have adverted to this confusion which has hardened to become worse confounded. If Cassirer's idea of mythical thought is already dated, as are his sources, one must expect the survey courses of the future to go farther in the same direction with sociological psychology and anthropological sociology, until all traces of the past have been wiped out. The masses will then have a culture of commonplaces reared on the common ideas of the last two centuries of history. Even the gifts of a Cassirer, who could discern the links between language and thought in modern science, left him defenseless when he accepted the most jejune reports of missionaries, and the most naïve intuitions of the obvious from the specialists of his own time. This makes his work "passé." Those are the wages of the sin of intellectual pride. In the very process of establishing an identity between non-discursive symbolism and "primitiveness," he cut himself off from the Kantian synthesis.

Where are the snows of yesteryear? In the very beginning of *Myth and Language*, a curious equivocation, quite unintentional, moves in with the words of Plato from the *Phaedrus*, a pleasant raillery at the intellectual exercises of the oversubtle with myth and "mythologemes." Clearly Professor Cassirer intends to take the reader into camp, and remind him with the authority of the Master that sober thought is in order, even concerning this "rustic science." Does he expect one to forget about the *Timaeus?* For in this, among his last Dialogues, Plato deals gravely and solemnly with first and last things, with the universe and the fate of the soul. And yet the *Timaeus* is, openly, explicitly, one great myth and nothing else. Is it then "unserious," as Plato perversely would like to have certain scholars believe? They have walked into the trap. For Plato not only has put into his piece all the science he can obtain, he has entrusted to it reserved knowledge of grave import, received from his archaic ancestors, and he soberly adjures the reader not to be too serious about it, nor even cultural in the modern sense, but to understand it, if he can. The scholar is already in a hopeless tangle, and Lord help him.

A simple way out would be to admit that myth is neither irresponsible fantasy, nor the object of weighty psychology, or any such thing. It is "wholly other," and requires to be looked at with open eyes. This is what we have tried to do.

II

Wandelt sich schnell auch die Welt
Wie Wolkengestalten,
Alles Vollendete fällt
Heim zum Uralten.*

R. M. RILKE, *Sonette an Orpheus*

In order to find bearings, one can go back for a moment to the thought of two fundamental moderns: Tolstoy, the last great epic writer, and Simone Weil, the last great saint of Christendom, even if a Gnostic one. Tolstoy, in his later years, was tormented with the question whether a way could be found to make some sense out of the events of history as he knew them. He concluded despairingly that sense there is none, that whatever the justifications of philosophers, the so-called makers of history are the puppets of fate. The reality of war destroyed all semblance of rationality, and left only a dreadful confusion. The modern consciousness is brought to face the stark events, from which one can draw only pragmatic inferences, starting from what is ascertained as the *fait accompli*.

And here, maybe, we find ourselves facing one of the Tolstoyan paradoxes driven to a point well-nigh unbearable. In his memorable and desperate letter of 1908 to Gandhi, then an obscure lawyer, which started the latter on his way to the teaching of nonviolence, *satyagraha*, Tolstoy denounces the various forms of violence, murder, and fraud on which society is based, which perpetuate education and class distinctions as a whole. In it he included all the official religions.

And then he pointed to science as the arch-culprit, because it teaches man to do violence to himself and to nature essentially. Of course, Tolstoy is thinking of the arrogant spirit of *scientism* with its heartless, un-understanding doctrines. It would never have oc-

---

* So quickly the world doth change/Like shapes in the clouds/Only the Achievèd remains/Cradled in Timeless Antiquity.

curred to him that science is really something else, with its spirit of pure research and serene dispassionateness. We would say now that technology is the culprit. But the finger is pointed unequivocally at our modern and vulgarizing idea of "science for the masses" and the consumer society. Against that, Tolstoy holds the one thing, Christian love—pure and simple—as wholly spontaneous, natural, and compelling. We might say, keeping away from Tolstoy and his illuminations, that what he asserts is respect for life and spontaneity, a holy restraint for the arcane ways of the cosmos itself, embodied in the community of beings with a conscience. He forgot perhaps, also, his own striving for harmony, which makes of him, in *War and Peace*, the legitimate successor of the great K'wei, that singular "Master of Music" in the new Empire of Letters. A reserved knowledge, that too, and far from our cliché of the "common man," for Christ addressed himself to "those who have ears to hear."

Simone Weil, lost in the turmoil of the Second World War, thought she saw a retrospective answer in the Greeks, in Homer himself, who had been called the Teacher of Greece. She called the *Iliad* the "Poem of Force" because it showed Force at the center of human history, a powerful and clear mirror of man's condition —with no soothing nonsense added. Death for the vanquished, nemesis for the conqueror—these are two members of the equation. The strictly geometrical atonement that comes with the abuse of force was the principal theme of Greek thought. It persisted wherever Greek thought had reached. And yet, Western man, the heir of the Judeo-Christian tradition, has lost it—so utterly that in no Western language is there a word to express it. The notions of limit, of measure, of commensurability, which guided the thought of sages have survived only in Greek science and in the catharsis of tragedy. This seemed to draw the boundary of understanding. It is a strange truth, notes Simone Weil, that men today should be geometers only with respect to matter. But Plato's famous lost lecture on the Good is known to have been based on geometrical demonstration. It had been so from the beginning in Greece. Not only Anaximander's ethical statics of the cosmos, but the whole

Pythagorean theory had been based on those three mathematical sciences: number, music, astronomy. Here lay the undeviating heart of truth on which the Good can rest, and the rest concerned engineers. Even in Thucydides, there is a kind of *reductio ad absurdum.* And it shows that if the Greeks were no less miserable than we are in life, still the great epic idea remains: no hate for the enemy, no contempt for the victim. The measures of the cosmos unfolded the facts. Force, Necessity, must be conceived as within an order. The crucial word remains that of the *Timaeus:* "Reason overruled Necessity by persuading her to guide the greatest part of the things that become towards what is best" (48A). There is a great idea here. This is how far the mind could read, and yet be able to make sense of reality. This is what the Greeks had accepted as the boundary of thought. However original their minds, one might say that the inheritance of archaic Measures had built up their patrimony, indestructibly.

Man has moved beyond that, and has brought the marvelous power of mathematics to the conquest of matter, as deep down as the core of the atom, as far out as the outer-galactic nebulae. But it is just as Simone Weil remarked, men are geometers only with respect to matter and energy. The rest one has to leave to events, and probabilities, to the physics of the dust. Man remains staring at what in his own frame is the denial of thought, the *fait accompli.* One dares not even examine the consequences of this geometry; men feel their way apprehensively around such fate-laden corollaries as "information" or "overkill." They turn under one's eyes into *faits accomplis.* The historical view of the past does not lend itself to contemplation. But as man tries by means of contrast to build up his experience of the true, he finds that truth is at odds with his ancient faith in continuity. Scientific prediction moves away from "instant catastrophes," on the subatomic level, breaks against the perpetual resurgence of falsifiability. Whatever is authentic expression in art, cleansed of context, scatters into the unceasing variety of styles, of responses, of happenings and discoveries; not even the specious present, but the fractured instant is for us the Now of Time.

III

> History is a nightmare from
> which I am trying to awake.
>
> James Joyce, *Ulysses*

In contrast with the present world, the archaic past has much to speak for it. It was based on a very high culture, an artistic one of a high order as everyone knows, and on a scientific culture too. It brought the first technological Revolution, on which so-called antiquity was to rest for millennia. Yet it lived on and flowered and let the world live. People like to ignore this archaic science because it started from the wrong foundations and drew any number of wrong conclusions, yet historians know that wrongness is not a test for relevance, that a course of reasoning may be scientifically important independent of its endpoints. Our forebears built up their world view from the idea which today would be called geocentric; they concluded with speculations about the fate of man's soul in a cosmos in which present geography and the science of heaven are still woven together. Worse, maybe, they built them up on a conception of time which is utterly different from the modern metric, linear, monotone conception of time. Their universe could have nothing to do with ours, derived as it was from the apparent revolutions of the stars, from pure kinematics. It has taken a great intellectual effort on the part of many great scholars to transfer themselves back to that perspective. The results have been astonishingly fruitful. For those forebears did not only build up time into a structure, *cyclic time:* along with it came their creative idea of Number as the secret of things. When they said "things are numbers," they swept in an immense arc over the whole field of ideas, astronomical and mathematical, from which real science was going to be born. Those unknown geniuses set modern thought on its way, foreshortened its evolution. But their ideas were at least as complicated as our own have come to be.

Cosmological Time, the "dance of stars" as Plato called it, was not a mere angular measure, an empty container, as it has now become, the container of so-called history; that is, of frightful and meaningless surprises that people have resigned themselves to calling the *fait accompli*. It was felt to be potent enough to control events inflexibly, as it molded them to its sequences in a cosmic manifold in which past and future called to each other, deep calling to deep. The awesome Measure repeated and echoed the structure in many ways, gave Time the scansion, the inexorable decisions through which an instant "fell due."

Those interlocking Measures were endowed with such a transcendent dignity as to give a foundation to reality that all of modern physics cannot achieve; for, unlike physics, they conveyed the first idea of "what it is to be," and what they focused on became by contrast almost a blend of past and future, so that Time tended to be essentially oracular. It presented, it announced, as it were; it oriented men for the event as the Chorus was later to do in a Greek tragedy. Whatever idea man could form of himself, the consecrated event unfolding itself before him protected him from being the "dream of a shadow."

Again and again, in the course of this essay, we have insisted on the vanity of any attempt to give an "image" of the archaic cosmos, even were it such an image as Rembrandt drew of the cabalistic apparition to the Initiate, or as Faust suddenly saw in the sign of the Spirit of Earth. Even as a magic scheme, it would have to be a design of insoluble complexity. Far worse did our own scholarly predecessors fare when they tried for a model, conceived mechanically, an orrery, a planetarium maybe, such as Plato suggests teasingly in his deadpan way with his whorls and spindles and frames and pillars. A real model might indeed help, he goes on without batting an eye, and one realizes it would come into the price range of a Zeiss Planetarium, still true to the kinematic rigor of the Powers of heaven, but blind to their moving soul in its action—and Plato's machinery promptly dissolves into contradictions, no real "model" at all. Plato will never yield on his "unseriousness," which for him is a matter of principle, a way of leaving mystery alone while respecting reason as far as it can go. Another image suggested by the past, still older than planetary models, would be the "tissue" woven in the skies, the Powers working at the whirring loom of Time, says Goethe, as they weave the living mantle of the Deity. But as in all those images, the real terms are life and harmony, many harmonies, such as Pythagoreans went on forever investigating. Our own "reconstruction," whatever it is, would come as close to a harmony as our cat achieves by stretching out full length on a keyboard. Kepler's mad attempt at writing out the notes of the "Harmony of the Spheres" was bound to fail abysmally to express the true lawfulness: what Plato called the Song of Lachesis.

Men have learned to respect it without thinking. Even today, as one celebrates Christmas, one invokes the unique gift of that cyclic time—the gift of not being historical; its opening into the timeless, the virtue of mapping the whole of itself into a vital present, laden with ancestral voices, oracles, and rites from the past. With what sincerity is left to them, men invoke the remission of ancient sins, the rebirth of the Soul even as was done many millennia ago. People beg from that Time the renewed strength to carry on against a senseless reality—and still ask their children to aid their unbelief.

True history goes by myths. Its forces are mythical. As Voltaire remarked coolly, it is a matter of which myth you choose.

The name of Revolutions is a true technical term of astronomic knowledge and myth: that which ever returns to the same point. It became insistently identified with the idea of the Great Change. As soon as men began to misunderstand it, it set History on the march with irreversible changes. But in the Middle Ages it still promised a return to the undefined Origins, to the Golden Age, when Adam delved and Eve span, or, more Christianly, to when the Lord still walked on earth. Joachim of Flora (c. 1200) was still the prophet of the Great Change that was to be a true accomplishment of ancient prophecies. After the ages of the Father and the Son men expected the Age of the Holy Ghost to follow immediately, when all men would be brothers—a great revolutionary moment sparked by the order of St. Francis. It lived on in the shrunken horizon of Enlightenment, which set the span back to the Greeks and Romans as semi-gods. And yet, in those classical times, that dream was already there. It was of a return far back to the birth of a Miraculous Child. And back far beyond that, to the clearer idea of cosmic configurations such as they were when time had not yet been set in motion. Here came the *Timaeus*.

The idea lingered on. In the Apocalypses and Cosmogonies, in the Vedic poems, time is scrambled artificially and deliberately into elements, lunar stations, or proto-chess, to restore the vision, the prophetic or sibylline vision. Out of this thoughtful scrambling of elements came the Alphabet. A prodigious conquest, like the making of iron, and a grievous gift unto men, as Hesiod might say. There is no doubt that one is dealing with the creations of genius, even if they were flashes in the darkness, which had found a way to perpetuate and propagate themselves.

It stands to reason that the actual chronicle of the archaic ages is full of "barbaric" events. What such migrations as those of the Cimmerians, of the Mongols, of the Peoples of the Sea achieved in the way of destruction and dispersal is beyond our imagination. One calls this the primitive way of life, and blithely conjectures extermination in the biological sense, forgetting what biology has to say of real conflicts among animal tribes. It is only man, more especially modern man, who knows the art of total kill, the quick and the slow. But archaic cultures, devoid of history but steeped in myth, did not find in events the surprise of the *fait accompli*, stunning and shattering to the mind in the way Auschwitz is to us. Mythical experience has its own ways of meeting catastrophe. Men were able to see things nobly. Narration became epic.

The great epic of the Fall of the Nibelungen mirrors in its own way the invasion of Attila and his Huns, the "Scourge of God." Official history might counter the Mongol hordes with the Roman victory on the Catalaunic fields, but the Attila of legend, chief of Gog and Magog, remains more imposing, even as he passes silently out of the scene, than Jenghiz or Tamerlane with their historic conquests and pyramids of skulls. He has little to act, he is the typical emperor of myth. Like Theodoric, like Arthur, like Kai Khusrau, he is the unmoved chess king around whom figures move. The Nibelungen story shows how mythical thought dealt with the crisis. It is Nemesis who destroys the German warriors at the last. Attila, "king Etzel," suffers in his turn, without losing the authority of the conqueror. His child dies at the hands of Hagen, last of the sinful brood who is cut down as a captive by the infuriated mother, destroyed in turn by Hildebrand, reconciled to the conqueror, who brings the drama to a catharsis. Attila the Hun and Theodoric the Goth, joined in the tale as allies, are left to weep together the death of great heroes. No hatred, no terror left, except at the working of Fate.

From the last night of Troy, extinguished in slaughter, what remains in living myth is the flight of the few survivors toward new shores. There they become mythical founding heroes in their turn, contended for by the great cities of the West. This is how myth deals with its own, and Nemesis is felt at last to catch up with the Roman Empire. The spirit of Homer's epic impassiveness led the ancient mind all the way up to the end of the classic world, purged of resentment and hatred, but nowhere more impressively than in Virgil's soul-stricken invocation, a vision of doom at a time when Rome fancied itself established forever: "Please, gods, have mercy. Have we not atoned enough for the original perjury of Troy?" *Iam satis luimus Laomedonteae periuria Troiae* . . . But there is no atonement in full measure within the unceasing rhythm of cycles and megacycles, which builds up a living dialectic of mythical imagination. The conquests and subversions which reshaped the world with Alexander were surely more important than any feats attributed to the legendary king of Uruk; but the latter's otherworldly sheen reverberated on the Macedonian, and tradition forced him into the pattern of another Gilgamesh, still bent on the discovery and conquest of all earth, water, and air, down to the end of the world and beyond, still questing in vain after immortal life. The molding capacity of established myth created the historic episodes that he needed to fit himself into the role, went beyond him to build up the "two-horned" demigod, Dhul-Karnein, he who erected a brazen wall against the path of destruction from the East, the peril of Gog and Magog, a fable that even the later glory of the Roman Empire could not imitate. For that kind of time always tends to move off into the forms of timelessness.

Let us go back to the end of the wonderful adventure of Dante's Odysseus, as he moves out of the straits of Gibraltar:

> *And having turned our poop towards the morning,*
> *Our oars we turned to wings in crazy flight*
> *Always gaining to the left-hand side.*

That is, he has "turned his poop to the east," and his prow directly west; he proceeds "always gaining to the left-hand side." In other words, it looks as if he were trying to circumnavigate Africa, not as Columbus but as Vasco da Gama did, going to India. The general direction of his "crazy flight" is actually south, across the equator and then the Tropic of Capricorn, just as he has already done in Homer under Circe's sailing directions: "follow the wind from the North." He is still looking for the "experience, beyond the sun, of the world without people." But in Dante's world scheme, he is clearly making for the Antipodes, which means, vaguely, the unknown South Seas.

> And, in fact, all the stars of the other pole had come into sight,
> and those of ours had sunk so low that they did not rise from the sea;
> five times the light of the moon had waxed and waned, when we
> described a tall mountain, dim from the distance, so tall that I
> had never seen any. We rejoiced, and soon it turned to tears . . .

for it was, as we already know (see chapter XIV, "The Whirlpool"), the mount of Purgatory, denied to the living. Hence, Providence decreed a whirl that swallowed the ship and all its hands, and that was the end.

What was Columbus' discovery? Hardly more.

Dante's description was not really an invention; it was derived from texts of his own time, and we find it, bodily transcribed, in Columbus' own extracts and notes, made in the years of waiting in Spain, from his favorite readings: "subtle shining secrecies, writ in the glossy margent of such bookes."[1] It is still the land of Eden.

A long distance by land and sea from our habitable land; it is so high that it touches the lower sphere, and the waters of the Flood never touched it . . . . The waters which descend from this very high mountain form an immense lake. The fall of such waters makes such a noise that the inhabitants are born deaf. From that lake as the one source flow the four rivers of Paradise: Physon which is the Ganges, Gyon which is the Nile, Tigris and Euphrates . . . . A fountain there is in Paradise which waters the garden of delights and which is diffused in the four rivers. According to Isidore, John of Damascus, Bede, Strabo and Peter Comestor . . . Pliny and Solinus, Marinus of

[1] Shakespeare, *The Rape of Lucrece.*

Tyre's corrections of Ptolemy show that the sea can be crossed with favorable winds in a few days, going down *per deorsum Africae*, along the back of Africa . . . for the earth extends from Spain to the Indies more than 180°. And the proof is that Ezra says that 6/7ths of the globe are land, Ambrosius and Augustine holding Ezra for a prophet . . . the degree being equal to 52 2/3 Roman Miles . . . .

The sources of Columbus are well known, one of them being Pierre d'Ailly's famous *Imago Mundi* of the 14th century, and another Aeneas Sylvius' *Historia Rerum ubicumque gestarum* of the 15th. Pierre d'Ailly differs even more from Ptolemy by ruining his celestial coordinates, whereas Aeneas Sylvius is no more than a compilation, a vague *miroir historial*, and yet these are the books in which Columbus put his trust, much more than in his maps, and rightly so. Even Toscanelli's famous letter to Martius does no more than emphasize Marco Polo's Cipango (Japan) and set it a thousand miles east; which at least encouraged the lonely Genoese, who to the end never suspected the existence of the Pacific, and made him look for the golden homes of Cipango while he was discovering Cuba. His never-never land, his own Island of St. Brandaen, must have been in his mind somewhere between the Canaries and the Empire of Prester John, along "the back of Africa," and this was enough impetus for him to discover America, or rather invent it out of his mythical enthusiasm, still bent on the Garden of Eden and its nightingales. As for Toscanelli, the "cosmographer," his impulse lay not so much in his geographical expertise on China as in his vaticination of a new world-age. Columbus' and Toscanelli's clear and very modern intention was "to search for the East by way of the West"; but what did it amount to? One of the authorities assured that Aryim, *umbilicus maris*, wherever it may be, was not "in the middle of the habitable earth," but further, 90° off. Another said that the distance between Spain and the eastern edge of India was "not much." Once out in the Atlantic, Columbus had to rely on his faith in timeless myth, from Gilgamesh to Alexander. To be sure, he had the compass, but his cosmography had lost the very idea of the heavens; and, like his Odyssean and medieval forebears, he had to keep searching for the Islands of the Blessed.

> *It may be we shall find the Happy Isles,*
> *and meet the great Achilles, whom we knew . . .*

What led him to his discovery was his wonderful skill as a navigator, which allowed him to ride out equinoctial storms and never lose a ship as he threaded his caravels through the tricky channels of the Indies. America was the reward for Paolo Toscanelli's[2] and Christopher Columbus' Archaic faith.

The relation of myth to history is very important indeed, but the influence of one on the other often goes counter to the interpretations of most debunkers. The famed nightingale from Eden that Columbus wrote he heard when he landed on Watling Island is only one striking counterexample. But there are more. For instance, myth had influence on the geopolitics of great Eastern conquerors like Tamerlane and Mohammed II. These two men of action, and decisive action, were far from illiterate. They had the cultures of two languages at their disposal, Turkish and Persian, and their inquisitive minds liked to dwell on great plans of adventure toward the West. Yet although they were obsessed with the destruction of the Empire of Rūm (Rome), it has been shown (von Hammer) that they had never so much as heard of Caesar and his great successors. Their historical information did not go beyond the "Romaunt of Alexander" in the Persian version. One is back again with Gilgamesh as the prime source. The comparison is all in their favor. While the potentates of Europe were loosing themselves in miserable quarrels, frittering away their possibilities, and even seeking an alliance with the Turk, only Pope Aeneas Sylvius, sick and dying, was finding the words for the occasion: "The barbarians have murdered the successor of Constantine together with his people, desecrated the temples of the Lord, overthrown the altars, thrown to the swine the relics of the martyrs, killed the priests, ravished their women and daughters, even the virgins consecrated to God; they have dragged along the camp the image of our crucified Savior, to the cry of 'there goes the God of the Christians' and have defiled it with filth and spit—and *we* seem to care for nothing." It was indeed the final tragedy of Christendom, vanishing first West, then East. At that point only the conquering Sultan found the words for the occasion. "The ruler of the world"—writes

[2] Giorgio de Santillana, "Paolo Toscanelli and His Friends," in *Reflections on Men and Ideas* (1968), pp. 33–47.

Tursum Beg, his chronicler—moved up like a spirit, to the summit of Saint Sophia; he watched signs of the already coming decay, and formed elegiac thoughts: "The spider serves as watchman in the porticoes of the cupola of Khusrau. The owl sounds the last post in the palace of Afrasiyab. Such is the world, and it is doomed to come to an end."

IV

What time span did the archaic world embrace within our own frame? Its beginning has already been placed in the Neolithic, without setting limits in the past; let prehistoric archaeologists decide. The astronomical system seems to conceive of the Golden Age, the Saturnian Era, as already mythical, in the proper sense. One can then say that it took shape about 4000 B.C.,[3] that it lasted into proto-history and beyond.[4] The fearful loss of substance that tradition suffered in the Greek Middle Ages (the same happened in Egypt too, before the Middle Kingdom) has created an almost complete gap with what we call Classical Antiquity, but enough did remain to ensure a continuity with those ancestors whom Plato and Aristotle liked to call "the men close to the Gods" and who were thought of in this way even to our Renaissance. In Plato's philosophy, Archaic Time stayed intact; it was resolutely understood as "wholly other" from extension, wholly incompatible with what Parmenides had already grasped in his Revelation, with what Democritus coldly theorized. But archaic time is the universe, like it circular and definite. It is the essence of definition, and so it continues to be throughout Classical Antiquity, which did not believe in progress but in eternal returns. In that world it was Space which, if taken by itself, brought in indefiniteness and incoherence. Ultimately, in Plato, space was identified with the nature of Non-Being. Plato called space the "Receptacle." This idea, so puzzling for us who think in spatial terms and cut up reality, as Bergson said, along dotted lines drawn in space, must have been for Plato an easy and natural conclusion. He had inherited the idea that reality, or rather Being, was

[3] See W. Hartner, "The Earliest History of the Constellations in the Near East and the Motif of the Lion-Bull Combat," JNES 24 (1965), pp. 1–16, 16 plates.

[4] But we do not know. These people could compute backward as well as forward.

defined in terms of Time above all. It was Space which brought in confusion, multiplicity, the resistance to Order, what Plato called the Unruly and Irregular which always resist the mind. In the beginning, it would seem, space even resisted the mind of the Creating Demiurge, for it presented him with the original chaos, a kind of foreign body intractably *plemmelōs kai atakteōs*, unorganized, devoid of any rhythm. Even the Demiurge must struggle to force it into shape, within the limits of his power.

When did the archiac world come to an end? There are many testimonials of the bewildering change. Plutarch, a true pagan, pondered about A.D. 60, why it was "that oracles had ceased to give answers." It is on this occasion that he told the tale of the voice that came from the sea, telling the pilot that "the Great Pan was dead." Recounting it on a previous occasion (above, chapter XX, p. 277), it was noted that the experts of the Emperor Tiberius decided that it must be Pan no. 3. Another world-age must be passing, together with the gods who belonged to it. For traditionalists it was indeed one more sign of the passing of the Age of Aries and the advent of the Age of Pisces. Historically, it is known as the advent of the Christian revolution, marked in so many ways by the sign of the Fish. It may have taken place with the Edict of Theodosius in A.D. 390. It was to be a change so profound that it would have caused Plutarch to lose his bearings. It was the end of the Parcae, the goddesses who lived Fate. The Song of Lachesis had been silenced. In a few centuries, it was as if new stars were shining over the heads of men brought up in a classical culture. The introduction of new gods from the East was certainly a contributory cause of the rapid conversion of the Roman elite, which appeared to the Christians a miracle in itself. Oracles and omens had been part of the texture of circular Time, using the sibylline language which continuously threw a rainbow bridge from the past into the future.

As later developments were to show, the great web of cyclical time suffered irreparable harm from the doctrine of the Incarnation, but did not snap asunder all at once. For a long time the belief in the Second Coming among Christians kept time together. But as it became established that the supramundane advent of Christ into the world had cleft time into an absolute Before and After, that it was a unique event not subject to repetition, then duration became simple extension, a waiting for the day of judgment, increasingly dependent on the vicissitudes of belief.

I tried to determine once, from the testimony of artistic experience, the period when the time frame of reality came to be felt and described in terms of three-dimensional space.[5] This first sign of the Scientific Revolution, I suggested, coincided with the invention of perspective in the 15th century. It arrived, as it were, surreptitiously, originating in the minds of great artists and technologists (artist being then the word for artisan). What is clear is that by the end of the Renaissance time and space had become what *we* mean by them.

Newton conceived of the frame of the universe as made up of absolute space and time. The mode of thought became natural, and not until Einstein did new and greater difficulties arise to resist the imagination. Today one should begin to appreciate the divine simplicity of the archaic frame, which took time as the *one* frame, even at the terrible price of making the cosmos itself into the "bubble universe." It was a decisive option. The choice went deep to the roots of man's being. It conditioned Aristotelian theory and conditioned Christian imagination. It constrained even Copernicus and Kepler. They both recoiled from unboundedness. That is why one sees Aristarchus, Bruno and Galileo not simply as bold generalizers or investigators of regularities, but as souls of superhuman audacity.

Aristarchus remained a loner, neglected in his time even by the sovereign mind of Archimedes. Twenty centuries later, Bruno was less a thinker than an inspired prophet of God's infiniteness, identical with the Universe itself. Galileo, the true scientist, still remained sufficiently dominated by the circularity he needed in his cosmos so that he did not dare to formulate the principle of rectilinear inertia which was already present to his mind. He held passionately to the circular cosmos. The circle was to him a metaphor of Being that he was still willing to accept even at the price of unacceptable epi-

[5] G. de Santillana, "The Role of Art in the Scientific Renaissance," in *Reflections on Men and Ideas* (1968), pp. 137–66.

cycles. However much he supported perfect circularity by sober and prosaic reasons, it remained to him first and last a "symbolic form," much as the Seven-ringed cup of Jamshyd, the magic Caldron of Koridwen, as the Cromlech of Stonehenge. The Untuning of the World, the dissolution of the Cosmos, was to come only with Descartes.

It was said earlier concerning the Mayan astronomers that the connections were what counted. In the archaic universe all things were signs and signatures of each other, inscribed in the hologram, to be divined subtly. This was also the philosophy of the Pythagoreans, and it presides over all of classical language, as distinct from contemporary language. This was pointed out perceptively by a modern critic, Roland Barthes, in *Le degré zero de l'Écriture*. "The economy of classical language," he says, "is rational, which means that in it words are abstracted as much as possible in the interest of relationship . . . No word has a density by itself, it is hardly the sign of a thing, but rather the means of conveying a connection." Today, the object of a modern poem is not to define or qualify relations already conventionally agreed; one feels transported, as it were, from the world of classical Newtonian physics to the random world of subatomic particles, ruled by probabilistic theory. The beginning of this was felt in Cézanne, in Rimbaud and Mallarmé. It is "an explosion of words" and forms, liberated words, independent objects—discontinuous and magical, not controlled, not organized by a sequence of "neutral signs." The interrupted flow of the new poetic language, Barthes remarks, "initiates a discontinuous Nature, which is revealed only piecemeal." Nature becomes "a fragmented space, made of objects solitary and terrible, because the links between them are only potential." More, they are arbitrary. They are supposed to be of the nature of the ancient *portentum*. The only meaning to be drawn from those links is that they are congenial to the mind that made them. The mind has abdicated, or it shrinks in apocalyptic terror. In the arts we hear of Amorphism, or "disintegration of form," of the "triumph of incoherence" in concrete poetry and contemporary music. The new syntheses, if any are still possible, are beyond the horizon.

CONCLUSION

Honneur des Hommes, saint Langage
Discours prophétique et paré . . .

VALÉRY, *La Pythie*

STARTING FROM one theme chosen among many, this self-styled first reconnaissance over uncharted ground has come a long and tortuous way since the early introduction of the Hamlet figure. The discussion has touched immense areas of myth whose probable value was indicated by previous discoveries. The treasures of Celtic tradition, of Egypt, China, tribal or megalithic India, and Oceania have barely been sampled. Nevertheless, the careful, inductive application of critical standards of form along the belt of High Civilizations has been enough to show the remnants of a preliterate "code language" of unmistakable coherence. No apologies are needed for having followed the argument where it led, but it is very much to be hoped that what has been uncovered will eventually prove sufficiently self-correcting to amend the inevitable errors of this essay.

What can the initial universe of discourse have been, that insensate scattering of dismembered and disjointed languages of the remote past, of earthy jargons and incommunicable experiences, from which, by a stroke of luck, scientific man was born? Clearly, man's capacity for attention, for singling out certain unattainable objects in the universe, must have overcome the convolutions and horrors of his psyche. There were some men, surely exceptional men, who saw that certain wondrous points of light on high in the dark could be counted, tracked, and called by name. The innate knowledge that guides even migratory birds could have led them to realize that the skies tell—yea, recount—the glory of God, and then to conclude that the secret of Being lay displayed before their eyes.

Their strange ideas, inscrutable to later ages, were the beginnings of intellect, and in the course of time they grew into a *koinē*, or *lingua franca*, covering the globe. This common language ignored local beliefs and cults. It concentrated on numbers, motions, measures, overall frames, schemas, on the structure of numbers, on geometry. It did all this although its inventors had no experience to share with each other except the events of their daily lives and no imagery by which to communicate except their observations of natural lawfulness.

Ordered expression, that is expression in accordance with laws or rules, comes before organized thought. After that, the spontaneous creation of fables occurs when there is a fund of direct experience to draw upon. For example, a prehistoric "tech talk," expressing only lawfulness in the grammatical or natural sense, may have begun by using terms that came from the earliest technology. Later, as experience increased, the same kind of talk using the same terms may have been extended to include alchemy and other imageries. In form, these exchanges would be transmuted tales, but because of their terminology they would possess an ordering and naming capacity that would diffuse stimuli over a sea of diversity. Ultimately, they could produce a sign code whereby the stars of Ursa became a team of oxen, and so on.

It is now known that astrology has provided man with his continuing *lingua franca* through the centuries. But it is essential to recognize that, in the beginning, astrology presupposed an astronomy. Through the interplay of these two heavenly concepts, the common elements of preliterate knowledge were caught up in a bizarre bestiary whose taxonomy has disappeared. With the remnants of the system scattered all over the world, abandoned to the drift of cultures and languages, it is immensely difficult to identify the original themes that have undergone so many sea-changes.

The language of the *Vedas*, for instance, which transposes a dazzling wealth of metaphysics into the discourse of hymns, is as remote from all other aspects of mythical thought as the stars of

Ursa themselves are from the verses which, as Masters, the *Vedas* suppose those stars to have written. In these verses, the notion of the way to transcendence and the absurdity and wild luxuriance of the imagery are certain to confound the Western mind and lead it far away from the subject of astronomical origins into a mystical dialectic.

And yet the original life of thought, born of the same seeds as the *Vedas*, worked its way in darkness, sent its roots and tendrils through the deep, until the living plant emerged in the light under different skies. Half a world away it became possible to rediscover a similar voyage of the mind which contained not a single linguistic clue that a philologist could endorse. From the very faintest of hints, the ladder of thought leading back to proto-Pythagorean imagery was revealed to the preternaturally perceptive minds of Kircher and Dupuis. The inevitable process became discernible, going from astronomical phenomena to what might be beyond them. Finally perhaps, as Proclus suggested, the sequence leads from words to numbers, and then even beyond the idea of number to a world where number itself has ceased to exist and there are only thought forms thinking themselves. With this progression, the ascensional power of the archaic mind, supported by numbers, has reestablished the link between two utterly separate worlds.

The nature of this unknown world of abstract form can also be suggested by way of musical symbols, as was attempted earlier. Bach's *Art of the Fugue* was never completed. Its existing symmetries serve only as a hint of what it might have been, and the work is not even as Bach left it. The engraved plates were lost and partly destroyed. Then, collected once more, they were placed in approximate order. Even so, looking at the creation as it now is, one is compelled to believe that there was a time when the plan as a whole lived in Bach's mind.

In the same way, the strange hologram of archaic cosmology must have existed as a conceived plan, achieved at least in certain minds, even as late as the Sumerian period when writing was still a jealously guarded monopoly of the scribal class. Such a mind may have belonged to a keeper of records, but not of the living word, still less of the living thought. Most of the plan was never recorded. Bits of it reach us in unusual, hesitant form, barely indicated, as in the wisdom and sketches of Griaule's teacher, Ogotemmeli, the blind centenarian sage. In the magic drawings of Lascaux, or in American Indian tales, one perceives a mysterious understanding between men and other living creatures which bespeaks relationships beyond our imagination, infinitely remote from our analytical capacity.

"From now on," said Father Sun, grieving over Phaethon, his fallen child, "you shall be Mink." What meaning can this have for us? For such an understanding between men and men, and other living creatures too, we would need the kind of help King Arthur had at hand: "Gwryr Interpreter of Tongues, it is meet that thou escort us on this quest. All tongues hast thou, and then canst speak all languages of men, with some of the birds and beasts." This ability was also attributed to Merlin and Gwyon, those masters of cosmological wisdom whose names resound through the legends of the Middle Ages. In general, all fabulous communication was conceived as having such a range, not merely the Aesopian fable with its flat, all-too-worldly wisdom.

Much of this book has been peopled with the inhabitants of a Star Menagerie of profoundly meaningful animal characters. The forms of animal life have varied from the Fishes who turned into hairy Twins to the remarkable succession of doglike creatures occurring around the world from Ireland to Yucatán. All of these animals have been of great significance, and each was invested with key functions in cosmological myth.

It would be possible, for example, to prepare a most informative edition of the *Romance of Reynard Fox* illustrated entirely with reproductions from Egyptian and Mesopotamian ritual documents. For it is likely that these documents represent the last form of international initiatic language, intended to be misunderstood alike by suspicious authorities and the ignorant crowd. In any case, the language forms an excellent defense against the kind of misuse which Plato speaks about with surprising earnestness in *Phaedrus* (274D–275B). At the point in question, Thot/Hermes is feeling very proud of himself for having invented letters, and he claims that the alphabet will make the Egyptians wiser and improve their memory. Plato has the god Thamus, "king of all Egypt," speak to him:

"Most ingenious Theuth," said the god and king Thamus, "one man has the ability to beget arts, but the ability to judge of their usefulness or harmfulness to their users belongs to another; and now you, who are the father of letters, have been led by your affection to ascribe to them a power the opposite of that which they really possess. For this invention will produce forgetfulness in the minds of those who learn to use it, because they will not practise their memory. Their trust in writing, produced by external characters which are no part of themselves, will discourage the use of their own memory within them. You have invented an elixir not of memory, but of reminding; and you offer your pupils the appearance of wisdom, not true wisdom, for they will read many things without instruction and will therefore seem to know many things, when they are for the most part ignorant and hard to get along with, since they are not wise, but only appear wise."[1]

Now that Plato's apprehensions have become fact, there is nothing left of the ancient knowledge except the relics, fragments and allusions that have survived the steep attrition of the ages. Part of the lost treasure may be recovered through archaeology; some of it —Mayan astronomy, for instance—may be reconstructed through sheer mathematical ingenuity; but the system as a whole may lie beyond all conjecture, because the creating, ordering minds that made it have vanished forever.

[1] Translation by H. N. Fowler, LCL. The Jowett translation reads: "The specific which you have discovered is an aid not to memory but to reminiscence, and you give your disciples not truth, but only the semblance of truth" (oukoun mnêmês all' hypomnêseôs pharmakon hêures; sophias de tois mathêtais doxan, ouk alêtheian porizeis).

L'Envoi

From harmony, from heavenly harmony,
    This universal frame began:
    When nature underneath a heap
        Of jarring atoms lay,
        And could not heave her head,
The tuneful voice was heard from high,
    "Arise, ye more than dead!"
Then cold, and hot, and moist, and dry,
    In order to their stations leap,
        And Music's power obey.
From harmony, from heavenly harmony,
    This universal frame began:
    From harmony to harmony
Through all the compass of the notes it ran,
The diapason closing full in Man. . . .
        As from the power of sacred lays
            The spheres began to move,
        And sung the great Creator's praise
            To all the Blest above;
        So when the last and dreadful hour
        This crumbling pageant shall devour,
        The trumpet shall be heard on high,
        The dead shall live, the living die,
        And Music shall untune the sky!

DRYDEN, *"A Song for St. Cecelia's Day"*

APPENDICES

Appendix 1

The only master of this kind of observation hitherto has been Marcel Griaule (d. 1956) but he left an impressive cohort of disciples. They have renewed the understanding of African studies, showing that such systems are still alive with the Dogon, whom Griaule "discovered," in the true sense of the word.

As Germaine Dieterlen writes: "The smallest everyday object may reveal a conscious reflection of a complex cosmogony . . . Thus for instance African techniques, so poor in appearance, like those of agriculture, weaving and smithing, have a rich, hidden content of significance . . . The sacrifice of a humble chicken, when accompanied by the necessary and effective ritual gestures, recalls in the thinking of those who have experienced it an understanding that is at once original and coherent of the origins and functioning of the universe.

"The Africans," she continues, "with whom we have worked in the region of the Upper Niger have systems of signs which run into thousands, their own systems of astronomy and calendrical measurements, methods of calculation and extensive anatomical and physiological knowledge, as well as a systematic pharmacopoeia. The principles underlying their social organization find expression in classifications which embrace many manifestations of nature. And these form a system in which, to take examples, plants, insects, textiles, games and rites are distributed in categories that can be further divided, numerically expressed and related one to another. It is on these same principles that the political and religious authority of chiefs, the family system and juridical rights, reflected notably in kinship and marriages, have been established. Indeed, all the activities of the daily lives of individuals are ultimately based on them."[1]

It goes without saying that we need not subscribe to the author's opinion that the Mande peoples invented "their own systems of astronomy . . ."

[1] Introduction to *Conversations with Ogotemmêli*, Marcel Griaule (1965), p. xiv.

Appendix 2

The father of Saxo's Amlethus was *Horvandillus,* written also Orendel, Erentel, Earendel, Oervandill, Aurvandil, whom the appendix to the *Heldenbuch* pronounces the first of all heroes that were ever born. The few data known about him are summarized by Jacob Grimm:[1]

> He suffers shipwreck on a voyage, takes shelter with a master fisherman Eisen,[2] earns the seamless coat of his master, and afterwards wins frau Breide, the fairest of women: king Eigel of Trier was his father's name. The whole tissue of the fable puts one in mind of the Odyssey: the shipwrecked man clings to a plank, digs himself a hole, holds a bough before him; even the seamless coat may be compared to Ino's veil, and the fisher to the swineheard, dame Breide's templars would be Penelope's suitors, and angels are sent often, like Zeus's messengers. Yet many things take a different turn, more in German fashion, and incidents are added, such as the laying of a naked sword between the newly married couple, which the Greek story knows nothing of. The hero's name is found even in OHG. documents: Orendil . . , Orentil . . . a village Orendelsal, now Orendensall, in Hohenlohe . . . But the Edda has another myth, which was alluded to in speaking of the stone in Thor's head. Groa is busy conning her magic spell, when Thorr, to requite her for the approaching cure, imparts the welcome news, that in coming from Jötunheim in the North he has carried her husband the bold Örvandill in a basket on his back, and he is sure to be home soon; he adds by way of token, that as Örvandil's toe had stuck out of the basket and got frozen, he broke it off and flung it at the sky, and made a star of it, which is called *Örvandils-tâ.* But Groa in her joy at the tidings forgot her spell, so the stone in the god's head never got loose (Snorri's Skaldskap. 17).

Powell,[3] in his turn, compares the hero to Orion in his keen interpretation:

> The story of Orwandel (the analogue of Orion the Hunter) must be gathered chiefly from the prose Edda. He was a huntsman, big enough and brave enough to cope with giants. He was the friend of Thor, the husband of Groa, the father of Swipdag, the enemy of the

[1] TM, pp. 374f. See also K. Simrock, *Der ungenähte Rock oder König Orendel* (1845), p. ix.

[2] Also written Ise or Eise, and derived from Isis, by Simrock; considering that the fisherman's modest home has seven towers, with 800 fishermen as his servants, Ise/Eisen looks more like the Fisher King of Arthurian Romances.

[3] In his introduction to Elton's translation of Saxo, p. cxxiii.

giant Coller and the monster Sela. The story of his birth, and of his being blinded, are lost apparently in the Teutonic stories, unless we may suppose that the bleeding of Robin Hood till he could not see, by the traitorous prioress, is the last remains of the story of the great archer's death. Dr. Rydberg regards him and his kinsfolk as doublets of those three men of feats, Egil the archer, Weyland the smith, and Finn the harper, and these again doublets of the three primeval artists, the sons of Iwaldi, whose story is told in the prose Edda.

It is not known which star, or constellation, Örvandils-tâ was supposed to be. Apart from such wild notions as that the whole of Orion represented his toe[4]—to identify it with Rigel, i.e., beta Orionis, would be worth discussing—even Reuter tries to convince himself that Corona borealis "looks like a toe,"[5] because he could not free himself from the fetters of seasonal interpretation of myth, nor dared he attack the romantic authority of Ludwig Uhland who had coined the dogma that Thor carried the sign for spring in his basket; accordingly a constellation had to be found which could announce springtime, and Reuter, choosing between Arcturus and Corona, elected the latter.

It is not his toe alone, however, which grants to Hamlet's father his cosmic background: some lines of *Cynewulf's Christ* dedicate to the hero the following words:

> *Hail, Earendel, brightest of angels thou,*
> *sent unto men upon this middle-earth!*
> *Thou art the true refulgence of the sun,*
> *radiant above the stars, and from thyself*
> *illuminest for ever all the tides of time.*[6]

The experts disagree whether Earendel, here, points to Christ, or to Mary, and whether or not Venus as morning star is meant, an identification which offers itself, since ancient glosses render Earendel with "*Jubar*,"[7] and Jubar is generally accepted for Venus on the presupposition that "morning star" stands every single time for Venus, which is certainly misleading: any star, constellation or planet rising heliacally may act as morning star. With respect to *juba*, i.e., literally "the mane

---

[4] R. H. Allen, *Star Names* (1963), p. 310.
[5] *Germanische Himmelskunde* (1934), p. 255.
[6] See TM, p. 375; I. Gollancz, *Hamlet in Iceland* (1898), p. xxxvii; Reuter, p. 256.
[7] O jubar, angelorum splendidissime . . . See R. Heinzel, *Über das Gedicht von König Orendel* (1892), p. 15.

of any animal," *jubar*, "a beaming light, radiance," we have, however, Varro's clear statement: "iuba dicitur stella Lucifer."[8] Nonetheless, several experts are against the equation Orendel/Earendel = Venus.[9] Gollancz abstains from precise identifications, but he procures the one more existing piece of evidence concerning the word *Earendel:*

> In Anglo-Saxon glosses "earendel" . . . or "oerendil" is interpreted jubar, but "dawn" or "morning star" would probably be a better rendering, as in the only other passage known in old English literature, viz. the Blickling Homilies (p. 163, 1. 3): "Nu seo Cristes gebyrd at his aeriste, se niwa eorendel Sanctus Johannes; and nu se leoma thaere sothan sunnan God selfa cuman wille"; i.e., And now the birth of Christ (was) at his appearing, and the new day-spring (or dawn) was John the Baptist. And now the gleam of the true Sun, God himself, shall come.[10]

Orendel/Earendel, then, seems to be the foremost among those which announce some "advent," not unlike the passage in the *Odyssey* (*13*.93f.) dealing with Odysseus' arrival in Ithaca: "When that brightest of stars (*astēr phaäntatos*) rose which comes to tell us that the dawn is near, the travelling ship was drawing close to an island." That might point, again, to Venus, but there are reasons to think of Sirius, the brightest of all fixed stars, as will come out later.

Another subject of discussion has been the etymology of the name, and since the identity of Orendel might depend on its etymology, we have to look into the matter, at least superficially. Jacob Grimm admitted freely:

> I am only in doubt as to the right spelling and interpretation of the word: an OHG. ôrentil implies AS. eárendel, and the two would demand ON. aurvendill, eyrvendill; but if we start with ON. örvendill, then AS. earendel, OHG. erentil would seem preferable. The latter part of the compound certainly contains entil = wentil.[11] The first part should be either ôra, eáre (auris), or else ON. ör, gen. örvar (*sagitta*). Now, as there occurs in a tale in Saxo Grammaticus . . . , a Horvendilus filius Gervendili, and in OHG. a name Kêrwentil . . . and

---

[8] See W. Gundel, *De stellarum appellatione et religione Romana* (1907), p. 106; Reuter, pp. 256, 295ff.

[9] E.g., A. Scherer, *Gestirnnamen* (1953), pp. 79–81.

[10] Gollancz, p. xxxvii*n*.

[11] In a footnote, Grimm asks (and we wish we knew the answer!): "Whence did Matthesius [in Frisch 2, 439[a]] get his 'Pan is the heathens' *Wendel* and head bag-piper?' Can the word refer to the metamorphoses of the flute-playing demigod? In trials of witches, Wendel is a name for the devil, Mones anz. 8, 124."

Gêrentil . . , and as *geir* (hasta) agrees better with *ör* than with *eyra* (auris), the second interpretation may command our assent; a sight of the complete legend would explain the reason of the name. I think Orentil's father deserves attention too: *Eigil* is another old and obscure name . . . Can the story of Orentil's wanderings possibly be so old amongst us, that in Orentil and Eigil of Trier we are to look for that Ulysses and Laertes whom Tacitus places on our Rhine? The names show nothing in common.

Scherer (p. 179) states shortly: "Earendel does not belong to *âusôs* 'dawn,' nor to OE. *éar* 'ear' (Ähre), but to OE. *ae, éar* m. 'wave, sea,' ON. *aurr* 'humidity.'" Gollancz, who is inclined to connect Earendel with Eastern (*ushas, eos, aurora,* etc.), mentions more current derivations, among which also that from *aurr* "moisture," and from the root signifying "to burn" in Greek, *euo,* Latin *uro,* Ves-uvius, etc. Decisive seems to us the derivation from *ör* = arrow, suggested by Grimm, and by Uhland, who explained Orendel as the one "who operates with the arrow" (in contrast to his grandfather, Gerentil, who worked with the *ger* = spear), and Simrock gives the opinion that the very gloss "Earendel Jubar" designates Earendel explicitly as "beam" (or "ray"), "which still in MHG. and Italian means 'arrow.'"[12]

Simrock did more. Taking into consideration that in the *Heldenbuch* Orendel is spelled *Erendelle,* and at some other place *Ernthelle,* he thinks it probable that "*Ern*" was dropped as epitheton ornans,[13] and he concludes from there that the story of *Tell* shooting the apple from the head of his son was once told of Orendel himself. That the historical (?) Tell was not the inventor of this famous shot, or even performed it, seems rather certain. As Grimm aptly stated:

> The legend of Tell relates no real event, yet, without fabrication or lying, as a genuine myth it has shot up anew in the bosom of Switzerland, to embellish a transaction that took hold of the nation's inmost being.[14]

Now there is no arrow to be found that could contest with Sirius in mythical significance. We know [mul]KAK.SI.DI, the "Arrow-Star" from

---

[12] *Handbuch der Deutschen Mythologie* (1869), § 82, p. 233.

[13] *Ibid.* See also Simrock, *Die Quellen des Shakespeare* (1870), pp. 129f.: "Dies ward aber wohl in Tell gekürzt, weil man die erste Silbe für jenes vor Namen stehende 'Ehren' ansah, das nach dem d. Wörterb, III 52 aus 'Herr' erwachsen, bald für ein Epitheton ornans angesehen wurde."

[14] TM 3, p. xxxiv.

Sumer, as well as "Tishtriya," the arrow from Ancient Iran—it is shot from a bow built up by stars of Argo and Canis Major (Sumerian: ^{mul}BAN). The very same bow is to be found in the Chinese sphere, but there the arrow is shorter and *aims* at Sirius, the celestial Jackal, whereas the same Egyptian arrow is aimed at the star on the head of the Sothis Cow, as depicted in the so-called "Round Zodiac" of Dendera—Sirius again. In India, Sirius is the archer himself (Tishiya), and his arrow is represented by the stars of Orion's Belt. And about all of them manifold legends are told. Thus, "Earendel, brightest of angels thou," might well point to the brightest among the fixed stars, Sirius.

But even the derivation from the root *aurr* = moisture, *ear* = sea, would not exclude Sirius. Quite the contrary. The Babylonian New Year's ritual says: "Arrow Star, who measures the depth of the sea"; the *Avesta* states: "Tishtriya, by whom the waters count." And as Tishtriya, "the Arrow," watches Lake Vurukasha (see p. 215), so Teutonic Egil is the guardian of Hvergelmer, the whirlpool, and of Elivagar, south of which "the gods have an 'outgard,' a 'saeter' which is inhabited by valiant watchers—*snotrir vikinger* they are called in *Thorsdrapa*, 8—who are bound by oaths to serve the gods. Their chief is Egil, the most famous archer in the mythology. As such he is also called Orvendel (the one busy with the arrow)."[15]

We had better stop getting diffuse concerning Sirius the Arrow and his role as guardian and as "measurer of the depth of the sea"; the few hints that were given here must suffice to show the level at which to look for the father of Hamlet.

Since, however, we can never resist the temptation to quote beautiful poems, we have still to confess our suspicion that the "Stella Maris" is Sirius too. Enough is known about Isis/Sirius as guardian-deity of navigators, to whom belongs the "carra navalis," and was it not "Mary or Christ" who was addressed with "Hail, Earendel"? In the same manner, the hymn "In Annunciatione Beatae Mariae" begins with the verses:

> *Ave, maris stella*
> *Dei mater alma*
> *atque semper virgo*
> *felix caeli porta*

[15] V. Rydberg, *Teutonic Mythology* (1907), pp. 424ff., 968ff.

> *Sumens illud Ave*
> *Gabrielis ore*
> *funda nos in pace*
> *mutans nomen Evae.*

And there is another hymn which was sung, according to the Roman Breviary, after Compline during Advent and Christmastide, and which has been ascribed to Herimanus Contractus of Reichenau (d. 1054), who would appear to have lived and died a cripple in his monastery:

> *Alma redemptoris mater, quae pervia caeli*
> *porta manes et stella maris, succurre cadenti,*
> *surgere qui curat, populo, tu quae genuisti*
> *(natura mirante) tuum sanctum genitorem,*
> *Virgo prius et posterius, Gabrielis ab ore*
> *sumens illud Ave, peccatorem miserere.*

"What I have been attempting to suggest," says the interpreter of this hymn,[16] "is that the attraction of this charming mediaeval prayer and hymn would seem to come, in large measure, from the intentional ambiguity, the different levels of meaning, and the sunken imagery . . . The 'nourishing mother' is perhaps pictured as a fixed constellation in the heavens, or perhaps as the morning star, guiding those on the sea. She is a celestial passage-way, always passable and ever accessible . . . The falling and rising has now (besides the constantly falling sinners) perhaps the further overtones of heavenly bodies rising and falling, perhaps of ships rising and falling on the sea, and lastly of tottering children who need their mother's help to walk . . . The poem . . . is a very striking one, and its force derives, in my view, from the subtle imagery of the first three lines . . . They offer us a symbol, a verbal icon, of the entire situation of man on earth in his struggle to rise to the stars, of his need of an otherworldly force which is at once strong and loving."

[16] H. Musurillo, S.J., "The Medieval Hymn, Alma Redemptoris," *Classical Journal* 52 (1957), pp. 171–74.

Appendix 3

Now, apart from the circumstance that the snowy burial ascribed to the followers of Kai Khusrau, Enoch, and Quetzalcouatl could hardly be claimed to be an "obvious" feature, the fate of Quetzalcouatl's companions might further our understanding; more correctly, the topos where this event is supposed to have happened might do so. The "five mountains" of Mexican myth, their "gods" respectively, the *Tepictoton*,[1] appear to represent the five Uayeb (= Maya; with the Aztecs: Nemontemi), the Epagomena, those days gained by Mercury from the Moon during a game of checkers, in order to help Rhea/Nut to days "outside of the year," when she could bring forth the five planets. As a matter of fact, in his chapter on the clothes and emblems of the gods, Sahagún puts the "Mountain-Gods" at the end of the list.[2]

Worth mentioning might be two more traits which Quetzalcouatl shares with his old-world brethren: Quetzalcouatl and Uemac, like Kai Ka'us and Kai Khusrau, are said to have ruled together, and Quetzalcouatl is accused of incestuous relations with his sister, as were Hamlet, Kullervo, Yama and, we might add, King Arthur.[3]

Appendix 4

It is as yet too early in the day to deal with "Uncle Kamsa," whom lexicographers make a *"mūra-deva,"* allegedly a "venerator of roots" (*mūla/mūra* = root). In his *Kleine Beiträge* (p. 11), Jarl Charpentier earnestly wishes us to accept as fact that "among the Indian natives fighting against the invading Aryans, there were such," namely, "venerators of roots" (and venerators of worms as well). Although we do not doubt that the species Homo sapiens is capable of any "belief," we cannot

[1] See E. Seler, *Gesammelte Abhandlungen*, vol. 2, p. 507, for an Aztec drawing of the *Tepictoton*.

[2] See T. S. Barthel, "Einige Ordungsprinzipein im Aztekischen Pantheon," *Paideuma 10* (1964), pp. 8of., 83. In this paper, Barthel has established, in a rather convincing manner, the presence of decans in Mexican astronomy.

[3] See W. Krickeberg, "Mexicanisch-peruanische Parallelen," in *Festschrift P. W. Schmidt*, p. 388.

perceive any cogent reason for subscribing to Charpentier's view. *Mūla/mūra*, the "root," is a Nakshatra, a lunar mansion woven around with tales: it is the sting of Scorpius, serving as Marduk's weapon in the "Babylonian Genesis," and as Polynesian Maui's fishhook; with the Copts it is "statio translationis Caniculae . . . unde et Siôt vocatur," i.e., the Coptic table of lunar stations takes lambda upsilon Scorpii as the precise opposite of Sirius/Sothis, as we are informed by Athanasius Kircher, whereas Indian tables ascribe the role of exact opposition to Betelgeuse, ruled by "Rudra-the-destroying-archer." Although we cannot pursue these and other tales further here, we think it at least appropriate to mention the concrete problems arising with such characters as "Uncle Kamsa," instead of accusing a true Asura of "veneration of roots."

Appendix 5

Sem Snaebjoern krad:
Hvatt kveda hraera Grotta
hergrimmastan skerja
ut fyrir jardar skauti
Eyludrs niu brudir;
thaer er, lungs, fyrir laungu
lid-meldr, skipa hlidar
baugskerdir ristr bardi
bol, Amloda mólu
Her er kallat hafit Amloda Kvern.

Gollancz (*Hamlet in Iceland*, p. xi) retranslated his translation into Old Norse so that the original and the *nolens volens* interpreting translation might be compared. The retranslation runs thus: kveda niu brudir eyludrs hraera hvatt hergrimmastan skerja grotta ut fyrir jardar skauti, thaer er fyrir longu molu Amloda lid-meldr; baugskerdir ristr skipa hlidar bol lungs bardi. Elton translates the passage:

"Men say that the nine maidens of the island-mill (the ocean) are working hard at the host-devouring skerry-quern (the sea), out beyond the skirts of the earth; yea, they have for ages been grinding at Amlodi's meal-bin (the sea)."[1]

Rydberg, too, offers a translation:

"It is said, that Eyludr's nine women violently turn the Grotte of the skerry dangerous to man out near the edge of the earth, and that those women long ground Amlode's *lid*-grist."[2]

In spite of the trickiness and the traps of the text Gollancz tries to solve the case; in fact, he tries too frantically (p. xxxvi): "The compound ey-ludr, translated 'Island-Mill,' may be regarded as a synonym for the father of the Nine Maids. *Ludr* is strictly '*the square case within which the lower and upper Quernstones rest*,' hence the mill itself, or quern."

With this we wish to compare O. S. Reuter's explanation: "*ludr* = Mühlengebälk (dän. *Luur* = das *Gerüst zu einer Handmühle*)" (*Germanische Himmelskunde*, p. 239; he also includes a drawing of the mill). On p. 242, note, he renders the lines of *Skaldskap. 25:* "Neun Schärenbräute rühren den Grotti des Inselmühlkastens (eyludr) draussen an der Erde Ecke (ut fyrir jardar skauti)," adding: "Das (kosmische?) Weltmeer ist als 'Hamlets Mühle' gesehen." At least he thought, even if within brackets and with a quotation mark, of "cosmic"—Rydberg is the only one who has grasped this point completely.

"Ey-ludr," Gollancz continues, "is the 'island quern,' i.e., 'the grinder of islands,' the Ocean-Mill, the sea, the sea-god, and, finally, Aegir. 'Aegir's daughters' are the surging waves of the ocean; they work Grotti 'grinder,' the great Ocean-Mill (here called 'skarja grotti,' the grinder of skerries, the lonely rocks in the sea), 'beyond the skirts of the earth' or perhaps, better, 'off yonder promontory.' The latter meaning of the words 'ut fyrir jardar skauti' would perhaps suit the passage best, if Snaebjörn is pointing to some special whirlpool." Non liquet: neither Aegir = eyludr, nor the nine maidens = waves, whether surging or not.

As concerns "off yonder promontory" which sounds ever so poetical and indistinct, see J. de Vries:[3] *skaut* n. Ecke, Zipfel, Schoss, Kopftuch, eig. "etwas Hervorragendes" ... Dazu *skauti* m. "Tuch zum Einhüllen," ae. sceata "Ecke, Schoss, Segelschote." *fyr.* praep. praef. "vor," durch,

---

[1] Saxo Grammaticus, *Danish History*, p. 402.
[2] *Teutonic Mythology*, § 80, p. 568.
[3] *Altnordisches Etymologisches Wörterbuch* (1961).

wegen, trotz, für . . . —lat. prae "voran, voraus," lat. *prior* "der frühere" —which tells us either nothing at all or, if we take "prior" for the proper translation, tells us the whole "story" by means of one single word; in the same manner as the mere fact that the pillars of Hercules were "fyr," called the pillars of Briareos, and before that time, the pillars of Kronos.

We stick, however, to Gollancz for some more lines. "The real difficulty," he says, "in Snorri's extract from Snaebjoern is . . . in its last line; the arrangement of the words is confusing, the interpretation of the most important of the phrases extremely doubtful. 'Lid-meldr' in particular has given much trouble to the commentators: 'meldr,' at present obsolete in Icelandic, signifies 'flour or corn in the mill'; but the word '*lid*' is a veritable crux. It may be either the neuter noun 'lid,' meaning 'a host, folk, people,' or ship, or the masculine '*lidr*,' 'a joint of the body.' The editors of the Corpus Poeticum Boreale read 'meldr-lid,' rendering the word 'meal-vessel'; they translate the passage, 'who ages past ground Amlodi's meal-vessels = the ocean'; but '*mala*,' 'to grind,' can hardly be synonymous with '*hraera*,' 'to move,' in the earlier lines, and there would be no point in the waves grinding the ocean. There seems, therefore, no reason why meldr-lid should be preferred to lid-meldr, which might well stand for 'ship-meal' (sea-meal), to be compared with the Eddic phrase 'graedis meldr,' i.e., sea-flour, a poetical periphrasis for the sand of the shore. Rydberg [*Teutonic Mythology* (1907), pp. 570ff. = pp. 388–92 in the 1889 edition], bearing in mind the connection of the myth concerning the fate of Ymer's descendant Bergelmer, who, according to an ingenious interpretation of a verse in Vafthrudnismal 'was laid under the millstone,' advanced the theory that 'lid-meldr' means 'limb-grist.' According to this view, it is the limbs and joints of the primeval giants, which in Amlodi's mill are transformed into meal . . . Snorri does not help us. The note following Snaebjoern's verse merely adds that here the sea is called "Amlodi's kvern." '"

In a note Gollancz adds that in some other manuscript he found the version: "Here the sea is called 'Amlodi's meal' " (*Amloda melldur*), and concludes: "No explicit explanation is to be found in early Northern poetry or saga. 'Hamlet's Mill' may mean almost anything." It is not as bad as that. Moreover, Gollancz (p. xvii, note) detected more relevant figures of speech in the four lines cited below which he ascribes to Snaebjörn: "The island-mill pours out the blood of the flood goddess's sisters (i.e., waves of the sea), so that (it) bursts from the feller of the land: whirlpool begins strong."

> *svad or fitjar fjoetra,*
> *flods asynju bolde*
> (*roest byrjask roemm*) *systra*
> *rytr, eymylver snyter.*

To which he adds: "In no other drottkvoett verse does eymylver occur: cp. eyludr above."

Appendix 6

It is not as easy to dispose of *Mysing*, as the specialists pretend, e.g., by preferring to interpret his name as "mouse-gray" instead of the equally possible "son of a mouse." Olrik (pp. 459f.) proposes to identify straightaway "King Mysing who killed Frith-Frothi, and the cow that struck down Frothi the Peaceful . . . King Mysing is merely a rationalistic explanation of the ancient monster." (For the death of Frodi by means of a sea cow, see also P. Herrmann's commentary on Saxo, pp. 380–84. This "cow"—in Iceland they remain within the frame of zoology and make it a stag—was, according to Saxo, a witch, who was pierced through by Frodi's men. Afterwards they kept Frodi's death a secret for three years, in the same manner as told by Snorri in his *Heimskringla* about Frey.)

A. H. Krappe, more observant, compared Mysing with Apollon Smintheus, the old "mouse-god" (ARW *33* [1936], pp. 40–56). He had in his mind, however, only the connection—undeniable as it is—between mice and rats and the plague, and the dragging-in of Smintheus does not much further the understanding of Mysing. This state of things was changed with the publication of the work by Henri Grégoire, R. Goossens and M. Matthieu, "Asklepios, Apollon Smintheus et Rudra: Études sur le dieu à la taupe et le dieu au rat dans la Grèce et dans l'Inde," although they do not even mention our Mysing, and although they loudly praise (p. 157) the merit of "Meillet . . . d'avoir fait descendre la mythologie du ciel sur la terre"; with Rudra, and with the rat of Ganesha (who, by the way, acquired his elephant's head because the planet Saturn, not being invited to the infant's "baptism," had looked upon the baby with his evil eye, thus destroying his head which was successfully replaced by that of an elephant), the mouse plot has got much deeper background. Nevertheless, the identity and the role of the mouse deity is hardly going to be settled without taking into account (1) "the tailed Mûs Parîk, arrayed with wings; the Sun fettered her to his own ray, so that she could not perpetrate harm; when she becomes free, she will do much injury to the world, till she is recaptured, having come eye-to-eye with the Sun"; this enigmatical winged mouse comes from the world horoscope in the Iranian *Bundahishn* (chapter V, A, Anklesaria translation, p. 63); (2) the colorful Polynesian myths dealing with the rat that gnawed through the "Nets of Makalii," i.e., Hyades and Pleiades; she could do so unpunished being Makalii's very own sister; (3) the warriors, in the guise of mice, of Llwyd, son of Cil Coed, "who cast enchantment over the seven cantrefs of Dyfed . . . to avenge Gwawl son of Clud," in the third branch of the Mabinogi. There are more items, to be sure, but we have to leave it at that.

Appendix 7

We want to stress the point that the haughty verdicts as given by Genzmer, Olrik, and others on Snorri's tale are not unknown to us. Their opinions run along these lines: "The last part of the story of Grotti and Mysing is 'How the sea grew salt.' This is a different motif, in no wise connected with the peace of Frothi."[1] Genzmer's wording is more arrogant still. The transportation of the mill by Mysing and the grinding of salt aboard the ship is "die Anschweissung einer zweiten selbständigen Sage; der grossartig einfache, ahnungsvolle Schluss unserer Dichtung wird durch ein solches Anhängsel tödlich geschädigt."[2]

It would be more adequate to state that the myth has been "fatally damaged" by the modern experts, and not by Snorri. When we come to

[1] A. Olrik, *The Heroic Legends of Denmark* (1919), p. 460.
[2] *Edda*, trans. F. Genzmer (1922), *Thule 1*, p. 181.

the little salt-mill of Kronos, the reader will understand the plot better. Olrik (pp. 457f.), however, has some pretty survivals to offer:

In 1895, Dr. Jakob Jakobsen, the well-known collector of the remnants of the ancient "Norn" language of the Western Islands, was informed by an old Shetlander, whose parents had come from the Orkneys (Ronaldsey) that near the most northerly of these islands there was an eddy called "the Swelki" [that is, Snorri's *svelgr*, "sea-mill, where the waters rush in through the eye of the mill-stone"]. On that spot a mill stood on the bottom of the sea and ground salt; and a legend of Grotti-Fenni and Grotti-Menni was connected with it. In the course of later investigations in the Orkneys themselves (South Ronaldsey) he learned about the sea mill in the Pentland Firth grinding salt. In 1909, Mr. A. W. Johnstone was told by a lady from Fair Isle that Grotti Finnie and Lucky Minnie were well known in her native island, being frequently invoked to frighten naughty children. Although the legend in those parts is in a fragmentary condition, reduced to incoherent survivals, the tenacity of the oral tradition shows how deeply rooted the legend is in these islands. Outside of the Orkneys neither Mysing nor his salt mill are known to tradition except in the songs of the Edda which themselves bear the stamp of Western provenience.

Appendix 8

Vafthrudnismal 35 is rendered by Gering: "Ungezählte Winter vor der Schöpfung / geschah Bergelmirs Geburt. / Als frühestes weiss ich, dass der erfahrene Riese / Im Boote geborgen ward." Simrock translates similarly, and he remarks (*Hdb. Dt. Myth.*, § 9): "Das dunkle Wort *ludr* für Boot zu nehmen, sind wir sowohl durch den Zusammenhang als durch die Mythenvergleichung berechtigt."

R. B. Anderson (*The Younger Edda* [1880], pp. 6of.) translates the verse—quoted by Snorri (*Gylf.* 7)—as follows: "Countless winters / Ere the earth was made, / Was born Bergelmer. / The first I call to mind / How the crafty giant / Safe in his ark lay."

Neckel and Niedner (*Die Jüngere Edda*, pp. 54f.) state that Bergelmer and his wife "stieg auf seinen Mühlkasten und rettete sich so." The lines above they render with the words: "Als frühestes weiss ich, dass der vielkluge Riese in die Höhe gehoben ward," adding in a footnote: "Das oben mit 'Mahlkasten' wiedergegebene Wort übersetzt man gewöhnlich mit 'Boot' oder auch mit 'Wiege,' ohne Begründung und gegen den Wortlaut der Prosa. Gegen den gewöhnlichen Wortsinn 'Mahlkasten' (Mühlsteinbehälter auf Pfosten) spricht nichts. Freilich kennen wir den angedeuteten Vorgang des Näheren nicht und wissen daher auch nicht, warum der Riese gehoben ('gelegt') werden musste, und wer ihn aufhob."

The ominous word *ludr* occurs again in *Helgakvida Hundingsbana* II, 2-4, where Helge—seeking refuge from king Hunding—works in a mill, disguised as a female, and almost wrecks the ludr.

In common with the mythologists who defend the "boat," in Vafthrudnismal 35, feeling entitled to it on account of comparative mythology (see Simrock, quoted some lines ago), and whom he fights explicitly, Rydberg upholds the notion that the Ark was a ship. It will come out later that this general notion is incorrect.

Appendix 9

As a matter of fact, Simrock already (*Handbuch der Deutschen Mythologie*, pp. 240f.) ventured to interpret Fengö (Amlethus' evil uncle) as "the grinding," and Amlethus as "the grain"; "wo selbst der Name mit Amelmehl [Greek *amylon*], Stärkemehl, Kraftmehl übereinstimmt." He even thought of the possibility (although taking this thought for audacious, "gewagt") to derive the family name of Thidrek's clan, i.e., the name of the Amelunge, from "Amelmehl." We shan't dwell upon the strange information given by Athenaeus (*Deipnosophistai* 3.114f.) about "Achilles, or very fine barley" (cf. Theophr. *8.4.2.* Aristoph. *Eq.* 819: Achilles cake), or on the surname of Ningishzida, namely Zid-zi "Meal of Life" (K. Tallqvist, *Akkadische Götterepitheta*, p. 406; cf. Riemschneider, *Augengott*, p. 133), and we point only to Ras Shamra texts, where the lady Anat ground Mot. (See C. Gordon, *Ugaritic Literature*, p. 45.) H. L. Ginsberg (ANET, p. 140) translates I AB, col. II:

> *She seizes the Godly Mot*
> *With swords she does cleave him*
> *With fan she does winnow him*
> *With fire she does burn him*

With hand-mill she grinds him
In the field she does sow him.
Birds eat his remnants
Consuming his portions
Flitting from remnant to remnant.

An astonished footnote states: "But somehow Mot comes to life entire in col. VI, and Baal even earlier." But there is absolutely nothing astonishing enough to shake the firm belief of experts in "chthonic" deities.

Appendix 10

For the first Irish harp (*cruit*), see Eugene O'Curry, *On the Manners and Customs of the Ancient Irish*, vol. 3 (1873), pp. 236f.; see also Rudolf Thurneysen, *Die irische Helden- und Königssage bis zum 17. Jahrhundert* (1921), pp. 264f.

There once lived a couple . . . And the wife conceived a hatred to him, and she was flying from him through woods and wilderness; and he continued to follow her constantly. And one day that the woman came to the sea shore of Camas, . . . she met a skeleton of a whale on the strand, and she heard the sounds of the wind passing through the sinews of the whale on the strand, and she fell asleep from the sounds. And her husband came after her; and he perceived that it was from the sounds the sleep fell upon her. And he then went forward into the wood, and made the form of a *Cruit*;[1] and he put strings from the sinews of the whale into it; and that was the first Cruit that was ever made.

Marbhan's legend about the beginnings of instruments and verses continues:

And again Lamec Bigamas had two sons, Jubal and Tubal Cain were their names. One son of them was a smith, namely, Jubal; and he discovered from sounds of two sledges (on the anvil) in the forge one day, that it was verses (or notes) of equal length they spoke, and he composed a verse upon that cause, and that was the first verse that was ever composed.

[1] "The word Cruit . . . signifies literally, a sharp high breast, such as of a goose, a heron, or a curlew" (O'Curry, *loc. cit.*).

The legend goes on to report, why the *timpan*—another stringed instrument, different from the cruit—was called Timpan Naimh (or saint's Timpan), because "at the time that Noah, the son of Lamech, went into the Ark, he took with him a number of instruments of music into it, together with a Timpan, which one of his sons had, who knew how to play it." When they finally left the ark, Noah caused his son to name the instrument after his own name, and only under this condition would he give it to him. "So that Noah's Timpan is its name from that time down; and that is not what ye, the ignorant timpanists, call it, but Timpan of the saints."

We introduce this legend for several reasons; first, because we felt reminded at once, as O'Curry did (p. 237), "of Pythagoras, who is said to have been led to discover the musical effect of vibrations of a chord by observing the sound of various blows on an anvil, though the Irish legend . . . does not appear to bear on the tones so much as on the rhythm of music." Second, because here we learn again about two successive stringed instruments, separated, so to speak, by a flood; Vainamoinen lost his Kantele when going to steal the Sampo, and had to construct a new one from wood, afterwards. These traditions must be thoroughly compared, one day, with the different lyres of Greece; we know that one was destroyed by Apollon—allegedly in a fit of repentance—after he had flayed Marsyas, and that Hermes made another one and presented it to Apollon; pike and whale of the northern seas have apparently replaced the turtle of Greek myth. We also know that the Pleiades, called the Lyre of the Muses by the Orphics, existed side by side with Lyra. And Michael Scotus still knew about a turtle figuring, so to speak, as prow of Argo, and "out of which the celestial lyre is made."[2] But before being trapped between the devil and the deep sea, we prefer to stop, although this turtle seems to be placed exactly there, where it "should" be, considering that upon its back the Amritamanthana was accomplished. We shall hear more about that considerable and mysterious man, Michael Scotus, later (see p. 258).

The long and the short of the various traditions is that with a new age new instruments, new strings, or, as in the case of Odysseus, a new peg are called for: a new "Harmony of the Spheres."

[2] Testudo eius (navis) est prope quasi prora navis . . . de qua testudine facta est lyra caeli. Cf. F. Boll, *Sphaera*, p. 447.

Appendix 11

Christensen, in his work on the Kayanids,[1] states: "La tradition nationale fait grand cas du forgeron Kâvag, qui s'insurgeait contre l'usurpateur Dahâg (le Dahâka des Yashts) et hissait son tablier de cuir sur une lance, ce qui fut l'origine du drapeau de l'empire sassanide, appelé *drafš ê kâvyân*, 'drapeau de Kâvag.' Cette légende, née d'un malentendu, la vraie signification du nom de drafš ê kâvyân étant 'le drapeau royal,' est inconnue dans la tradition religieuse."

By means of such statements—apart from "modestly" insinuating that Firdausi spun whole chapters of his *Shahnama* out of "malentendus"—the way to relevant questions is effectively blocked. The story of the smith Kâvag—also written Kâweh,[2] or Kawa—is told by Firdausi in the book dealing with the 1000 years' rule of Dahâk, that fiendish tyrant out of whose shoulders grew two serpents[3] that had to be fed with the brains of two young men every day. The predestined dragon-slayer, and much expected savior, Faridûn—Avestan Thraethona—a true predecessor of Kai Khusrau, had been saved from the snares of Dahâk as a baby, and hidden away in the mountains. When the archdevil Dahâk claimed the sacrifice of the last son of Kaweh—seventeen sons had already been fed to the dragon-heads—the smith started the revolution for the sake of Faridûn:

> He took a leathern apron, such as smiths
> Wear to protect their legs while at the forge,
> Stuck it upon a spear's point and forthwith
> Throughout the market dust began to rise . . .
>
> He took the lead, and many valiant men
> Resorted to him; he rebelled and went
> To Faridûn. When he arrived shouts rose.
> He entered the new prince's court, who marked

---

[1] A. Christensen, *Les Kayanides* (1932), p. 43.

[2] F. Justi, *Iranisches Namenbuch* (1895), p. 160. In the most recent translation of the *Shahnama* (Firdousi: *Das Königsbuch* [1967]—so far, only Pt. 1, Bks. 1–5 have come out), H. Kanus-Credé boldly identifies the smith Kawa with "awestisch Kawâta," i.e., with Kai Kobâd, the first Iranian ruler.

[3] Dahâk with his two additional serpent heads is the same as the "powerful, raving Dasa with his 6 eyes and 3 heads" of RV *10.99.6.*: Visvarupa, son of Tvashtri, and "Schwestersohn der Asura"; cf. Mbh. *12.343* (Roy trans., vol. *10*, p. 572).

The apron on the spear and hailed the omen.
He decked the apron with brocade of Rûm
Of jewelled patterns on a golden ground,
Placed on the spear point a full moon—a token

Portending gloriously—and having draped it
With yellow, red, and violet, he named it
The Kawian flag. Thenceforth when any Shah
Ascended to the throne, and donned the crown,

He hung the worthless apron of the smith
With still more jewels, sumptuous brocade,
And painted silk of Chin. It thus fell out
That Kawa's standard grew to be a sun
Amid the gloom of night, and cheered all hearts.

Now, if there was only the "royal" flag to explain, why should Firdausi (or his sources) invent a *smith* with the name Kâweh (Kavag, Kawa), if there was no connection whatever between kingship and the smith? Even if we leave out of consideration the widely diffused motif of great smiths as foster-fathers and educators of the hero[4] as well as the Chinese mythical imperial smiths, and all the material collected by Alföldi in his article on smith as a title of honor among the kings of Mongols and Turks:[5] the very name of the dynasty of Iranian kings which is of the greatest interest for us, i.e., the Kayanides, is derived from *Kavi/Kawi*.[6] The most "kawian" Shah is Kai Ka'us, whose name even contains the relevant word twice, the "Kavi Kavi-Usan," who cannot be separated from Kavy Usa (or Usanas Kavya) of the *Rigveda* and the *Mahabharata*,[7] who shows several of the decisive char-

---

[4] To mention only Mimir, Regin, Gobann. Kâweh's son Karnâ, by the way, whose life was spared thanks to the rebellion, became a famous paladin of Faridûn, as Wittige/Wittich, son of Waylant the smith, became a strong paladin of Thidrek.

[5] Cf., for Turkish traditions, R. Hartmann, "Ergeneqon," in *Festschrift Jacob* (1932), pp. 68–79.

[6] For the word *kavi*, see H. Lommel, *Die Yäshts des Awesta* (1927), pp. 171f.; E. Herzfeld, *Zoroaster and His World* (1947), pp. 100–109.

[7] See Lommel's article "Kavy Uçan," in *Mélanges linguistiques offerts à Charles Bally* (1939), pp. 210f. That C. Bartholomae (*Altiranisches Wörterbuch* [1904], col. 405) confesses that he is "unable to find relations" between Iranian Kavi Usan and Rigvedian Kavy Usha is a precious gem in the collection of philological atrocities. "Falls meine Etymologie richtig ist, entfällt auch die Namensähnlichkeit." Similarity he calls it! It will come out in the course of this essay that his

acteristics of the *Deus Faber*. Not alone is he said to have forged the weapon for Indra[8]—instead of Tvashtri—and to have given Soma to Indra who, otherwise, has stolen (or has just drunk) the Soma in the "House of Tvashtri" (e.g., RV 3.48.2f.), but we are told that, during one of the never-ceasing wars between Asura and Deva for the "three worlds," the Asura elected Kavya Ushanas for their "priest" or "messenger,"[9] the Deva elected Brihaspati (or Vrihaspati, i.e., Jupiter, in *Taittiriya Sanhita* Agni). Many warriors were slain on both sides, but—so the *Mahabharata* tells—"the open-minded Vrihaspati could not revive them, because he knew not the science called Sanjivani (re-vivification) which Kavya endued with great energy knew so well. And the gods were, therefore, in great sorrow."[10] The *Bundahishn*, in its turn, gives the following report in chapter 32, dedicated to "the mansions which the Kayans erected with glory, which they call marvels and wonders,"[11] in verse 11: "Of the mansions of Kay Us one says: 'One was of gold where-in he settled, two were of glass in which were his stables, and two were of steel in which was his flock; there-from issued all tastes, and waters of the springs giving-immortality, which smite old-age,—that is, when a decrepit man enters by this gate, he comes out as a youth of fifteen years from the other gate,—and also dispel death." According to Firdausi, Kai Ka'us had a kind of balm by means of which he could have restored Shurab to life, but he did not give it to Shurab's father Rustem who implored him for this gift.[12] To which Lommel remarks

proposition to derive the name Usan from "usa- m. (1) Quelle, Brunnen; (2) Abfluss, Leck . . ," is no obstacle at all to the understanding of Kavy Usan. Kronos too has been derived from Greek *krounós*, i.e., "source," "spring" (see Eisler, *Weltenmantel*, pp. 378₂, 385₀, reminding us also of the Pythagorean formula concerning the sea: "Kronou dakryon, the tear of Kronos").

[8] RV *1.51.10*; 121.12, 5.34.2. It is particularly the Shushna-myth, where K. U. replaces Tvashtri.

[9] *Taittiriya Sanhita 2.5.8* (Keith trans., vol. *1*, p. 198).

[10] Mbh. *1.76* (Roy trans., vol. *1*, p. 185). For this role of Kavya Ushanas, cf. Geldner, in R. Pischel and K. F. Geldner, *Vedische Studien*, vol. *2* (1897), pp. 166–70; for a life-restoring lake or well, owned by the "wicked Dânavas," see Mbh. *8.33* (Roy trans., vol. 7, p. 83). In Ireland the Tuatha de Danann were able to revivify the slain (in the Second Battle of Mag Tured), the Fomorians were not.

[11] *Zand-Akâsîh: Iranian or Greater Bundahishn*, trans. by B. T. Anklesaria (1956), p. 271; cf. Christensen, p. 74.

[12] In the same manner, Lug—the strength and heart of the Tuatha De Danann as Krishna was that of the Pandava—denies the revivifying pig's skin to Tuirill who, by means of it, could have restored to life his three sons, Brian, Juchair, and Jucharba.

(*Mélanges Bally*, p. 212): "Und das ist der hässlichste Zug im Bilde des Kay Kâus, dass er die Herausgabe des Wunderheilmittels verweigert, da Rostem und Sohrab, wenn beide am Leben wären, vereint ihm zu mächtig wären." It is a rather idle occupation to look for "ugly traits" in the "character" of the Demiurge, even if he comes our way in the disguise of a Shah.

These few hints must suffice for now; it is bad enough that the burden of "proof" rests with the defenders of sense in our deteriorated century, whereas everyone who presupposes non-sense and "malentendus" can get away with the most preposterous claims. In other words: even if the individual Kâweh/Kavag *should* have been "invented" by Firdausi, the notion of the Deus Faber and Celestial Smith as the disposer and guardian of kingship,[13] as the original and legitimate owner of the "water of life,"[14] is by no means an accidental fancy,[15] and the significance and meaning of the smith's apron as "Kawian flag" would have been understood from China to Ireland.

Appendix 12

It should be stressed that the disinclination of philologists to allow for the "essential" connection of Chronos and Kronos rests upon the stern belief that the "god" Saturn has nothing to do with the planet Saturn, and upon the supposition that an expert in classical philology has nothing whatever to learn from Indian texts. Were it not so, they might have stumbled over Kāla, i.e., Chronos, as a name of Yama, i.e., Kronos, alias the planet Saturn.

[13] To repeat: the "Lord of the Triakontaeteris," the period of thirty years, i.e., the Egyptian and Persian "Royal Jubilee" (Saturn's sidereal revolution), is Ptah-Hephaistos.

[14] Also of the intoxicating beverage replacing it; Soma belonged to Tvasthri; Irish Goibniu brewed the ale which made the Tuatha De Danann immortal, and the beer of the Caucasian smith Kurdalogon played the same role. When Sumerian Inanna was almost lost in the underworld, it was Enki who gave to his messengers the life-restoring fluid with which to besprinkle the goddess. And, last but not least, it is Tane/Kane, the Polynesian Deus Faber, whose are "the Living Waters."

[15] Leo Frobenius, when accused—as happened sometimes—of having been deceived by African informants who "made up" any amount of fairy tales which were not "true," used to smile benevolently, and to point to what he called "*stilgerechte Phantasie.*"

Indians have indeed, written more about their Kāla—and the Iranians about their Zurvan—than the Greeks about Chronos, but with the translated *Vedas* being what they are, we won't claim the relevant texts to be transparent, nor the scholarly interpretations to be particularly elucidating, all of the experts starting, as they do, from the unfounded conviction that "astrology" must be a "late" phenomenon.

To throw "identifications" around, does not lead anywhere, in our opinion, so we do not mean to simplify by nailing down, once and for all, Kāla/Chronos as being the very same as Yama/Kronos/Saturn. To recognize Kronos/Saturn as *auctor temporum* is quite sufficient for the time being,[1] and so are the Indian notions, according to which Yama is often called Kāla; in other passages he is the commander of Kāla (and Kāla, in his turn, the commander of Mrityu, Death).[2]

Kāla plays his unmistakable role already in *Rigveda 164*, but the *Atharva Veda* dedicates to this "god" two whole hymns (*19.53* and *19.54*), and it is worth recalling Eisler's statement (*Weltenmantel*, p. 499): "Zu dieser Kāla-Lehres des Atharvaveda ist später nichts mehr dazugekommen; die jüngeren Quellen führen nur die Vorstellungen weiter aus."

Here are some verses from these two hymns dedicated to Kāla, without the numerous notes and comparisons with other translations, as treated by Bloomfield and Whitney (*Atharva Veda*, trans. by Bloomfield [1964], pp. 224f.):

19.53: (1) Time, the steed, runs with seven reins (rays), thousand-eyed, ageless, rich in seed. The seers, thinking holy thoughts, mount him, all the beings (worlds) are his wheels.

(2) With seven wheels does this time ride, seven naves has he, immortality is his axle. He carries hither all these beings (worlds). Time, the first god, now hastens onward.

(3) A full jar has been placed upon Time; him, verily, we see existing in many forms. He carries away all these beings (worlds); they call him Time in the highest heaven.

---

[1] We do not think it is an "accident" that this originator of time begins with the letter X, representing the obliquity of the ecliptic in Plato's *Timaeus*.

[2] See J. Scheftelowitz, *Die Zeit als Schicksalsgottheit in der indischen und iranischen Religion* (1929), pp. 18ff. See also Burgess (*Surya Siddhanta*, p. 5), who generalizes: "To the Hindus, as to us, Time is, in a metaphorical sense, the great destroyer of all things; as such, he is identified with Death, and with Yama, the ruler of the dead."

(4) He surely did bring hither all the beings (worlds), he surely did encompass all the beings (worlds). Being their father, he became their son; there is, verily, no other force higher than he.

(5) Time begot yonder heaven, Time also (begot) these earths. That which was, and that which shall be, urged forth by Time, spreads out.

(6) Time created the earth, in Time the sun burns. In Time are all beings, in Time the eye looks abroad. . . .

(8) . . . Time is the lord of everything, he was the father of Prajāpati.

(9) By him this (universe) was urged forth, by him it was begotten, and upon him this (universe) was founded. Time, truly, having become the brahma (spiritual exaltation), supports Parameshtin (the highest lord).

(10) Time created the creatures (prajāh), and Time in the beginning (created) the lord of creatures (Prajāpati), the self-existing Kashyapa and the tapas (creative fervour) from Time were born.

19.54: (1) From Time the waters did arise, from Time the brahma (spiritual exaltation), the tapas (creative fervour), the regions (of space did arise). Through Time the sun rises, in Time he goes down again.

(2) Through Time the wind blows, through Time (exists) the great earth; the great sky is fixed in Time. In Time the son (prajāpati) begot of yore that which was, and that which shall be.

(3) From Time the Rks (= the Rig Veda) arose, the Yajus (= the Yajur Veda) was born from Time; Time put forth the sacrifice, the imperishable share of the gods.

(4) Upon Time the Gandharvas[3] and Apsarases are founded, upon Time the worlds (are founded), in Time this Angiras and Atharvan rule over the heavens.

(5) Having conquered this world and the highest world, and the holy (pure) worlds (and) their holy divisions; having by means of the brahma conquered all the worlds, Time, the highest God, forsooth, hastens forward.

[3] See A. Weber (*Die Vedischen Nachrichten über die Nakshatras*, Pt. 2, p. 278, n. 3) about the Gandharvas as representing the days of the "year" of 360 days, according to the *Bhagavata Purana* 4.29.21 (Sanyal trans., vol. 2, p. 145); the Indians reckoned with several types of "years" at the same time, and so did the Maya.

Where we alternately read once "beings," and "worlds," the Sanskrit word is *bhuvana*, from the radical *bhū-* (= Greek *phyō-*) as discerned from the radical *as-*, *bhū-* meaning "to be" in the sense of perpetual change, "coming to be and passing away," *as-* being reserved for the changeless, timeless existence beyond the planetary "instruments of time," the *organa chronou* of Plato's *Timaeus*. As a matter of fact, Plato would have understood at once the verbs *bhū-* and *as-*, and he might well have applauded the utterance of the vanquished Daitya King Vali:[4]

> "O Indra! Why are you vaunting so much? All persons are practically urged on by Kāla in engaging themselves in an encounter. To the heroes, glory, victory, defeat and death gradually come to pass. This is the reason that the wise behold this universe as being guided by Kāla, and they therefore neither grieve nor are elated with joy."

Nor is there much "primitive belief" to be squeezed out of such statements as "many thousand Indras and other divinities have been overtaken by Kāla in the course of world periods."[5] But the classicists usually prefer to keep silent about the most revealing sentence of Anaximander, handed down to us by Cicero (*De Natura Deorum* 1.25): "It is the opinion of Anaximander, that gods are born in long intervals of rising and setting, and that they are innumerable worlds (or *the*—much discussed—innumerable worlds. Anaximandri autem opinio est, nativos esse deos longis intervallis orientis occidentisque eosque innumerabiles esse mundos)"; and if they do not keep silent, they claim it to be "much more natural" to understand these intervals as being in space than in time (Burnet), by which means every way to understanding is effectively blocked.

This much only for the time being: a broader discussion of Iranian Zurvan would wreck our frame; we do not think, however, that Zurvan/Chronos represents a "Zoroastrian Dilemma"; to style it thus (with Zaehner) is one more mistake: it is not the "beliefs" and "religions" which circle around and fight each other restlessly; what changes is the celestial situation.

[4] *Bhagavata Purana* 8.11 (Sanyal trans., vol. 3, p. 126).
[5] Quoted by Eisler, *Weltenmantel*, p. 501. What the author (pp. 385f.) has to say about "anthropomorphic, most primitive empathies" (?Einfühlungen), connected with Ouranos, Gē, Helios and Selene, which are, allegedly, miles away from the "step of highly abstract conceptions about eternal Time," is not only a *contradictio in adjecto*, but plain thoughtlessness.

Appendix 13

Some say, he bid his angels turn askance
The poles of earth, twice ten degrees and more
From the sun's axle, they with labour pushed
Oblique the centric globe: some say, the sun
Was bid turn reins from the equinoctial road . . .
. . . else had the spring
Perpetual smiled on earth with vernant flowers
Equal in days and nights, except to those
Beyond the polar circles; to them day
Had unbenighted shone; while the low Sun
To recompense his distance, in their sight
Had rounded still the horizon, and not known
Of east or west; which had forbid the snow
From cold Estotiland, and south as far
Beneath Magellan. At that tasted fruit
The sun, as from Thyestean banquet, turn'd
His course intended; else how had the world
Inhabited, though sinless, more than now
Avoided pinching cold and scorching heat?

MILTON, *Paradise Lost, 10*

Appendix 14

The name Mundilfoeri (Mundel-fere) raises a cluster of problems, and nothing is gained by evasive statements such as that given by de Vries (*Altnord. Etym. Wb.*, p. 395): "*Mundilferi.* Name of the father of the Moon . . . *Mundill.* Name of a legendary figure."

As concerns *mund*, feminine, it means "hand" (Cleasby-Vigfusson, s.v.), but *mund* comprises the meaning of tutelage, guardianship (cf. German Vor*mund*). *Mund* as a neutrum means "point of time, mood, humor, measure, and the right time" (de Vries, *loc. cit.*).

Mundill (Mundell) is an unknown "legendary figure," certainly; we should be glad to know what the name indicates precisely, but the specialists do not tell us. There is a small but promising hint: Gering, in his commentary on the *Edda* (vol. *1*, p. 168), remarked, "The name occurs again among the *saekonunga heiti* Sn. E. II, 154." *Heiti* are a kind of denominations (Neckel renders it "Fürnamen") which the skalds used side by side with *kenningar* (circumlocutions); the list of "heiti of sea-kings" is to be found in the *Third Grammatical Tract* contained in Snorri's *Edda* (ascribed to Snorri's nephew Olaf), and among the twenty-four *heiti*, no. 11 is *Mysingr*, no. 15 is *Mundill*.[1] Everyone who is familiar with the many names given to the cosmic personae—specific names changing according to the order of time—in Babylonian, Indian, Chinese, etc., astronomy, is not likely to fall for the idea that these *heiti* were names of historical kings.[2] The consequences resulting from the understanding of Mysing and Mundill (together with twenty-two more *heiti*) as representatives of the same cosmic function will not be worked out in detail here: he who keeps his eye on the different fords, ferrymen, pilots, personified divine ships, and kings of the deep sea that cross his path in the course of this essay may eventually work out his own solution. As for the word *fere* (in Mundelfere), Gering feels certain that it is the same word as OHG *ferjo*, MHG *verge*, i.e., ferryman, the name meaning "ferryman of Mundell." Gering refers to Finnur Johnsson who understood the *mund* in the name as "time," and "explained the name which he took for the name of the moon, originally, as 'den der bewaeger sig efter bestemte tider,'" i.e., somebody who moves according to definite times, let us say: according to his timetable (or schedule).

There is no reason at all to take Mundilföri for "originally" the name of the moon, this luminary not being the only timekeeper at hand. Vafthrudnismal 23 says of the Sun and Moon, the *children of Mundilföri*, that they circle around the sky serving as indicators of time.[3]

"Ferryman of Time" would make a certain sense, but not enough yet to enlighten us about Mundill "himself." The same goes for Sim-

[1] *Den tredje og fjaerde grammatiske afhandling i Snorres Edda*, ed. by Björn Magnússon Ólson (1884), III.15 (vol. *2*, pp. 154f.).
[2] Ólson, apparently a hardened euhemerist, stated in a note: "Hoc versu memoriali viginti quatuor nomina archipiratorum sive regulorum maritimorum continentur."
[3] Gering, *loc. cit.*: "himen hverfa . . . 'den Himmel umkreisen' . . . aldom at ártale, 'um den Menschen die Zeitrechnung zu ermöglichen.' Daher führt auch der Mond den Namen ártale 'Zeitberechner.'"

rock's rather imaginative Mundilfoeri = "Achsenschwinger," i.e., "axis-swinger," but Simrock has at least thought about a sensible meaning, and maybe he has hit the mark quite unbeknownst. Ernst Krause, too, racked his brains, modestly asking the experts to examine the relation of this *mundil* with Latin *mundus*.[4] We do not mean to meddle earnestly with this particular question, the less so as *mundus* translated into "the world" has become an empty and insignificant word altogether, but it certainly is depressing to watch the progressists working out their latest "solutions" for Latin *mundus*, namely, (1) "ornament," (2) "jewellery of women,"[5] without recalling Greek *kosmeo* which does mean also "to adorn," to be sure, but not "originally," and not essentially; to establish order, especially in the sense of getting an army into line, is what *kosmeo* means, whence kosmos. And we are not entitled to give the silliest of all imaginable meanings to such a central word as *mundus*.

We should like to approach the words in question by means of the common objective significance underlying the vast family of word-images engendered by the radical *manth, math*, whence also (Mount) Mandara, *mandala*, Latin *mentula* (penis), and also our *möndull*,[6] which is supposed to have replaced the older form *mandull*. True, mandull/möndull is not yet mundill, and mundus is not identical with mandala, yet the whole clan of words depends from a central conception sticking firmly to *mnt/mnd*, and these consonants connote a swirling, drilling motion throughout. We are, here, up to a veritable jungle of misunderstandings, and the closer we look into the "ars interpretandi" of professionals, the more impenetrable the jungle becomes. But let us try to get a shred of sense by laying bare the more or less "subconscious" blunders accomplished by the interpreters dealing with the radical *manth*, the heart and center of the Indian Amrita*manth*ana, the "Churning of Ambrosia," i.e., the Churning of the Milky Ocean in order to gain Amrita/Ambrosia, the drink of immortality. It is some sort of case history, the "case" being that *manth, math* appears to have two fundamentally different meanings (and some more), for which we

---

[4] *Tuisko-Land* (1891), p. 326; see also p. 321.

[5] I.e., (1) "Schmuck," (2) "Putz der Frauen"; see Walde-Hofmann, *Lat. Etym. Wb.*, vol. 2, pp. 126f.

[6] Cf. A. Kuhn (*Die Herabkunft des Feuers und des Göttertranks* [1886], p. 116) where he refers to Aufrecht: "*möndull* m., axis rotarum, cotis rotatilis et similium instrumentorum"; *ibid.* note 2, quoting Egilson: "*möndull* m. lignum teres, quo mola trusatilis circumagitur, mobile, molucrum; *möndultrè* m. manubrium ligneum, quo mola versatur."

quote Macdonell's Sanskrit dictionary (p. 218): "*manth*—á churning, killing, mixed beverage (= the Soma mixture); mantha-ka m. churning stick; manth-ana, producing fire by attrition." On page 214 we find s.v. *math, manth:* "whirl around (agnim), rub (a fire stick), churn, shake, stir up, agitate, afflict, crush, injure, destroy, . . . mathita bewildered, . . . strike or tear off, . . . uproot, exterminate, kill, destroy, . . . strike or tear off, drag away."[7]

So far, so good. But why insist on such misleading verbs as "striking" or "tearing off," etc.? Did not we hear about Fenja and Menja who "ground out a sudden host" for Frodhi, i.e., Mysing? And this is not an isolated instance. We know, for instance, of an extremely relevant Hittite prayer to the Ishtar of Nineveh who is asked "to grind away from the enemies their masculinity, power and health"[8]—the Hittites are quite respectable members of the Indo-European family of languages. Whether something is gained, or something is lost—peace, gold, health, heads, virility, and what else—it is *ground* out, or ground away, when the underlying image is a *mola trusatilis;* it is *drilled* out, or drilled away, when the motion of the cosmos is understood as *alternative* motion, as in the case of the Indian churn. We have sufficient reasons to take alternative motion for the older conception, but this is irrelevant right here and now; relevant is the general conception, expressed by the manifold words engendered by the radical *manth/math*, that every event is due to the rotary motion (whether "true" or alternate, compare appendix #17) of the celestial mill or churn,[9] i.e., of the combined motions of the planetary spheres and the sphere of fixed stars.

At the same moment, when we understand mill and churn as the celestial machinery, the stumbling stone of "to drill" versus "to rob, to destroy" becomes insignificant, and this is important enough, since it helps to clear the decent name of the hotly debated Prometheus.

[7] See also H. Grassmann, *Wörterbuch zum Rig-Veda* (1955), col. 976f.
[8] See L. Wohleb, "Die altrömische und hethitische evocatio," in ARW *25* (1927), p. 209, n. 5: "Ferner mahle den Männern (nämlich des feindlichen Landes) Mannheit, Geschlechtskraft(?) Gesundheit weg; (ihre) Schwerter, Bogen, Pfeile, Dolch(e) nimm und bringe sie ins Land Chatti."
[9] We touch only slightly the family of Amlodhi's *kvern;* it must be enough to state that *quairnus* means "millstone, mill" in Gothic, whereas Old Norse *kirna* is the churn. Jacob Grimm (*Geschichte der deutschen Sprache* [1848], p. 47) wanted to derive quairnus from žarna, žrno, Lith. girna, Latv. dsirnus = corn, kernel, but there seems to be no way from there to English churn, and kirna, the Old Norse churn. Kuhn (p. 104) calls attention to Sanskrit cûrna, ground powder, derived in the Petersburger Wb. from *carv*, to crush, to chew.

Adalbert Kuhn, surely a great scholar, has dealt broadly with the radical *manth*, with Mount Mandara, the churning stick used by the Asura and Deva for the churning of the Milky Ocean, and he tried hard to bring about a happy marriage between this *manthana* and Greek *manthano* "to learn," confronting us with his rather strange opinion of what is "natural." This is what he says (pp. 15ff.):

> Mit der bisher entwickelten Bedeutung der Wurzel *manth* hat sich aber schon in den Veden die aus dem Verfahren natürlich sich entwickelnde Vorstellung des Abreissens, Ansichreissens, Raubens entwickelt und aus dieser ist die Bedeutung des Griech. *manthánō* hervorgegangen, welches demnach als ein an sich reissen, sich aneignen des fremden Wissens erscheint. Betrachten wir nun den Namen des Prometheus in diesem Zusammenhang, so wird wohl die Annahme, dass sich aus dem Feuer entzündenden Räuber der vorbedächtige Titane erst auf griechischem Boden entwickelt habe, hinlänglich gerechtfertigt erscheinen und zugleich klar werden, dass diese Abstraktion erst aus der sinnlichen Vorstellung des Feuer*reibers* hervorgegangen sein könne. Was die Etymologie des Wortes betrifft, so hat auch Pott . . . dasselbe auf *manthano* in der Bedeutung von *mens provida, providentia* zurückgeführt . . . , aber er hätte, sobald er das tat, das Sanskritverbum nicht unberücksichtigt lassen sollen . . . Ich halte daher an der schon früher ausgesprochenen Erklärung fest, nach welcher Promētheús aus dem Begriff von pramātha, Raub, hervorgegangen ist, so dass es einem vorauszusetzenden Skr. pramāthyus, der Räuberische, Raub liebende, entspricht, wobei jedoch wohl auch jener oben besprochene pramantha—i.e. the upright drilling stick— auf die Bildung des Wortes mit eingewirkt hat, zumal Pott auch noch einen *Zeus Promantheus* . . . aus Lycophron 537 nachweist, so dass in dem Namen auch der Feueranzündende zugleich mit ausgedrückt wäre.

It goes without saying that we do not think it either "natural" or "obvious" to "develop" learning from robbing, or providence from learning: Prometheus (Lykophron's *Promantheus*)-pramantha drilled new fire, at a new place, at new crossroads of ecliptic and equator; the "gods" did not like that (about which more later).

Now, *pramantha*, alias the male fire stick, having the well-known naughty connotations, and with the Fecundity-Trust standing around the corner, classical philologists fought bitter battles against Kuhn's proposition, for the sake of noble Prometheus who simply should not be a fire stick or, worse, the fascinum. The highly emotional classicists remained victorious upon the battlefield until very recently, when we learn the newest tidings from Mayrhofer,[10] who rules firmly: *"manth,* 'quirlen' ist etymologisch von *math-, mathnāti* 'rauben' (offenbar nasallos) verschieden."* After having dealt with the different meanings of the words, already known to us, he continues: "An ausserindischen Nachweisen der Vorstufe von ai. *math-* 'rauben' . . . besteht vorerst nur die vorsichtig ausgesprochene, aber sehr glaubhafte Zusammenstellung von ai. pra- math- mit griech. Promētheús, dor. Promatheús (Narten)."

That is exactly what "progress" means nowadays: that we are offered as a brand-new, "cautiously uttered, but very credible connecting of Sanskrit pra-math with Greek Prometheus" in 1963, when Kuhn's second edition had been published in 1886. We do not wish to dwell upon the claimed "etymological difference" of the radicals manth and math: if philologists do not understand a subject, they invent different radicals, which are "mixed" in later times, allegedly, as here math- and manth "in post-Vedic times."[11]

Prometheus was a "pramantha," as were Quetzalcouatl, Tezcatlipoca, the four Agnis, and very many more, drilling or churning with "Mount Mandara," or with Möndull: why not call him Mundilfoeri, the axis-swinger? We have, indeed, Altaic stories about one or the other Mundilfoeri "begetting" Sun and Moon. Uno Holmberg states (*Die religiösen Vorstellungen der altaischen Völker* [1938], pp. 22, 63, 89f.):

In the myths of the Kalmucks the world mountain—Sumeru, Meru, alias Mandara—appears as the means of creation. The world came into being, when four powerful gods got hold of Mount Sumeru, and whirled it around in the primordial sea, just as a Kalmuck woman turns the churning stick when preparing butter. Out of the vehemently agitated sea came, among others, Sun, Moon, and stars. The same significance has, doubtless, the story of the Dorbots, according to which once upon a time, before Sun and Moon existed, some being began to stir the primordial ocean with a pole of 10,000 furlongs, thus bringing forth Sun and Moon. A similar creation is described in a Mongolian myth, where a being coming from heaven—a Lama it is supposed to have been, see Holmberg, *Finno-Ugric Mythology,* p. 328—stirs up the primeval sea, until part of the fluid becomes solid.

[10] *Kurzgefasstes Etymol. Wörterbuch des Altindischen,* vol. 2 (1963), pp. 567f., 578ff.

[11] The worst among the relevant cases is the Greek radical *lyk,* which the experts insist upon being two different ones, i.e., lyk = light, and lyk = wolf, without spending a thought on Pythagoras, who taught us: "The planets are the dogs of Persephone"; all mythical canines have just everything to do with light.

These "creation stories" are more or less deteriorated survivals of the Amritamanthana, "the incomparably mighty churn," in the course of which one constellation after the other emerged from the wildly agitated Milky Ocean.[12] And the same goes for the "creation" brought forth by the Japanese "parents of the world," who, standing upon the Celestial Bridge, stirred with the celestial jewel-spear the primordial sea until parts of it thickened and became islands. The Amritamanthana survived also in Greece, in the beginning of *Iliad* 8, and in the myth of Plato's Statesman, and Plutarch spotted it in Egypt: but this subject would make another book. The relevant point was, here, to place figures as Mundilfoeri, or some surviving Lama, or Vishnu Cakravartin on the cosmological stage, where their modes of "creation" make sense.

[12] The collector of merely funny survivals might enjoy the following yarn from Switzerland (Grimm, TM, p. 697): "In the golden age when the brooks and lakes were filled with milk, a shepherd was upset in his boat and drowned; his body, long sought for, turned up at last in the foamy cream, when they were churning, and was buried in a cavity which bees had constructed of honeycombs as large as town-gates."

Appendix 15

As concerns the removing of the Pole star, the most drastic version is told by the Lapps:

> When Arcturus (alpha Bootis, supposed to be an archer, Ursa Major being his bow) shoots down the North Nail with his arrow on the last day, the heaven will fall, crushing the earth and setting fire to everything.[1]

Other legends prefer to deal with the fate of circumpolar stars, the result being the same.

> The Siberian Kirghis call the three stars of the Little Bear nearest the Pole star, which form an arch, a "rope" to which the two larger stars of the same constellation, the two horses, are fastened. One of the horses is white, the other bluish-grey. The seven stars of the Great Bear they call the seven watchmen, whose duty it is to guard the horses from the lurking wolf. When once the wolf succeeds in killing

[1] U. Holmberg, *Finno-Ugric and Siberian Mythology* (1964), p. 221. See the drawing made by J. Turi in *Das Buch des Lappen Turi* (1912), plate xiv: Arcturus = Favtna, Polaris/North Nail = Boaje-naste, or Bohinavlle.

the horses the end of the world will come. In other tales the stars of the Great Bear are "seven wolves" who pursue those horses. Just before the end of the world they will succeed in catching them. Some even fancy that the Great Bear is also tied to the Pole Star. When once all the bonds are broken there will be a great disturbance in the sky.[2]

According to South Russian folklore, a dog is fettered to Ursa Minor, and tries constantly to bite through the fetter; when he succeeds, the end of the world has come.

Others say that Ursa Major consists of a team of horses with harness; every night a black dog is gnawing at the harness, in order to destroy the world, but he does not reach his aim; at dawn, when he runs to a spring to drink, the harness renews itself.[3]

A very strange and apparently stone-old story is told by the Skidi-Pawnee about the end and the beginning of the world.[4]

Various portents will precede: the moon will turn red and the sun will die in the skies. The North Star is the power which is to preside at the end of all things, as the Bright Star of Evening was the ruler when life began. The Morning Star, the messenger of heaven, which revealed the mysteries of fate to the people, said that in the beginning, at the first great council which apportioned to star folk their stations, two of the people fell ill. One of these was old, and one was young. They were placed upon stretchers, carried by stars (Ursa Major and Ursa Minor),[5] and the two stretchers were tied to the North Star. Now the South Star, the Spirit Star, or Star of Death, comes higher and higher in the heavens, and nearer and nearer to the North Star, and when the time for the end of life draws nigh, the Death Star will

---

[2] Holmberg, p. 425; cf. Holmberg's *Die religiösen Vorstellungen der altaischen Völker* (1938), p. 40.

[3] A. Olrik, *Ragnarök* (1919), pp. 309f. The author regards it as "ein neues Motiv, dass der Hund am Himmel angebracht ist und mit den Sternbildern zu tun hat. Sonst haben wir die Hunde in einem Berg am Ende der Welt . . ."

[4] H. B. Alexander, *North American Mythology* (1916), pp. 116f.

[5] The Sioux take Ursa Major for a coffin, accompanied by mourners. This picture is not too "obvious," so it is significant that Ursa is *banat na'sh* with the Arabs, i.e., the bier and its daughters; the bier is formed by the chest of the wagon, El-na'sh, the handle of the Dipper being the daughters. See Ideler, *Sternnamen*, pp. 19f. Kunitzsch, *Arabische Sternnamen in Europa*, p. 149, no. 71, adds that, according to Athanasius Kircher, christianized Arabs recognized in the constellation the coffin of Lazarus, followed by the mourners Maryam, Marta, and their maid (*al-ama*). See also Henninger, ZfE *79*, p. 81. Due to Islamic influence, the constellation is called *bintang al'nash*, star of the bier, by the people of Minangkabau, Southern Sumatra. (See H. Werner, "Die Verstirnung des Osiris-Mythos," IAfE *16* [1954], p. 154.)

approach so close to the North Star that it will capture the stars that bear the stretchers and cause the death of the persons who are lying ill upon those stellar couches. The North Star will then disappear and move away and the South Star will take possession of earth and its people. The command for the ending of all things will be given by the North Star, and the South Star will carry out the commands. Our people were made by the stars. When the time comes for all things to end our people will turn into small stars and will fly to the South Star where they belong."

To return to better known provinces, Proclus informs us that the fox star nibbles continuously at the thong of the yoke which holds together heaven and earth; German folklore adds that when the fox succeeds, the world will come to its end.[6] This fox star is no other than Alcor,[7] the small star g near zeta Ursae Majoris (in India Arundati, the common wife of the Seven Rishis, alpha-eta Ursae; see p. 301 about Arundati and Elamitic Narundi, sister of the Sibitti, the "Seven"), known as such since Babylonian times.[8]

The same star crosses our way again in the Scholia to Aratus[9] where we are told that it is Electra, mother of Dardanus, who left her station among the Pleiades, desperate because of Ilion's fall, and retired "above the second star of the beam . . . others call this star 'fox.'"

This small piece of evidence may show the reader two things: (1) that the Fall of Troy meant the end of a veritable world-age. (For the time being, we assume that the end of the Pleiadic age is meant; among various reasons, because Dardanos came to Troy after the third flood, according to Nonnos.); (2) that Ursa Major and the Pleiades figuring on the shield of Achilles, destroyer of Troy, have a precise significance, and are not to be taken as testimony for the stupendous ignorance of Homer who knew none but these constellations, as the

---

[6] (Proklos ad Hesiod, opp. 382) Boll and Gundel, in Roscher s.v. Sternbilder, col. 876.

[7] For the name Alcor, and its tradition, see Kunitzsch, pp. 125f.

[8] See F. X. Kugler, S.J., *Ergänzungsheft zum 1. u. 2 Buch* (1935), pp. 55f.; P. F. Gössmann, *Planetarium Babylonicum*: "The star at the beam of the wagon is the fox star: Era, the powerful among the gods. In astrological usage, it represents above all the planet Mars/Nergal." See also E. F. Weidner, *Handbuch Babyl. Astr.* (1915), p. 141; E. Burrows, S.J., "The Constellation of the Wagon and Recent Archaeology," in *Festschrift Deimel* (1935), pp. 34, 36. The said Nergal, i.e., Mars, to whom "belongs" Alcor in the Series mulAPIN, starts the first flood, as we learn from Utnapishtim—see p. 297—under the name of Era, he succeeds in starting a new one, according to the Era-Epos.

[9] 257; E. Maass, *Commentariorum in Aratum Reliquae* (1898), p. 391, ll. 3ff.

specialists want us to believe. There are, indeed, too many traditions connecting Ursa and the Pleiades with this or that kind of catastrophe to be overlooked. Among the many we mention only one example from later Jewish legends, some lines taken out of a most fanciful description of Noah's flood, quoted by Frazer:[10]

> Now the deluge was caused by the male waters from the sky meeting the female waters which issued forth from the ground. The holes in the sky by which the upper waters escaped were made by God when he removed stars out of the constellation of the Pleiades; and in order to stop this torrent of rain, God had afterwards to bung up the two holes with a couple of stars borrowed from the constellation of the Bear. That is why the Bear runs after the Pleiades to this day; she wants her children back, but she will never get them till after the Last Day.

[10] *Folk-Lore in the Old Testament* (1918), vol. *1*, pp. 143f.

Appendix 16

For *Hallinskidi* see Reuter, p. 237; Simrock, *Handbuch*, p. 277; Gering (*Edda* trans., p. 320): "gebogene Schneeschuhe habend." Much (in *Festschrift Heinzel*, p. 259), connecting -*skidi* with Celtic skêto, skêda (English: humerus, scapula) and taking *halle* for "stone," ventures to propose the reconstruction "he with the stone shoulder . . . which would presuppose a similar story as that about Pelops and his ivory shoulder."

As concerns *mjötvidr*, A. V. Ström renders vol. *2*:[1]

> *Ich erinnere mich neun Welten*
> *Neun im Baume (oder neun Heime),*
> *des ruhmvollen Massbaums*
> *unter der Erde.*

And he quotes Hallberg's statement: "Der Baum selbst ist das Mass für die Existenz der umgebenden Welt—in der Zeit."[2] The last remark goes

[1] "Indogermanisches in der Volüspa," *Numen 14* (1967), pp. 173.

[2] Why the author, in this excellent article, drags in "ecstatic visions," remains incomprehensible, unless we prefer to call every account of astronomical situations "ecstatic visions," which would be a true miötvidr to measure the vast abyss between sciences and humanities in our time.

without saying, mythic measures are time measures, generally, but this fact is so seldom recognized that this white raven has to be welcomed enthusiastically. The "localization under the earth" points to the (invisible) South of the world, as will come out later. By which we do not mean to say we understood the enigmatical picture of this measuring tree.

Now, Heimdal and Loke, perpetual enemies as they are, kill each other at Ragnarök, but Heimdal's death is accomplished by means of a very strange weapon, i.e., by a "head." Snorri's *Skaldskaparmal 8* (see also *69*) offers an ambiguous kenning: "Heimdal's head is the sword, or, the sword is Heimdal's head,"[3] or we learn that the sword was called "miötudr Heimdaler," and that is, according to Jacob Grimm,[4] "the measurer (sector, messor)." Thus, Heimdal measures—or is he measured?—by means of a sword that is also said to be his very own head. Strange goings-on, indeed. Ohlmarks[5] declared the sword to be the Sun—a pleasant change for once, otherwise everything and everybody is the Moon, with him—but although the measuring instrument, whether the "golden rope" or not, usually is the sun (see p. 154 on Varuna, and p. 246 on Theaethetus 153c [the latter is by Plato]), we have the suspicion that the case of Heimdal's head/sword is more complicated, and that it may not be settled until we know much more about Loke.

3 Heimdalur hoefut heitir sverdh; cf. Simrock, *Handbuch*, pp. 272f.

4 TM, p. 22 (see also p. 1290); the English translation says "the wolf's head, with which Heimdal was killed," but the original (*Deutsche Mythologie*, p. 15) does not mention a wolf.

5 *Heimdalls Horn* (1937), p. 151.

Appendix 17

To prevent rash critics from hurling into our faces the—maybe they will style it thus—"complete absence of technological knowledge," we hasten to assert that the relevant inquiries are not as foreign to us as they might assume.[1] Curwen might point to his enlightened sentence:

1 To mention only a few useful titles: Joseph Needham, *Science and Civilisation in China*, vol. *4*, Pt. II, (1965); Gordon Childe's chapter on "Rotary Motion," in Singer et al., eds., *A History of Technology*, vol. *1* (1954), pp. 187ff.; Hugo Theodor Horwitz, "Die Drehbewegung in ihrer Bedeutung für die Entwicklung der

We are, happily, emerging from that state of blissful ignorance of the subject which made possible such an anachronism as Décamps' well-known picture of "Samson grinding in the Prison-house," wherein Samson is seen turning a huge mill-stone by means of a long lever like a capstan-bar, after the fashion of the Roman slaves a thousand years later.[2]

There are, indeed, "a number of reasons for questioning the common belief that grainmills were rotary," as Moritz states (p. 53). And whereas Forbes (*Studies in Ancient Technology*, vol. 3, p. 155, n. 3) votes for "rotary querns . . . in Assyrian times," Lynn White (p. 108) says: "But while continuous rotary motion was used in this large mola versatilis and, of course, in the water mill which appears in the first century B.C., it is by no means clear how early such a motion was used with querns," which is certainly true. That true rotary motion was used with the potter's wheel much earlier is unquestionable, which is the more relevant, as the potter's wheel, too, belongs to the cosmological instrumentation, e.g., in the hands of Ptah and Khnum. Decisive is the Ancient Egyptian instrument for drilling out stone vessels, which was perhaps even cranked, but there is no unanimity among the historians of technology as to the real nature of this device. In this case and in that of the mill, the accent goes with "true" rotary motion, because there are two kinds of rotary motion, to which we quote Gordon Childe (Singer, p. 187) on the difference "between continuous, true and complete rotary motion, and partial or discontinuous rotary motion. For true rotary motion, the revolving part of the instrument must be free to turn in the same direction *indefinitely*. There are, however, a number of processes which involve a partial turn of the instrument, such as boring and drilling by hand. There are even machines like the bow-drill or the pole-lather which allow a number, but only a limited number, of complete revolutions of the revolving part. Partial rotary motion of this sort has been used by man much longer than true rotary motion."

Now, we do not wish to suppress White's footnote (p. 109), where he claims Fenja's and Menja's Grotte to have been an apparatus involving alternative motion, "no doubt." This might be the case, although we

---

materiellen Kultur," *Anthropos 28* (1933), *29* (1934). John Storck and Walter Dorwin Teague, *Flour for Man's Bread: A History of Milling* (1952); Lynn White, *Medieval Technology and Social Change* (1962)—this title is a grotesque understatement!

[2] "Querns," *Antiquity 11* (1937), pp. 133f. See also L. A. Moritz, *Grain-Mills and Flour in Classical Antiquity* (1958), p. 12—he makes it a medieval mill.

do not agree with the "no doubt": several doubts are permitted. We shall abstain, however, from discussing this and similar questions as long as we do not understand precisely and thoroughly how the "Churning of the Milky Ocean" was thought to work, in India, and in Egypt, where the specialists insist upon calling the celestial churn a "symbol of uniting the two lands," and in the survivals in Homer and Plato. For the time being we do think that the oldest technological device used in cosmological terminology was, indeed, a churn or a drill, implicating alternative motion.

The point is this: whether or not Samson, or Fenja and Menja, waited on an oscillating quern or on a true rotary mill is a cosmological question, and will hardly be decided by historians of technology. To illustrate this, we have a look at that "mill" of the Cherokee Indians, mentioned in the chapter on the Galaxy, where it is told that "people in the South had a corn mill," from which meal was stolen again and again; the owners discovered the thief, a dog, who "ran off howling to his home in the North, with the meal dropping from his mouth as he ran, and leaving behind a white trail where now we see the Milky Way, which the Cherokee call to this day . . . 'Where the dog ran.' " In his supplementary notes (p. 443), Mooney explains: "In the original version the mill was probably a wooden mortar, such as was commonly used by the Cherokee . . ." Well, in the "original version," *as told by the Cherokee*, we may rely on their talking of a mortar—but certainly not in the truly "original" myth. There is no possible way whatsoever of "developing" out of "primitive" mortars (or grindstones) cosmological imagery; in other words: the Cherokee mortar is a "deteriorated" mill (whether oscillating or not).

The cosmic machine (mill, drill, or churn) produces periods of time, it brings about the "separation of heaven and earth," etc. Along the way of diffusion into unfamiliar surroundings, particularly tropical ones (lacking grain, plow culture, etc.), the Mill (or churn) ceases to be understood, while the memory sticks to an instrument for crushing foodstuff. And, suddenly, we are told in several continents how Heaven, who once was lying closely upon Earth, withdrew in anger because of women who, busy with their mortars, kept bumping with their pestles against Heaven's body. An extremely pointless idea, the origin of which is only to be understood when we follow it back to the highly complicated machinery which stood at its beginning (historically as well as "sinngemäss"), and begot quite innocently such strange offshoots.

Although we do not like to apply strictly scientific models to historical phenomena, here we abuse the case of entropy: to derive Grotte (the Amritamanthana, etc.) from those utterly nonsensical females bumping their pestles against "Heaven" would be on the same level as to derive the original substances from the state of randomly mingled gases.

These minima only for the technological problem. We keep these questions under lock and key on purpose, and not because it has not dawned upon us that the technological aspect is a very important one. On the contrary, we nurse the suspicion that next to nobody has an idea of the huge difficulties that arise with churn, mill, and fire drill, if one understands them properly as machines which were meant to describe the motions of nested spheres.

Appendix 18

Compare *Popol Vuh: The Sacred Book of the Ancient Quiché Maya* (Eng. trans. by D. Goetz and S. Morley, [1951], pp. 99–102). As concerns the escape of Zipacna, compare the distribution map, given by Frobenius (*Paideuma 1* [1938], p. 8, map 3—"Der Lausbub im Hauspfeiler").

For the whole motif of pillars and houses pulled down, compare Eduard Stucken, *Astralmythen* (1896–1907): pp. 73f. for the death of Nebrôd, according to Cedrenus—of Cain, according to Leo Grammaticus Chron. p. 8 (Kain, hōs legei Mōysēs, tēs oikias pesousēs ep'auton eteleutēsen); pp. 329f. for the case of Susanowo; p. 348 for Turkish Depe Ghöz; pp. 402f. for Zipacna; there, he also wants to incorporate Job 1.18. Stucken's complete blindness to the mere existence of planets has prevented him from better understanding; thus, he claims for the case of Job 1.18: "Auch hier ist die Orion-Gottheit (Satan-Ahriman), welche den Hauseinsturz verursacht, um die Plejaden-Gottheit (Hiob) zu züchtigen." This blindness is the more astonishing as Stucken has read Eisenmenger's huge opus, "Entdecktes Judenthum" (1711), where he should have detected the identity (as claimed by rabbinical literature) of the planet Mars with the serpent in Paradise, with Kain, Esau, Abimelech, Goliath, Sammael, the Scape-Goat, and many others.

Appendix 19

A remarkable amount of information about submarine creatures is contained in Mansikka's inquiry into Russian magic formulae, already mentioned;[1] intermingled as the material is with the author's rather violent "interpretatio christiana," it is well-nigh impossible to lay one's hands on the bare facts. This much can be said, however: in the middle of the "Blue Sea" (or "the middle of the whole earth"), there is either (a) an island—most of the time called *Bujan*, from the same radical as buoy—"the center of celestial power," upon which there is a tree, or a stone, or a tree upon a stone, sometimes the cross or the "mountain of Zion" itself;[2] or there is (b) the "White Altar-Stone," which is a "fiery" one, lying in the navel of the sea without being supported by an island; under this stone, there is "a green fire, the king of all fires," or an "eternal, unquenchable fire" that "has to be procured from under the stone" (Mansikka, p. 188—we are not told for what purpose the fire has to be fetched from there; the text says only "for burning"). Sometimes it is said that upon this stone—regardless of its being "holy" and the "Stone of the Altar," and even "Christ's Throne"—was the "habitation of the Devil himself";[3] in other formulae the point is stressed that this fire "scorches and burns the decayed and impure power of the devil" (i.e., "die verfallene, unreine Macht des Teufels," where "verfallen" may mean either "decayed" or "forfeited"). As long as this unquenchable fire remains safely under a stone, nothing dangerous is going to happen; accordingly, a German formula (Mansikka, p. 37) says: "In Christ's Garden there is a well, in the well there is a stone, under the stone lies a golden snake." That snake can also be a scorpion, as we have just seen (footnote 3).

[1] *Über russische Zauberformeln* (1909), pp. 168–213: "The Sea, the Stone, the Virgin Mary."
[2] Thus it is said that "upon the mountains of Zion, upon the white stone stands the pillar and the altar of Christ," or, "a pillar from the earth to heaven." In a prayer Christ is addressed: "O, thou deadly stone pillar" (o, du tödliche Steinsäule, Mansikka, p. 187).
[3] Mansikka, p. 189; see also the formula on pp. 35f.: "Es gibt ein heiliges Meer Ozean, in seiner Mitte liegt ein weisser stein, aus dem weissen Stein kommt eine grimmige Schlange, der Skorpion, hervor . . . In dem teuflischen Sumpf liegt der weisse Stein Latyr; auf dem weissen Stein Latyr aber sitzt der leibhaftige Teufel."

The Mordvinians[4] have a long story to tell about God, Tsham-Pas, who was rocking to and fro upon a stone in the primordial sea, thinking deeply about how to create the world and how to rule it afterward, and complaining: "I have neither a brother nor a companion with whom to discuss the matter." Angrily he spat into the sea, the spittle turned into a large mountain from which emerged Saitan and offered himself as partner in the discussion. Tsham-Pas sent his new companion to the bottom of the sea to fetch sand, admonishing him to mention his (God's) name before touching the sand. Saitan did not do so, and was burned heavily by the flames which came out of the bottom of the sea; this happened twice, and Tsham-Pas warned Saitan that, should he not mention the divine name when diving for the third time, the flames would consume him completely. The bad companion obeyed and brought, finally, the sand necessary for the creation. But since he could not abstain from playing tricks, God chased him away, saying: "Go away to the bottom of the sea, to the other world, in that fire that burned you when you were too proud to mention the name of your creator. Sit there and suffer for all eternity."

In India, where the word "eternity" is not applied as thoughtlessly as in European lege.:ds, the *Harivamsa* tells us the following about the offspring of the sage Aurva (i.e., "born from the thigh," *uru*), as we hear from Dowson:[5]

> The sage was urged by his friends to beget children. He consented, but he foretold that his progeny would live by destruction of others. Then he produced from his thigh a devouring fire, which cried out with a loud voice, 'I am hungry; let me consume the world.' The various regions were soon in flames, when Brahmā interfered to save his creation, and promised the son of Aurva a suitable abode and maintenance. The abode was to be at Badavā-mukha, the mouth of the ocean; for Brahmā was born and rests in the ocean, and he and the *newly produced fire* were to consume the world together at the end of each age, and at the end of time to devour all things with the gods, Asuras, and Rākshasas. The name Aurva thus signifies, shortly, the submarine fire. It is also called Badavānala and Samvarttaka. It is represented as a flame with a horse's head, and it is also called Kāka-dhwaya, from carrying a banner on which there is a crow.

In the *Mahabharata*,[6] this story is told by the Rishi Vasishtha (zeta Ursae Majoris) in order to appease his grandson, who likewise wished

[4] O. Dähnhardt, *Natursagen* (1907–1912), vol. *1*, pp. 60–62.
[5] J. Dowson, *A Classical Dictionary of Hindu Mythology* (8th ed. 1953), pp. 32f.
[6] Mbh. *1*.180–82 (Roy trans., vol. *1*, pp. 410–14).

to destroy the whole world without delay: "Then, o child, Aurva cast the fire of his wrath into the abode of Varuna.[7] And that fire which consumeth the waters of the great Ocean, became like unto a large horse's head which persons conversant with the *Vedas* call by the name of Vadavamukha. And emitting itself from that mouth it consumeth the waters of the mighty ocean."

This fiery horse's head guides the curious straight into the mazes of the *Mahabharata* and the *Shatapatha Brahmana* where they are most impenetrable because they deal with the enigmatic story of the Rishi Dadhyañk, whose horse's head was dwelling in Lake Saryanāvant, after it had revealed the "secret of madhu" (madhuvidyâ; *madhu* = honey mead) to the Ashvins, the Dioscures,[8] and out of whose bones (the bones of the horse's skull) Tvashtri forged the thunderbolt for Indra, thus enabling him to slay "the 99 vritras"[9]—as Samson killed the Philistines with the jaw-bone of an ass—whereas Vishnu used this head to reconquer the *Vedas* that had been carried away by two Daityas during one of those time-swallowing "Yoga-sleeps" of Vishnu. Bereft of the *Vedas*, Brahma, to whom they served as "eyes," was unable to continue the work of creation, so that he implored the Lord of the universe to awake. "Praised by Brahma, the illustrious Purusha . . . shook off his slumber, resolved to recover the Vedas (from the Daityas that had forcibly snatched them away). Applying his Yoga-puissance, he assumed a second form . . . He assumed an equine head of great effulgence, which was the abode of the Vedas. The firmament, with all its luminaries and constellations, became the crown of his head . . . Having assumed this form endued with the equine head . . . the Lord of the universe disappeared then and there, and proceeded to the nether regions"[10]—to return with the *Vedas*, successfully, and resuming his sleep, as goes without saying.

In other words, the "equine head" is as important a "form" of Vishnu as an enigmatical one, so much so, in fact, that the more "popular" tradition seems to ignore it, although the Great Epic tells us the following:

[7] "The water from which the world took its origin," according to H. G. Jacobi, *Mahabharata* (1903), p. 20.

[8] Cf. RV *1*.116.12; SB *14*.1.1.18–25 (Eggeling trans., vol. *5*, pp. 444f.); Saunaka's *Brihad Devata 3.16.25* (Macdonell trans., vol. *2*, pp. 82–85).

[9] Cf. RV *1*.84.13; Mbh. *12*.343 (Roy trans., vol. *10*, p. 578). Compare for the whole tradition, K. Rönnow, "Zur Erklärung des Pravargya, des Agnicayana und des Sautrāmanī," in *Le Monde Oriental* (1929), pp. 113–73; see also A. Keith, "Indian Mythology," MAR *6* (1917), pp. 61, 64.

[10] Mbh. *12*.348 (Roy trans., vol. *10*, p. 605).

In days of yore, for doing good to the world, Narayana [Vishnu] took birth as the great Rishi Vadavamukha [see above, Aurva's son, the mouth of the ocean, Vadavamukha]. While engaged in practising severe austerities on the breast of Meru, he summoned the Ocean to his presence. The Ocean, however, disobeyed his summons [Greek Okeanos, too, was in the habit not to make his appearance, when Zeus summoned everybody to assemble.] Incensed at this, the Rishi, with the heat of his body, caused the waters of the Ocean to become as saltish in taste as the human sweat. The Rishi further said, 'Thy water shall henceforth cease to be drinkable. Only when the Equine-head, roving within thee, will drink thy waters, they will be as sweet as honey.'—It is for this curse that the waters of the Ocean to this day are saltish to the taste and are drunk by no one else than the Equine head.[11]

The translator, Pratap Chandra Roy, remarks in a footnote (p. 583), without referring to the first book of the epic:

The Hindu scriptures mention that there is an Equine-head of vast proportions which roves through the seas. Blazing fires constantly issue from its mouth and these drink up the sea-water. It always makes a roaring noise. It is called Vadava-mukha. The fire issuing from it is called Vadava-nala. The waters of the Ocean are like clarified butter. The Equine-head drinks them up as the sacrificial fire drinks the libations of clarified butter poured upon it. The origin of the Vadava fire is sometimes ascribed to the wrath of Urva, a Rishi of the race of Jamadagni. Hence it is sometimes called Aurvya-fire.

None of the authorities quoted hitherto thought it worth mentioning where this Vadava-mukha was supposed to be. Only when checking the word in Macdonell's *Practical Sanskrit Dictionary* (p. 267) did we learn —exactly as foreseen, although Macdonell means a terrestrial South Pole, presumably—that "vádabā, f. = mare; Vivasvat's wife, who in the form of a mare became the mother of the Ashvins . . . vadaba-agni, m. submarine fire (supposed to be situated at the south pole) . . . vadaba-mukha, n. mare's mouth = entrance of hell at the south pole."

We are not likely to change these dark plots into a lucid and coherent story by dealing, here and now, more closely with Dadhyañk, whose name is said to mean "milk-curdling," and who is a "producer of Agni," and by comparing the several characters who are accused of swallowing up the ocean: we only hope to guide the attention to one among the many unperceived concrete problems.

[11] Mbh. *12*.343 (Roy trans., vol. *10*, p. 583).

We might be suspected of proposing to identify the sea-swallowing horse's head with the equally thirsty Agastya-Canopus,[12] just to simplify the situation, and there are factors which invite such a "solution."[13] But the horse is the animal of Mars, and it is "the khshatriya Apām Napāt with the swift horses" who "seizes the hvarnah," hiding it in the "bottom of the deep sea, the bottom of the deep lake":[14] the *"nephew"* (*napāt*) of the waters (*apām*), and not the original (and highest) ruler of the "mouth of the ocean," alias *pī narātī*, "the confluence of the rivers," i.e., Canopus, which the Tahitians of old called "Festivity-from-whence-comes-the-flux-of-the-sea" (T. Henry, *Ancient Tahiti* [1928], p. 363). Aurva's frightening son is, moreover, a "newly produced fire," as we have heard, and Apām Napāt is by no means the one and only "Agni"; the *Rigveda* knows of four "fires," Agnis, allegedly consumed by the sacrificial service, one after the other. No valid insight is likely to be gained before we cease to disregard the only mythical dimension that counts: time.

Horses' heads not being connected with deep waters quite "naturally," we might close with some stories collected by Jacob Grimm (TM, pp. 597f.) which go to show that

Lakes cannot endure to have their depth gauged. On the *Mummelsee,* when the sounders had let down all the cord out of nine nets with a plummet without finding a bottom, suddenly the raft began to sink, and they had to seek safety in a rapid flight to land . . . A man went in a boat to the middle of the *Titisee,* and payed out no end of line after the plummet, when there came out of the waves a terrible cry: "Measure me, and I'll eat you up!" In a great fright the man desisted from his enterprise, and since then no one has dared to sound the

---

[12] See p. 263. Cf. also Varāhamihīra, *The Brihad Sanhita,* trans. by H. Kern, in JRAS 5 (1871), p. 24. For a related and very peculiar legend of the Maori, see *The Lore of the Whare-wānanga,* trans. by S. Smith, in *Mem. Polynesian Soc. 3* (1913), pp. 156f., 164, and M. Makemson, *The Morning Star Rises: An Account of Polynesian Astronomy* (1941), p. 157, for a summary. There, the heavenly waters of Rangi-tamaku (i.e., the sky which lies directly above the visible one) became overheated and evaporated, so that whole tribes of celestial fish had to emigrate by descending on the "Road of the Spider," where they met Tawhaki ascending on his expedition to avenge his father.

[13] E.g., Stephanus of Byzantium mentions a temple of Poseidon-Canopus; see P. Casanova, "De quelques Légendes astronomiques Arabes," in BIFAO 2 (1902), p. 11.

[14] Yasht *19.51*; see E. Herzfeld, *Zoroaster and His World* (1947), p. 571; to the Iranian conceptions one has to compare the Rigvedian hymn dedicated to Apām Napāt (RV 2.25), where he is said to "shine in the waters," blazing unquenchably, the driver of horses (2.35.5: "Er hat sich in den Gewässern—apsú—ausgestreckt" . . . 2.35.6: "Dort ist der Geburtsort des Rosses und dieser Sonne").

depth of the lake ... There is a similar story ... about *Huntsoe*, that some people tried to fathom its depth with a ploughshare tied to the line, and from below came the sound of a spirit-voice: "i maale vore vägge, vi skal maale jeres lägge!" Full of terror they hauled up the line, but instead of the share found an old horse's skull fastened to it.

Appendix 20

Such stories are no jokes, although they make this impression when they cross our way in Eurasian folklore. "Air" is a strictly astronomical and, therefore, also a "religious" term. Thus, we hear from Rabbi Eleazar b. Pedath (ca. A.D. 270): "Als der Pharaoh aus Agypten auszog, die Israeliten zu verfolgen, erhoben sie ihre Augen gen Himmel und sahen den Engelsfürsten Ägyptens in der Luft fliegen."

"That signifies the fall of Egypt," adds Bertholet, who mentions this case in his article on the "guardian angel of Persia" (*Festschrift Pavry*, p. 38), starting from Isa. xxiv.21 and its rabbinical interpretations. He also points to the utterance of Rabbi Chanina (ca. A.D. 225): "Nicht bestraft Gott eine Nation eher, als bis er zuvor ihren Engelfürsten im Himmel bestraft hat," to which he compares Ps. xxiv.21: "On that day the Lord will punish the host of heaven, in heaven, and the kings of the earth, on the earth."

These "guardian angels" will be identified sooner or later, insofar as this has not yet been accomplished in older literature which our contemporaries disdain as "obsolete"; one among them, the "angel-lord" of Esau/Edom, with whom, according to the Zohar, Jacob wrestled (Gen. xxxii.24–33), is the planet Mars.[1] How the whole system really works— e.g., these punishments first in "heaven," subsequently "on earth"—will not be understood before Plato's *Timaeus* is taken as earnestly as it was taken by the Pythagorean Timaios himself, whom Plato introduced as

---

[1] See J. Eisenmenger, *Entdecktes Judenthum 1* (1711), pp. 844–46; cf. *The Zohar*, 144a, 146a (trans. by H. Sperling and M. Simon, [1956], vol. *2*, pp. 63, 70f.): "For Jacob conquered the serpent with prudence and craft, but chiefly by means of the he-goat; and although the serpent and Sammael are the same, yet he also conquered Sammael by another method, as described in the passage, saying: and there wrestled a man with him until the breaking of the day (Gen. xxxii.25–26)." And: "Another blessing he [Jacob] received from that angel, the chieftain of Esau." A. Jeremias (ATAO, p. 324) maintains that the wrestling took place at "Nibiru," which he identifies, here, with the solstice, but see appendix #39. For angels as stars, see also M. Knapp, *Antiskia* (1927), pp. 33–36.

"astronomikōtaton hēmōn," i.e., the most astronomically-minded among us, and before it is accepted as the foundation from which to proceed further. (See below, chapter XXII, for a superficial touching on this cosmic system.)

Appendix 21

A faint, though rather pleasant, echo of such huge events, comes from an Esthonian story about the Lake Eim changing his bed (Grimm, TM, p. 599):

> On his banks lived wild and wicked men, who never mowed the meadows that he watered, nor sowed the fields he fertilized, but robbed and murdered, so that his bright wave was befouled with the blood of the slain. And the lake mourned; and one evening he called his fish together, and mounted with them into the air. The brigands hearing a din cried: "The Eim has left his bed, let us collect his fish and hidden treasure." But the fish were gone, and nothing was found at the bottom but snakes, toads and salamanders, which came creeping out and lodged with the ruffian brood.

> But the Eim rose higher and higher, and swept like a white cloud through the air; said the hunters in the woods: "What is this murky weather passing over us?" and the herdsmen: "What white swan is flying in the sky?" All night he hung among the stars, at morn the reapers spied him, how that he was sinking, and the white swan became as a white ship, and the ship as a dark drifting cloud. And out of the waters came a voice: "Get thee hence with thy harvest, I come to dwell with thee." Then they bade him welcome, if he would bedew their fields and meadows, and he sank down and stretched himself in his new couch. They set his bed in order, built dikes, and planted young trees around to cool his face. Their fields he made fertile, their meadows green; and they danced around him, so that old men grew young for joy.

In a note, Grimm quotes F. Thiersch's opinion on this lake:

> Must not *Eim* be the same as *Embach* (mother-beck, fr. emma mother . . .) near Dorpat, whose origin is reported as follows? When God had created heaven and earth, he wished to bestow on the beasts a king, to keep them in order, and commanded them to dig for his reception a deep broad beck, on whose banks he might walk; the earth dug out of it was to make a hill for the king to live on. All the beasts set to work, the hare measured the land, the fox's brush tailing

after him marked the course of the stream; when they had finished hollowing out the bed, God poured water into it out of his golden bowl.

How tough the life of tradition is! And how obvious—here, we mean it—that more is meant than the changing of the bed of a river or a lake; that rivers have their own method of establishing a new course, instead of flying, fish included, in the air and hanging among stars, is a fact that, we trust, was not unknown to our ancestors, whether Esthonians or not.

Appendix 22

A survival, vague as it is, and evidently mistaking a chariot for a wain, we find in India. The *Sūrya-Siddhānta* states: "In Taurus, the 17th degree, a planet of which the latitude is a little more than two degrees, south, will split the *wain of Rohini.*"[1]

According to Burgess (p. 214), Rohini's (= Aldebaran) wain "contains five stars, in the grouping of which Hindu fancy has seen the figure of a wain," i.e., the Hyades, containing epsilon, delta, gamma, nu, alpha Tauri. Burgess continues (p. 249): "The *Siddhanta* does not inform us what would be the consequences of such an occurrence; that belongs rather to the domain of astrology than of astronomy. We cite from the *Pancatantra* (vv. 238–241) the following description of these consequences, derived from the astrological writings of Varahamihira:

'When Saturn splits the wain of Rohini here in the world, the Mādhava rains not upon the earth for 12 years.

When the wain of Prajāpati's asterism is split, the earth, having as it were committed a sin, performs, in a manner, her surface being strewn with ashes and bones, the Kāpālika penance.

If Saturn, Mars, or the descending node splits the wain of Rohini, why need I say that, in a sea of misfortune, destruction befalls the world?

When the moon is stationed in the midst of Rohini's wain, the men wander recklessly about, deprived of shelter, eating the cooked flesh of children, drinking water from vessels burnt by the sun.'

By what conception this curious feature of the ancient Hindu astrology is founded, we are entirely ignorant."

[1] *Sūrya-Siddhānta*, trans. by E. Burgess (1860; repr. 1935), 8.13, pp. 248ff.

The bad experiences which Saturn had with Auriga's vehicle—whether beta zeta Tauri, or the Hyades—seem to have left a trace in the memory of Indian astrologers.

Appendix 23

See J. Kepler, "De Stella Nova in Pede Serpentarii et qui sub ejus exortum de novo iniit Trigono Igneo," in *Opera Omnia*, ed. C. Frisch (1859), vol. 2, p. 636. See also J. Kepler, "De vero anno quo Aeternus Dei Filius humanam naturam . . . assumsit," in *Opera Omnia* (1863), vol. 4, pp. 346ff.

Kepler was less interested in the revolution of one angle of the trigon through the whole zodiac than in the span of time which the conjunctions needed to pass through all four "elements," particularly between conjunctions in the "fiery triplicity." The zodiac is divided into four "elementary" trigons or triplicities in the following manner:

Fire: Aries, Leo, Sagittarius
Earth: Taurus, Virgo, Capricornus
Air: Gemini, Libra, Aquarius
Water: Cancer, Scorpio, Pisces

The "great conjunction" of Saturn and Jupiter, occurring every twenty years, remains about 200 years within one triplicity; it moves through all four "elements" in 800 years (more exactly: in 794⅓ years). By means of the average of 800 years which it took the conjunction to pass from one "fiery triplicity" to the other, Kepler reconstructed history:

4000 B.C.	Adam	Creatio mundi
3200	Enoch	Latrocinia, urbes, artes, tyrannis
2400	Noah	Diluvium
1600	Moses	Exitus es Aegypto. Lex
800	Isaiah	Aera Graecorum, Babyloniorum, Romanorum
0	Christ	Monarchia Romana. Reformatio orbis
800 A.D.	Carolus Magnus	Imperium Occidentis et Saracenorum
1600	Rudolphus II	Vita, facta et vota nostra, qui haec disserimus

As concerns the—faraway—2400 A.D., Kepler remarks: "Ubi tunc nos et modo florentissima nostra Germania? Et quinam successores nostri? An et memores nostri erunt? Siquidem mundus duraverit." ("Florentissima Germania": this was written before the Thirty Years' War started.)

Compare H. H. Kritzinger (*Der Stern der Weisen* [1911], pp. 35, 44, 59), who deals broadly with the significance of "great conjunctions," and who adds: "The same table was repeated, with more precise data, by Riccioli in his *Almagestum Novum* (Tom. 1, 672–75), beginning with the verses:

> Ignea Triplicitas, coniunctio Maxima dicta
> Saturniq. Jouisque, annis redit Octingentis."

What is called here "great conjunction"—occurring every twenty years—has been styled in earlier times, i.e., in Sasanian and Arabian astrology, "small conjunction," as we learn from E. S. Kennedy:[1]

> After about 12 such small conjunctions the next conjunction will pull forward into the next triplicity. This event, called the shift or transit (intiqâl al-mamarr) is also known as the *middle conjunction* . . . Four middle conjunctions carry the phenomenon through all the triplicities and make up a *big conjunction*. But in order that the entire cycle recommence from a particular initial sign, taken as Capricorn, three big conjunctions are required, these making up a *mighty conjunction*.

A "mighty conjunction" thus corresponds to the revolution of one angle or corner of the trigon of Jupiter-Saturn conjunctions—built up in sixty years (more correctly: 59.6 years)—through the whole zodiac, completed in 2400 years (2383 years, respectively).

For one particular reason why the "big conjunction" of 800 years should be multiplied by 3, see Oscar Marcel Hinze's article:[2] within the frame of archaic "Gestalt-Astronomie," it was the revolution of the trigon as a whole that "counted." (Hinze deals also with the hexagon, i.e., the "Gestalt" of Mercury—revolution of one corner about twenty years—and with the famous "Pentagramma Veneris.")

As concerns the role of Saturn-Jupiter conjunctions in Iran and India, cf. also D. Pingree ("Astronomy and Astrology in India and Iran,"

---

[1] "The Sasanian Astronomical Handbook Zîj-i Shâh, and the Astrological Doctrine of 'Transit' (Mamarr)," JAOS 78 (1958), p. 259.

[2] "Studien zum Verständnis der archaischen Astronomie," in *Symbolon, Jahrbuch für Symbolforschung* 5 (1966), pp. 162–219, esp. pp. 203ff.

ISIS *54* [1963], p. 244), and the forthcoming paper by B. L. van der Waerden on "the conjunction of 3102 B.C."—this very conjunction introduces the flood of the *Mahabharata*. Allegedly, there is no trace of big conjunctions in Hindu and Hellenistic astrology. Astrology, however, is not found in texts only which are recognizable as such at first glance. Apart from Greece, where we have—besides the wrestling of Kronos and Saturn at Olympia—also the *Daidalia*, held in the interval of sixty years—sixty-year cycles in India, or in the West Sudan, are not likely to be understood, if the scholars prefer to inhibit the trigon of the Saturn-Jupiter conjunction; this inhibition being the logical outcome of the persistent refusal to recognize Saturn and Jupiter as Saturn and Jupiter.

The decisive conjunction of 6 B.C. (that "opened" our age of Pisces) having been near zeta Piscium, it is slightly surprising to learn from Burgess (*Sūrya-Siddhānta*, p. 14) the following—he explains the Indian notion of *nutation* (also called libration): "The vernal equinox librates westwards and eastwards from the fixed point, near zeta Piscium, assumed as the commencement of the sidereal sphere," the "libration" moving in eastern and western directions for twenty-seven degrees from this fixed point. On p. 230 he states about zeta Piscium that "it coincided in longitude with the vernal equinox in the year 572 of our era."

Appendix 24

Eduard Stucken (*Astralmythen*, pp. 190ff.) and, later, F. W. Albright (JAOS *40*, pp. 329f.) drew attention to the very same method employed when Rishyasringa, son of Vibhandaka (son of Kashyapa) and a hind, was lured by a courtesan, ordered by King Lomapada, into the latter's town, because only with Rishyasringa present would the country have rain. (Compare H. Lüders, "Die Sage von Rsyasrnga," in *Philologica Indica* [1940], pp. 1–42; also Lüders, "Zur Sage von Rsyasrnga," *Philologica Indica*, pp. 43–73.)

The major difference between GE and the story told in the *Mahabharata* 3.110–13 (Roy trans., vol. 2, pp. 242–48) is that Father Vibhandaka is the one "whose body was covered with hair down to the tip of the nails . . . and whose life was pure and was passed in religious meditation"; seduced is the son, not a hairy one, apparently, but "there was a horn on the head of that magnanimous saint." "Saints" they were both —those Indians of "high and far-off times" were in the habit of building up *tapas*, "ascetic heat," an instrument of the utmost cosmic "efficiency," if we may style it thus.

Appendix 25

It is not yet securely established what the word *sippu* means. (See W. Baumgartner, "Untersuchungen zu den akkadischen Bauausdrücken," ZA *36* [1925], pp. 27, 63; A. Schott, "Zu meiner Übersetzung des Gilgamesh-Epos," ZA *42* [1934], pp. 105f.) For the style of this battle, characterized by Cyrus Gordon as "Beltwrestling" (JNES 7, p. 264), see A. Oppenheim, Or. *17*, pp. 29f. "They seized each other (by their girdles), like experts/ they wrestled./ They destroyed the doorpost/ The wall shook." See also E. A. Speiser, "Akkadian Myths and Epics," ANET, p. 78. This "doorpost" is no *quantité négligeable*, because some similar "object" comes our way again at the "entrance" of the Cedar Forest, and does the most devilish things to poor Enkidu. (Compare J. Friedrich, "Die hethitischen Bruchstücke des Gilgamesh-Epos," ZA *39* [1929–30], pp. 48f., dealing with the Hittite fragments; he established at least that it was not the bolt.) In fact, were we to have started from GE, instead of paying it a casual visit, the several "doors" with their "posts," or "pillars," with their "fillings" and "thresholds" would have had as much of a paralyzing effect upon us as the eye of Medusa. Meanwhile, detrimental translations are quite enough to turn the reader to stone.

Appendix 26

See P. Gössmann, *Planetarium Babylonicum* (1950), 99: "ᶦᶦDapinu, 'the prevalent, the strong,' surname of *Nusku* (passim), of *Nabu*, of *Marduk* . . . As star-god ᶦᶦDapinu is the Marduk-star Jupiter, identical with ᵈSUL. PA.E₃ . . . , ᵐᵘᶦUD.AL.TAR . . . Since UD.AL.TAR can

also mean the fixed star *Procyon, also* [il]Dapinu should have this significance (Jensen, "der Furchtbare, Gewaltige (= Humbaba)," ZDMG
67, S. 517)." (For the identification of Nusku with Mercury, see H. and
J. Lewy, "The God Nusku," Or. *17* [1948], pp. 146–59.) See also
Gössmann, 137 s.v. [mul]UD.AL.TAR: "I. Akkadian as much as umu
dapinu . . . the full name of Jupiter, II. Procyon. Procyon seems to have
been counted with Jupiter's hypsoma, Cancer." See also E. Weidner,
Handbuch der Babylonischen Astronomie (1915), p. 25. (For Procyon
as part of Cancer, see RLA *3*, p. 77; for al. lu$_5$, representing sometimes
the zodiacal sign Cancer, otherwise Procyon, see B. van der Waerden,
"The Thirty-Six Stars," JNES *8* [1949], p. 21.)

Langdon (*Semitic Mythology* [1931], p. 268) mentions the identification Humbaba = Procyon, without giving the source, and without paying heed to such notion.

As concerns Humba with the determinant [mul](Babylonian [kakkab], respectively), Weidner (RLA *2*, p. 389) informs us of the existence of
two lists dealing with "7 astralen Enlil-Gottheiten." List 1 states—we
give it according to Weidner, since it is not essential, right here, to establish whether or not his identifications are right throughout: "Perseus
is the Enlil of Nippur, g Ursae Majoris is the Enlil of Enamtilla, alpha
Cassiopeiae is the Enlil of Hursag-kalama, Columba is the Enlil of Kul-
lab, Taurus is the Enlil of Aratta, [k]Humba (=?) is the Enlil of Šuba
(?)-Elam, Arcturus is the Enlil of Babylon." List 2 omits [mul]Humba
(compare also Weidner, *Handbuch*, pp. 58–60). Gössmann 188 states,
pointing to F. Boll–C. Bezold (*Antike Beobachtungen farbiger Sterne*
[1916], p. 121), that, according to VAT 0418 III 3, "[mul]HUMBA replaces [mul]APIN." The latter, the "plow constellation," is triangulum and
gamma Andromedae (see van der Waerden, JNES *8*, p. 13).

Now it is of considerable interest to learn from Hüsing (*Die einheimischen Quellen zur Geschichte Elams* [1916], pp. 11, 95) that "the
highest god of Elam . . . Humban (Hanubani, Hamban—Umman,
Imbi)" is (supposedly) the same as Hanuman, the monkey-god, the
crafty adviser of Rama (Hüsing also takes Humban for a monkey);
and from Charles Dupuis (*Origine de tous les cultes et toutes les religions* [1795], vol. *3*, p. 363) the following: "Dans l'explication
des Fables Indiennes, nous avons toujours trouvé que *Procyon* étoit le
fameux singe *Hanuman*. Il fixe le lever du Sagittaire, avec lequel le
singe est en aspect (Kircher: Oedipus 2 II, p. 201)."

Considering that Procyon has been counted among the stars of Cancer, a constellation which had the name Nangar = Carpenter, the

Twelfth Tablet of GE, of pure Sumerian origin, might gain a completely new significance. Gilgamesh does, there, a lot of "wailing" and "lamenting" about some objects that he left (or failed to leave) there, where they might have been in safety, in "the house of the Carpenter," *nangar*. Apart from Procyon, the fixed representative of Jupiter and Mercury, once Humbaba is purged from his "ogrish" reputation, the time will have come to approach Kombabos and his doubles in Iranian and Indian mythology.[1] The story of young Kombabos, who castrated himself as a precaution when he was appointed the traveling companion of "Caesar's wife," has been hitherto incompatible with the "monster" of the cedar forest, although the scholars agree that the names Humbaba and Kombabos are identical. It would be worth investigating whether or not the proposed equation Humbaba = Mercury might also fit Kombabos. F. K. Movers, however, was inclined to take Kombabos for Saturn.[2]

Appendix 27

See A. Oppenheim, "Mesopotamian Mythology," *Or. 17* (1948), p. 40: "After Enkidu tossed towards her . . . what is euphemistically termed the 'right thigh' of the bull, the goddess and her devotees performed age-old rites over the part of the bull."

True as this statement certainly is, it does not explain much—nor is it even asked why it must be the right thigh (*imittu;* compare H. Holma, *Die Namen der Körperteile im Assyrisch-Babylonischen* [1911], pp. 131f. See for the "euphemism" Holma, pp. 96f.).

The consensus of the experts, in overlooking that the GE talks explicitly of the celestial bull, keeps them from asking relevant questions, and their conviction that Mesopotamians and Egyptians had not much in common prevents them from recognizing the "bull's thigh" when

---

[1] Lucian, "De Dea Syria," in *Lucian,* trans. by A. M. Harmon, vol. *4,* cols. 19–27, LCL. Lucian claims that Kombabos was the prototype of the *galloi,* i.e., that after his example the priests of the Great Goddess castrated themselves and put on female garments. See also F. Liebrecht, *Des Gervasius von Tilbury Otia Imperialia* (1856), pp. 216f.; Ganschinietz, in RE *11,* cols. 1132–39; E. Benveniste, "La Légende de Kombabos," in *Mélanges Syriens offerts à René Dussaud* (1939), pp. 249–58.

[2] *Die Phönizier* (1841/1967), vol. *1,* pp. 154, 306–09, 686–89.

they see it. Yet it is there: Maskheti, the thigh of the bull, Ursa Major, depicted on the astronomical ceilings in the tombs of Senmut, Seti, in the Ramesseum, etc. In Altaic mythology, Ursa turns into the leg of a stag; in Mexico we find it as the lost "foot" of Tezcatlipoca.

The constellations are named according to a system, and if we meet "incomplete" or mutilated characters among them, we have to ask for the sufficient reason, e.g., why the ship *Argo* is a stern only, why Pegasus is barely half a horse—apart from its standing on its head and having wings—and why Taurus is the head and first third of a bull, his "thigh" turning around in the circumpolar region. Thus, it might be something to think about that in the Round Zodiac of Dendera (Roman period), the circumpolar "thigh" shows a ram sitting on it, looking back, moreover, as befits the zodiacal Aries (see F. J. Lauth, *Zodiaques de Denderah* [1865], p. 44). G. A. Wainwright, in "A Pair of Constellations," *Studies presented to F. L. Griffith* (1932), p. 373, with reference to Bénédite, mentions a thigh with the head of a ram from Edfu, called the "Foreleg of Khnum" (cf. *Monumenti dell'Egitto e della Nubia*, Ippolito Rosellini, ed. [1844], vol. 3, plate 24).

Appendix 28

In the GE Enkidu appears later on the stage of events than Gilgamesh. This does not entitle us to take him for the prototype of the "younger brother" (see, e.g., W. Albright, "Gilgamesh and Engidu," JAOS *40* [1920], pp. 312, 318). Actually, the hairy partner of the Twins, the "Dog," is the prototype of the older one who is cheated out of his primogeniture in various ways. Esau, the hairy, is the first born; so is Hono-susori no Mikoto (*Nihongi*, trans. by W. G. Aston [repr. 1960], pp. 92–108; K. Florenz, *Die historischen Quellen der Shinto-Religion* [1919], pp. 204–21) who, together with his offspring, after having been passed by the Japanese "Jacob," had to serve as "dogs," as clowns, playactors, guardians of the imperial palace for eighty generations; at New Year and during coronation ceremonies these Hayahito had to bark three times.

Particularly obvious is the case in Egypt, where we learn from H. Kees (*Der Götterglaube im Alten Ägypten* [1956], p. 193, n. 3): "wtw means 'jackal' and 'the eldest,' " and it happens that Kees made this remark when dealing with a classical case of cheating: when Geb/Kronos declared Horus the eldest, cutting out Seth/Typhon completely, as reported in the Shabaka Inscription. Actually Geb claims Horus to be Upuaut, the Opener of the Way—Upuaut being the Upper Egyptian Jackal or Wolf. The complex of the "Dog-Twin" is, however, of such a size and weight that it cannot be attacked here.

A particularly relevant and revealing case of inseparable "twins" comes our way in Cherokee mythology, where the thunder-boys are called "Little Men." At the beginning we hear of one boy only, born in proper wedlock.by "The Lucky Hunter" and "Corn," but soon the boy "finds" his "Elder Brother" in the river, and the latter has the name "He-who-grew-up-wild." These two arrange the world and human life as it is now, model cases of what ethnologists call "heroes of culture." Gilgamesh and Enkidu all over, they were asked to give "verdicts," alias oracles, after they had finally left the "earth."[1]

Appendix 29

We might call it Lethe, and feel happy about it, were it not for the deplorable uncertainty of Lethe's localization, with respect to the celestial itinerary of the soul particularly. The Milky Way being as large as it is, it does not help to state that one has to look for a galactical section. Worse, it remains unclear at which occasion the souls were supposed to drink from the water of this river of forgetfulness, whether they did so shortly after having arrived in Hades or before their reincarnation, or at both times. Although the supposition of an intake of Lethe right at the entrance of Hades would deprive the underworld jurisdiction, together with the good or bad recompenses for former conduct, of its significance, both views were upheld. (See Stoll, in Roscher s.v. Lethe, col. 1957; O. Gruppe, *Griechische Mythologie und Religionsgeschichte* [1906], pp. 403–405, 1036–41. On p. 760, n. 8, Gruppe quotes a passage, according to which a soul which has not yet crossed the river Lethe comes back to molest the living.)

Our most competent witnesses for Orphic-Pythagorean tradition take Lethe for the last "station" before rebirth, e.g., Plato in the myth of Er

[1] J. Mooney, "Myths of the Cherokee," 19th ARBAE 1897–98 (1900), pp. 243–50.

(*Republic 10.620*), and Virgil in the sixth book of the *Aeneid* (748–51), but only Macrobius (*Commentary on the Dream of Scipio*, trans. by W. Stahl [1952], *1.12.8*) pretends to know the source of the drink: the constellation Crater, the "bowl of Bacchus." This does not make sense[1] but, anyhow, he makes the souls descend through the northern intersection of Galaxy and zodiac, taking the southern crossroads, between Scorpius and Sagittarius for the entrance, which fits the "Hades-constellations" of the Sphaera barbarica. Yet we have observed, in other parts of our globe (see pp. 242f.), some uncertainty concerning entrance and exit: the Nicaraguan "Mother Scorpion at the end of the Milky Way" receives the souls of the dead, and takes care of the babies going to be reborn, whereas the Cherokee appear to assume the entrance at the "Northern End" of the Milky Way (Gemini-Taurus), from where the souls migrate to the "Spirit-Star" in Scorpius. We are not informed precisely whether the souls follow the Milky Way for a whole half-circle, either turning to the north or to the south, or whether they go first in one direction and return later on the same way. The latter seems to be expressed in the *Vishnu Purana* which restricts the "Way of the Fathers" to the region on the north of Canopus, and south of three lunar mansions in Sagittarius and Scorpius; the "Road of the gods" (*devayana*) runs north of three lunar stations in Taurus and Aries, and south of the Seven Rishis, the Big Dipper. *Vishnu Purana 2.8* (Wilson trans. [1961], p. 186) reads:

> On the north of Agastya, and south of the line of the goat [Ayavithi, i.e., the said three nakshatras in Scorpius and Sagittarius] lies the road of the Pitris. There dwell the great Rishis, the offerers of oblation with fire, reverencing the Vedas, after whose injunctions creation

[1] Macrobius' "uranography" is most embarrassing. He claims that "so long as the souls heading downwards still remain in Cancer they are considered in the company of the gods, since in that position they have not yet left the Milky Way. But when in their descent they have reached Leo, they enter upon the first stages of their further condition . . . The soul, descending from the place where the Zodiac and the Milky Way intersect, is protracted in its downward course from a sphere, which is the only divine form, into a cone . . ." We have remarked already (p. 242) that Macrobius, in calling the "Gate of Cancer" the crossroads of Galaxy and zodiac, talks of *signs*, not of constellations. And so he does, when pinning down the "bowl of Bacchus"—Crater—"in the region between Cancer and Leo": Crater is "between" Leo and Virgo, i.e., south of these constellations. How the souls, coming "down" from those crossroads of Galaxy and ecliptic, i.e., between Taurus and Gemini, should get hold of Lethe in Crater, south of Leo and Virgo, remains a mystery. Macrobius was, apparently, not in the habit of looking at the sky, and in this respect he was a very modern character.

commenced, and who were discharging the duties of ministrant priests: for as the worlds are destroyed and renewed, they institute new rules of conduct, and reestablish the interrupted ritual of the Vedas. Mutually descending from each other, progenitor springing from descendent, and descendent from progenitor, in the alternating succession of births, they repeatedly appear in different houses and races along with their posterity, devout practices and instituted observances, residing to the south of the solar orb, as long as the moon and the stars endure.

In a similar direction might point the report given by Pausanias about the oracle of Trophonios in a deep cave (9.39.8): the visitor comes first to "*fountains of water very near to each other*.[2] Here he must drink water called the water of forgetfulness (*Lēthēs hydōr*), that he may forget all that he has been thinking of hitherto, and afterward he drinks of another water, the water of memory (*hydōr mnēmosynēs*) which causes him to remember what he sees after his descent." Not enough, after the oracle has been given, and the inquirer ascended from the chasm (9.39.13), "he is again taken in hand by the priests, who set him upon a chair called the *chair of memory* (epi thronon mn.) and they ask of him, when seated there, all he has seen or learned. After gaining this information they then entrust him to his relatives. These lift him, paralyzed with terror and unconscious both of himself and of his surroundings . . . Afterwards, however, he will recover all his faculties, and the power to laugh will return to him."[3]

Nor does this "chair of memory" remain without its partner: Apollodorus (*Epit. 1.24*) tells us of the "Chair of Forgetfulness," to which Theseus and Pirithous "grew and were held fast by coils of serpents." That we learn also of "houses" of Lethe (Plutarch, *Consolatio ad Appolonium*, ch. 15, 110E, quoting an unknown poet) does not make this quarter more lucid. On the Etruscan Bronze Liver of Piacenza, *letham*, the river, divides the lower—otherwise empty—side into approximately equal parts—the invisible southern arch of the Milky Way?

[2] So are the rivers of lust and mourning (Hēdonē and Lypē) of Theopompus (Book 8 of his *Philippika*) which have been compared to our rivers by E. Rohde ("Zum griechischen Roman," Rh. Mus. *48* [1893], pp. 123f.). In Polynesia we meet near together the "water of life" and the "water of death" (see R. Williamson, *Religious and Cosmic Beliefs of Central Polynesia* [1924], vol. *1*, pp. 334, 344; vol. *2*, pp. 169f.).

[3] Of considerable interest are several terrestrial rivers called Lethe, mentioned by Gruppe (*Griechische Mythologie* [1906], p. 817): they are flowing at the foot of several "White Rocks" (*Leuketēs skopelos*), one among which has the name *agelastos petrē*, the laughterless rock.

Considering this state of confusion and uncertainty, we abstain from calling it rightaway either the drink of forgetfulness or the drink of memory, although one or both of them could very well be found upon the shelves of Ishara tamtim, alias Mother Scorpion.

Appendix 30

See P. F. Gössmann, *Planetarium Babylonicum* (1950), 94: "^{mul}GIR$_2$-TAB ^{d}Ishara tam-tim. Anton Deimel (*Pantheon Babylonicum* [1914], pp. 148f.) takes ^{mul}GIR.TAB for beta delta alpha Scorpii only: 'Ishara est dea quaedam partus, quae relationem habet ad Gestin anna, Adad.' " See also W. J. Hinke, *A New Boundary Stone of Nebuchadnezzar I from Nippur* (1907), pp. 223, 243; A. Jeremias, HAOG (1929), pp. 223, 385; F. Hommel, *Ethnologie und Geographie des Alten Orients* (1926), pp. 563, 770–74, 783; and D. O. Edzard, "Die Mythologie der Sumerer und Akkader," in *Wörterbuch der Mythologie*, vol. *1*, p. 9.

We might be accused of a clumsy contradiction because of having claimed Sirius to be the "Sea-Star" in appendix #2, when *here* it is evident that Ishara tamtim, the goddess of Scorpius, is entitled to this dignity. We are not only aware of this apparent "contradiction," but we also hope to unravel the mystery in the future. It is a mysterious scheme, but not a hopeless case. Clue number one is contained in the Coptic list of lunar mansions, already mentioned in appendix #4 (cf. A. Kircher, *Oedipus Aegyptiacus* [1653], vol. 2, pt. 2, p. 246), where it is stated with respect to the twentieth lunar mission, the sting of the scorpion (lambda upsilon Scorpii): "Aggia, Sancta, Arabice al-Sa'ula [i.e., "the sting"]; *statio translationis caniculae* in coelum, unde et *siot* vocatur . . . Longitudo huius stationis est a quarto Sagittarii usque ad decimum septimum eiusdem. Haec statio ab Aegyptiis quoque vocatur *soleka* sive *Astrokyon* . . . statio venationis." Eduard Stucken (*Der Ursprung des Alphabets und die Mondstationen* [1913], p. 7) identified this *soleka* immediately with Egyptian Selket/Serqet, the Mesopotamian Ishara tamtim, the Scorpion goddess. Whether or not this is permissible under the stern laws of linguists, it is a fact that we find regularly on the Egyptian astronomical ceilings Selket standing above, i.e., beyond, the bull's thigh (Big Dipper), which means that Selket represents the opposition to the perpetual center of attention: Sirius/Sothis. (Yes, we are aware of the circumstance that fourteen degrees is no ideal opposition to one star.) Clue number two are the stories spun around Indian *mura*, "the root" (or "tearer out of the root"), again lambda upsilon Scorpii—compare appendices #4 and #39—which have to be combined with the ocean of most atrocious yarns dealing with *Mandragora* (Alraun), the famous root that can be pulled out only by a *dog* that dies immediately after having completed this feat. Clue number three is carefully hidden away in the Mexican traditions concerned with the hunting festival Quecholli (*statio venationis*, and Quecholli is not to be separated from the "hunt" for *hikuli*, the peyote, as undertaken by Huichol and Tarahumare), which rehearses the great "fall" of the gods who had plucked the forbidden flowers, in Tamoanchan, "the house of descending."

Appendix 31

These unknown factors, crucial as they are, resist successfully every decoding for the time being. Šu-ut abnē, "those of stone," represent "an expression which recurs and has not been explained" (S. Langdon, *Semitic Mythology* [1931], pp. 213f., 405). Alexander Heidel (*The Gilgamesh Epic and Old Testament Parallels* [1963], p. 74, n. 157) remarks: "The Hittite Version has 'two images of stone.' These images may perhaps have been idols of an apotropaic character enabling Urshanabi to cross the waters of death." Speiser ("Akkadian Myths and Epics," ANET, p. 91, n. 173) makes it "apparently stone figures of unusual properties . . . "

According to Speiser (Assyrian version, Tabl. 10, col. 3, 37f., ANET, p. 92; cf. Heidel, p. 76) Urshanabi states: "Thy hands, Gilgamesh, have hindered [the crossing]: Thou hast broken the Stone Things . . . ," which can hardly be correct, since they do cross, after all.

F. M. Th. de Liagre Böhl, in his translation of GE, seems to have boldly claimed that the "stone objects" were "part of the fence of Siduri's yard," to which I. M. Diakonoff (Review article on the GE translations of F. M. Th. Böhl and P. L. Matous, *Bibliotheca Orientalis* *18* [1961], p. 65) remarked: "The sūt abnē cannot have any connection with Siduri's yard (indeed, no such yard is mentioned)."

Luckenbill (AJSL *38* [1922], pp. 96–102) seems to have voted for anchors (see *Gilgamesh et sa légende*, ed. by P. Garelli [1958], p. 17, item 146). Orally, three years ago, Florence Day proposed "load stones." For further keen propositions, see A. Salonen, *Die Wasserfahrzeuge in Babylonien* (1939), pp. 131f.

Some new light falls upon these objects through a Neo-Babylonian fragment published by D. J. Wiseman (*Gilgamesh et sa légende*, pp. 128–30), but the author himself states that the new reading (u šu-ut NA₄.MES) "appears at present to help little towards the understanding of this much discussed term. The restoration of parts of ll.35–41, now possible, shows that the end of this column describes the manner in which Gilgamesh met Ur-shanabi and obtained the boat and its equipment for his journey over the 'waters of death.'

> When Gilgamesh heard this,
> he took up the axe in his hand,
> drew the dagger from the belt,
> crept along and went down . . .
> Like a lance he fell among them . . .
> within the forest he sat down and . . .
> Ur-shanabi saw the flashing of the dagger,
> heard the axe and . . .
> Then he smote his head . . . Gilgamesh
> seized the wings . . . its breast
> and the sūtabnē . . . the boat . . ."

More annoying still, these stone-things are not the only vexing items to be found in the neighborhood of Urshanabi. Heidel simply drops them, and renders line 29 of the Assyrian version (Tabl. 10, col. 2, p. 74): "With him are the stone images (?), in the woods he picks . . . ," and accordingly he deals with column 3, 38f.: only the stone-things are mentioned. Speiser (ANET, p. 91) continues after the "Stone-Things": "In the woods he picks ['*urnu*'-snakes]." And column 3 he renders: "Thou hast broken the Stone Things, hast picked [the 'urnu'-snakes]. The Stone Things are broken, the 'urnu' is not [in the woods]."

In note 174 Speiser refers to Landsberger (*Die Fauna des Alten Meso-potamien* [1934], p. 63), who "points out that the urnu snake has long been supposed to be a favorite with sailors. At all events, whatever the meaning of the term may be in the present connection, its properties seem to be on a par with those of the Stone Things."

Now, let us first express our disapproval of Urshanabi's lack of "fair-ness," just in case this translation might be correct: Siduri states it as well known that "Urshanabi, with whom are the stone-things, picks urnu-snakes" in the woods, and here he accuses Gilgamesh of having done so, taking it, evidently, for an improper thing to do! In the second line, B. Landsberger (*Fauna*, p. 63; cf. pp. 45f., 52, 60) identified tenta-tively the "urnu-snake" (maybe also "the yellow (green) snake," muš. sig₇. sig₇) with the waran, and considers, since even today warans are eaten, that the urnu were collected in order to serve as roast meat for the sailors. He thinks it possible that in later times "urnu" was meant as "land-crocodile." If urnus belonged to the usual travel provisions, why should the picking of these animals be an impediment for the crossing of the waters of death? Although one should not criticize others, least of all scholars of the rank of Landsberger, if one has no positive propo-sitions to offer, reading through this learned work, it becomes less and less comprehensible how he could misapprehend these animals, particu-larly the snakes, for a veritable terrestrial fauna, these seven-headed, one-eyed, unicorned creatures belonging to Anu, Nergal, Ningishzida, etc.

Appendix 32

Considering that removed posts or pegs, pulled-out pins, wrecked axles, and felled trees have accompanied this whole investigation as a kind of *basso ostinato*, we cannot pass in silence over these superimpor-tant posts; considering, on the other hand, that technical details are not likely to make pleasant reading, we prefer to deal with this specimen outside the main text, although we deem it essential indeed.

The object that Irragal is tearing out is called *tarkullu*, Sumerian DIM.GAL, which has been translated into "(Anchor-)post," "ship's mast," "mooring-post" (Heidel), also "anchor" itself, and even "steering-oar" (Jensen).[1] In the Era Epic, Era (=Irragal=Nergal), when announcing a new catastrophe, threatens that he is going to tear out the tarkullu, that he will make the ship drift off, break the steering oar so that the ship cannot land, and remove the mast and all that belongs to it.[2]

We meet the word also in names given to temples, as we learn from Burrows,[3] who considers "the evidence for the relation of the temples to (1) heaven, (2) earth, (3) underworld," and tells us what follows:

(1) The idea of the Bond of Heaven and Earth is given explicitly. Dur-an-ki, was the name of sanctuaries at Nippur, at Larsa, and probably at Sippar. Also in Semitic markas šamē u irsiti, Bond of Heaven and Earth, is used of the temple E-hursag-kur-kur-ra and of Babylon.

(2) Idea of Bond of the Land. Probably by extension of religious use the royal palace of Babylon is called *markas* (bond) of the Land. An ancient Sumerian temple-name, which probably expresses an analogous idea, is *"dimgal* of the Land." This was the name of the temple of Der, an old Sumerian center beyond the Tigris; a name given to Gudea's temple at Lagash; a temple of Šauška of Niniveh; and probably the temple of Nippur was another *"dimgal* of the Land." The pronunciation and meaning of *dimgal* are disputed. "Great binding-post" is perhaps a fair translation. The religious terms *"dimgal* of the Land" and the like perhaps indicate the temple as a kind of towering landmark which was a center of unity by its height.

(3) Idea of the bond with the underworld. Gudea uses *dimgal* also with reference to the *abzu*, i.e., the waters of the underworld: he laid two temens, ritual foundations—the temen "above" or "of heaven" and the temen "of the *abzu*," and the latter is called "great *dimgal*." The idea may be that the temple is as it were a lofty column, stretching up to heaven and down to the underworld—the vertical bond of the world. The same passage mentions, it seems, a place cf libation

[1] See P. Jensen, *Die Kosmologie der Babylonier* (1890), pp. 377, 422f.; K. Tallqvist, *Akkadische Götterepitheta* (1934), p. 244 (see also p. 283; Dim gul-an-na "Himmelspfahl" = Ninurta, and Dim gulkalam-ma "Weltpfahl" = Ninurta). See C. Bezold, *Babylonisch-Assyrisches Glossar* (1926), p. 296: "Pfahl, Prügel, Schiffspfahl, Mast"; A. Salonen, *Nautica Babyloniaca* (1942), p. 85; "(Anker)pfahl." On p. 104 Salonen explains *tarkulla* as "the mast," and it is the mast of Ea's ship: "sein (des Ea-Schiffs) Mast ist in der Schiffsmitte aufgestellt, schwebt am Himmelsband." See also R. Labat, *Manuel d'Epigraphie Akkadienne* (1963), no. 94, p. 81: DIM *riksu*, lien; dimmu, colonne; DIM-GAL *tarkullu*, mât; no. 122a, p. 93: DIM GUL *tarkullu*, mât. Cf. B. Meissner, *Beitrage zum Assyrischen Wörterbuch* I (1932), pp. 58f., and A. Schott, *Das Gilgamesch-Epos* (1958), p. 90, n. 19: "Das Weltenruder?"

[2] For the explanation of the several termini, see P. F. Gössmann, *Das Era-Epos* (1956), p. 55; see also Ebeling, AOTAT, p. 227.

[3] Eric Burrows, S.J., "Some cosmological patterns in Babylonian religion," in *The Labyrinth*, ed. by S. H. Hooke (1935), pp. 46ff. (That we do not share the author's too-simple opinions goes without saying.)

to the god of the underworld. Drains or pipes apparently destined for libations to the underworld have been discovered at Ur. Thus, if these interpretations are right, the temples expressed not only, in their height, the idea of the bond with heaven but also, in their depth, that of union with the netherworld.

Were we to hear less of "towering landmarks" and "lofty columns," for the sake of being presented with one single thought dedicated to the fact that these alleged "temples" and "columns" were torn out in order to start a deluge, we would be better off. Much more astonishing, however, is the circumstance that nobody seems to have taken the trouble of looking for relevant enlightenment in Egypt, i.e., of dealing with the Egyptian *mnj.t*.

According to Erman-Grapow (*Wörterbuch der Aegyptischen Sprache* [1957], vol. 2, pp. 72ff.) the word is used as (1) symbolical expression for the king (als Lenker des Staatsschiffes); (2) symbolical expression for Isis and Nephthys who fetched Osiris from the water. It is a constellation, the instrument for impaling, the post to which a person to be punished is bound. The transitive verb (*mnj*) means to bind to a post, to tether (anpflocken); the intransitive verb means to land, from persons, and from ships, and to die, sometimes supplemented "at Osiris" (bei Osiris landen).

This *mnj.t wr.t*—Mercer writes it *min.t*—the "great landing stick,"[4] is said "to mourn" for the soul of the dead in the Pyramid Texts,[5] and Mercer comments[6] that "the great stake . . . is personified as a 'mourning woman' in reference here to Isis." The "mooring-post" being a constellation, as even the *Wörterbuch der Aegyptischen Sprache* has to admit, the question is where to look for this *mnj.t*. The constellation—transcribed *menat* by Brugsch,[7] *mnit* by Neugebauer[8]—occurs in two categories of astronomical monuments, namely (1) in the Ramesside

[4] See W. Max Müller, *Egyptian Mythology* (1918), p. 376, n. 79.
[5] *Pyramid Texts*, ed. by S. Mercer (1952), p. 794c: "The great min.t (-stake) mourns for thee"; cf. 876c, 884b ("the great min.t laments for thee, as for Osiris in his suffering"), and 2013b.
[6] *Pyramid Texts*, vol. 2, p. 399; see also p. 361. See pp. 371, 398 for *mini* "to pasture, to land (i.e., to die)," and for *min.w*, derived from *mini*, as an epithet of Anubis 793c: "he who is upon the *min.w*"). "The min.w here seems to indicate a cask for the limbs of Osiris."
[7] H. Brugsch (*Thesaurus Inscriptionum Aegyptiacorum* [1883–91; repr. 1968], pp. 122, 130, 188) takes it for a "knife" or "sword"; later (*Die Aegyptologie* [1891], p. 343) he spelled it "ship's peg" ("Schiffspflock" and "Doppelpflock").
[8] O. Neugebauer and R. Parker, *The Ramesside Star Clocks* (1964), p. 7.

Star Clocks,[9] and (2) in the ceiling pictures of royal tombs, in the zodiacs of Dendera, etc. In every case the peg or post rests in the hands of Isis disguised as a hippopotamus; fastened to the mooring-post is a rope or chain, to the other end of which is tied *Maskheti*, the bull's thigh, i.e., the Big Dipper, and in one of the texts it is stated (Brugsch, *Thesaurus*, p. 122) that "it is the office of Isis-Hippopotamus to guard this chain."

According to the Ramesside Star Clocks, *mnj.t* included six different parts,[10] and only after these six parts follow *rrt* "female hippopotamus," comprising eight positions. Boll (*Sphaera* [1903], p. 222) remarks that this constellation must be thought of as being parallel to either the equator or the zodiac, and as being rather "long," because otherwise it could not need more than four hours of ascending.

Most of the scholars dealing with the Egyptian astronomical ceilings took it for granted that the main scenery represented the northern circumpolar constellations, because the Big Dipper, Maskheti, holds the "determinant" position upon the stage, and they tried their hardest to identify Isis-Hippopotamus holding the mooring-post, and carrying upon her back a crocodile, with a constellation very near the Pole. Now, we do not mean to go into details of the Egyptian sphere as represented in these ceiling decorations, which is an extremely difficult task, and nothing has been gained in the past by the different efforts to settle the affair by simply looking at the sky (worse, at sky-maps) trying to imitate Zeus by "catasterizing" on one's own account, and giving keen verdicts. Let us say only this much: (1) as yet no single proposition concerning the Hippopotamus holding the mooring-post is satisfying;[11] (2) that the determinative group of the ceiling pictures

---

[9] Formerly they were called "Theban hour-tables" (*Thebanische Stundentafeln*, or *Thebanische Tafeln stündlicher Aufgänge*).

[10] Neugebauer and Parker, "The Ramesside Star Clocks," p. 7: (1) the "predecessor," or the "front of the mooring post," (2) "is not translatable," (3) "follower of the front of the mooring post," (4) "mooring post," (5) "follower of the mooring post," (6) "follower which comes after the mooring post."

[11] We hope for enlightenment to be contained in the third volume of Neugebauer's *Egyptian Astronomical Texts*. In vol. 2, p. 7 he states, with respect to the hour-stars: "To what extent, if at all, the constellations of the lion, the mooring post, the hippopotamus, and perhaps others, can be identified with similar figures in the *so-called 'northern'* constellations as depicted on many astronomical ceilings . . . is a problem into which we do not intend to enter until all the evidence can be presented in our final volume. That the problem is more complex than would appear at first glance—at least in so far as the two hippopotami are concerned— is sufficiently indicated by the fact that on the ceilings the hippopotamus is *never*

show decisive factors of the "frame": Leo, Scorpius, Taurus,[12] serving thus as a kind of "key" of the whole presentation.[13] But, if our "frame" is meant, i.e., the structure of colures, where is the southern celestial landscape? We do not dare to molest the reader with the impenetrable text (Brugsch, *Thesaurus*, p. 122), out of which we quoted only one sentence which states that Isis-Hippopotamus is guarding the chain; this much at least is recognizable, that this text jumps from the Big Dipper— via "the middle of the sky"—to positions "*South* of Sah-Orion."

And here Casanova[14] comes in quite handy with his proposition to understand *mnj.t* (he writes it *menat*) as *Menouthis*, the wife of Canopus, steersman of Menelaus, whom we know from late Greek texts (also written Eumenouthis). Epiphanius[15] talks of the tomb of both, i.e., Canopus and his wife, in Alexandria. Stephanus of Byzantium knows of a village "at Kanobos" which had the name Menouthis.[16] It would lead us too far to deal with Canopus-steersman-of-Menelaus, and the Canopic mouth

---

named *rrt, never* is shown with two feathers as a headdress, and very frequently has a crocodile on its back." (We are only too grateful for everybody who recognizes that the problems are "more complex"—a hundred times more complex, indeed—"than would appear." The underlining of "so-called 'northern'" is ours; that of the two "never's" is Neugebauer's.)

[12] That the Dipper is said to be the thigh of a bull indicates Taurus clearly enough; we have mentioned that there is also a "foreleg of Khnum" available, i.e., that of a ram, and that in Dendera a ram is sitting on the Ursa-Leg: whose leg it is depends from the constellation marking the vernal equinox. To the objection that the constellation as depicted in Egyptian pictures clearly shows the *hindleg of an ox*, we have to answer that the texts insist on talking about the bull's foreleg; in other words, the real resemblance does not count so much, apparently (cf. appendix #27).

[13] Even if we had no other evidence, the Ramesseum would be good enough, showing in the center, precisely below Maskheti, the baboon sitting upon the Djed-pillar—we know from Horapollo (*1.16*) that the squatting baboon indicates the equinoxes; whereas the third, lowest register shows the sitting dogs at both ends, and we know from Clemens Alexandrinus (*Strom. 5.7, 43.3*) that these represent the Tropics.

[14] P. Casanova, "De quelques Légendes astronomiques Arabes," BIFAO 2 (1902), p. 18.

[15] Quoted by P. E. Jablonski, *Pantheon Aegyptiorum* (1752), vol. 3, pp. 141f.

[16] Casanova, p. 153. Cf. H. Kees in RE s.v. Menuthis, cols. 968f., who also mentions a dedication to "Eisidi Pharia, Eisin tēn en Menouthi," and who points to a sanctuary of Menouthis famous as "sanatorium" and replaced, later, by a monastery. W. Max Müller, in his turn (*Egyptian Mythology* [1918], p. 397, n. 94), informs us thus: "In the Greek period the name *Menuthias* ('Island of the Nurse') was given to a mythical island in the South as being the abode of the divine nurse [of Horus], and later this was identified with Madagascar as the most remote island in the south, i.e., the lower world." Müller seems to take Menouthis for the

of the Nile: the modern *Homo occidentalis* is bound to shrink back from the mere idea that the Nile represented a circle, where "source" and "mouth" meet, so that there is nothing preposterous in the notion that a Canopic mouth can be found in the geographical North, and here it is not necessary to discuss the question. It is sufficiently striking to see the mooring-post "married" to Canopus in a similar manner as Ur-shanabi is "married" to Nanshe, Enki's daughter, to whom is consecrated the holy stern of the ship.

Admittedly, we know as little as before where precisely the *mnj.t* of the star clocks has to be looked for,[17] but we have at least made it more plausible that DIM.GAL/*tarkullu*/*mnj.t* must be the decisive plumb line connecting the inhabited world with the celestial South Pole or, let us say, with the orbis antarcticus: Osiris being depicted as a circle (see Brugsch, *Religion und Mythologie*, plate facing p. 216), the verb *mnj.t*, "to land (at Osiris)," points in this direction. (We recall once more Virgil's statement that the "shades infernal" and Styx see the South Pole.) It has not escaped our attention that GE *11.101* seems to talk of posts, in the plural: as, in some Egyptian texts, we have the "double mnj.t." We do not know yet why: the Era Epic uses the singular, but Era is going to pull out a different post from the one he had torn out previously in GE under his name Irragal. There are possible solutions, but we leave alone this question as well as the next difficult problem arising with the suspicious similarity of the ship's peg with the nose-bone of the Horus-Eye (numerical value 1/64), however tempting this problem is.

---

same as Thermouthis, the daughter of that Pharaoh who found Moses in the Nile (cf. Josephus, *Jewish Antiquities* 2.9.5–7, 224; Bk. Jub. XLVII.5: Tharmuth), without giving sources or reasons for doing so. We should very much like to know whether or not *mnj.t* is identical, or has something to do at all with "Menât or Heliopolis," whom Brugsch identified with Satit of Elephantine (of all deities!); it would be decisive to know it. (Cf. Brugsch, *Religion und Mythologie* [1891], p. 301; Brugsch, *Thesaurus* [1883–91; repr. 1968], p. 107.)

[17] Some years ago, a mathematician in Frankfurt, who had invested much computer time in the star clocks, felt sure that *mnj.t* must end in alpha Centauri. As concerns the astronomical ceilings, we have presumably to mind the manner in which the late zodiacs of Dendera and Esne (Roman time) "project" the Big Dipper/*Maskheti*, the bull's thigh (together with Isis-Hippopotamus and the chain) into the zodiac, namely, between Scorpius and Sagittarius (Esne), and between Sagittarius and Capricornus (Dendera). There is, moreover, a remarkable Arabian survival (R. Böker, quoting Chwolson [1859], in A. Schott's translation of Aratus, *Sternbilder und Wetterzeichen* [1958], p. 119) stating to Sagittarius degree 30: "To the right of the degree is *Meshkedai*, the moulder of divine images."

Appendix 33

The mere notion of the emperors sleeping makes it clear that they are expected to awake and to return one day;[1] be it Quetzalcouatl (in the heart of the sea), Ogygian Kronos himself, or Arthur, "ruler of the lower hemisphere," who announces in a fictitious letter "that he has come, with a host of antipodean subjects"[2]—according to Étienne de Rouen (c. 1169; see R. S. Loomis [ed.], *Arthurian Literature in the Middle Ages* [1959], p. 69); that Geoffrey of Viterbo placed Arthur straightaway into the depth of the sea has been mentioned on p. 299, n. 35.

Few scholars only, among them Franz Kampers and Robert Eisler, have recognized the awe-inspiring age of such traditions, and even they have been incapable of calling the much-expected "redeemer" and "kosmokrator" by his very own name: Saturn. Says Kampers, concerning the apocryphal Apocalypsis of Daniel:[3]

> Alexander wird hier . . . nicht mit seinem Namen genannt, sondern er wird als Johannes eingeführt. Nach all dem Gesagten wird es nicht mehr allzu kühn erscheinen, in diesem Namen Johannes eine prophetische Chiffre zu erkennen. Wenn Nimrod in einer altslawischen Sage auch Johannes heisst, wenn der erdichtete Erretterkönig der Kreuzfahrer, wie wir sehen werden, Johannes genannt [=Prester John] und auch in Beziehung gesetzt wird zu dem Weltenbaum, so dürfte die Annahme, dass hier fortlebende altorientalische *Oannes*-Erwartungen sich äussern, nicht von der Hand zu weisen sein.

And right here, he refers to Robert Eisler's chapter, "John-Oannes?" which states:[4]

> We should not hesitate even to presuppose that the same syncretism of John and Oannes, which seems so natural with Neo-Babylonian Gnostics [the Mandaeans are meant], existed also among the more immediate Jewish followers of the Baptist, seeing that an influence

[1] See for the rich theme of "heroes inside hills," J. Grimm, TM, pp. 951–62; Axel Olrik, *Ragnarök* (1922), pp. 353–62.

[2] This role is otherwise ascribed to Beli (or Bilis), brother of Bran, "the dwarf King of the Antipodes"—later he had the name Pelles. "In Welsh poetry the sea is referred to as Beli's liquor and the waves as Beli's cattle" (R. S. Loomis, *The Grail* [1963], pp. 110–12). "Elsewhere he is implored as 'victorious Beli . . . that will preserve the qualities of the honey-isle of Beli' " (McCulloch, in ERE 3, p. 290).

[3] F. Kampers, *Vom Werdegange der abendländischen Kaisermystik* (1924), p. 109; Kampers, *Alexander der Grosse und die Idee des Weltimperiums in Prophetie und Sage* (1901), pp. 145–48.

[4] *Orpheus the Fisher* (1921), pp. 151–62, esp. p. 153.

of the Babylonian belief in ever new incarnations of the primeval Oannes—Berossos knows as many as six such reincarnations in past times—on the Messianic hopes of the later Jews is far from credible. In ch. 12f. of IV Esra (temp. Domitian, 81–96 A.D.), the redeemer of the world, the celestial "Man" is expected to rise from the "heart of the Ocean" before his coming, as Daniel (7.13) says, with the clouds of the sky, for: "As no man can search or discover that which is in the depths of the Ocean, even so no mortal can see the Son of God nor his hosts except in the hours of His day."

Accordingly, we find in 4 Ezra XIII.3 (in E. Kautzsch [ed.], *Pseudo-epigraphen des Alten Testaments* [1900]) the sixth vision of the prophet: "Ich schaute, siehe da führte jener Sturm aus dem Herzen des Meeres etwas wie einen Mann hervor." In a note (p. 395) the Latin translation of the Syriac version is quoted: "Et vidi et ecce ipse ventus ascendere faciebat de corde maris tanquam similitudinis hominis."

We know well enough that the Oannes of Berossos is Ea, i.e., Saturn, whose "town" is Eridu/Canopus, the very depth of the sea. That Ogygian Kronos is unmistakably the planet Saturn is not to be overlooked by anyone who reads Plutarch's report (*De facie quae in orbe lunae apparet* 941) of the "servants" of Kronos who—every thirty years, when Saturn is standing in Taurus—sail to Ogygia to remain there in service for thirty years, after which they are free to go; but most of them prefer to stay, because there, in Saturn's island, the Golden Age lasts on and on. The servants spend their whole time on mathematics, philosophy, and the like, and there is no reason to worry about food, it is all conveniently at hand.

The reluctance at recognizing the almost uncanny power of the oldest traditions is a very modern invention. Kampers still knew very well that the "type" of the medieval emperor was coined in the most ancient Near East, Alexander being a "repetition" of Gilgamesh, and the emperor repeating Alexander again and again. (Cf. Kampers, *Vom Werde-gange*, pp. 21f., 35, and *passim*.)

Appendix 34

Actually, we are up against a completely incomprehensible narrative of events which occurred during a sea voyage. The plant, according to Albright (AJSL *36*, p. 281, n. 2) literally "thorny grapevine," is supposed to grow in the apsu, and to be accessible by way of a "water-pipe." This pipe, *rātu*, however, is a conjecture right here: the word occurs only later when, after his bath in a well, and the following loss of the plant, Gilgamesh complains bitterly about his frustration, i.e., about having obtained a boon for the "earth lion" instead of for himself. The "earth lion," identified with the thievish serpent, is assumed in its turn to live "in a well which communicated with the apsu" (Albright, AJSL *35*, p. 194). It is then (GE *11*.298) that the hero says: "When I opened the water-pipe and [. . .] the gear, I found that which has been placed as a sign for me: I shall withdraw and leave the boat on the shore" (Speiser trans., ANET, pp. 96f.). Heidel makes it: "When I opened the . . . I have found something that [has been s]et for a sign unto me; I will withdraw!" Instead of that "sign," Albright (RA *16*, pp. 175f.) recognized a flood rising out of the *pipe* (if so, why does Gilgamesh talk about it only after his bath in the well?): "When I opened the water-pipe, I overturned the cover (?). Let not the sea rise to my side, b[efo]re (it) let me retire"; and so did Ungnad-Gressmann (pp. 63f.) and Schmoekel (p. 111). From this passage the translators derive the occurrence of the word *rātu* in the earlier passage, where Gilgamesh dives for the plant. Speiser alone[1] refers to another occasion where the word is used, and it is a decisive occasion; namely, in the (wrongly called) "Eridu creation story" (v. 11), where it is told that before anything was created and when all lands were sea (tamtim), then "the spring which is in the sea was a water pipe; then Eridu was made, Esagila was built" (Heidel, *The Babylonian Genesis* [1963], pp. 61f.). Sayce (ERE *4*, p. 129) makes it a "current" within the sea; with Jensen (*Assyrisch-Babylonische Mythen und Epen* [1909], p. 41) it is a "Wasserbecken"; with Ebeling (AOTAT, pp. 130f.) a "Schoepf-rinne." Considering that Eridu is Canopus, and Esagila is "l-Iku"—the Pegasus-square between the two Fishes that ruled the hibernal solstice during the Age of Gemini—this particular *rātu* seems to have been the connection between the two depths of the sea, between Pisces as the depth of the salt sea and Canopus as the depth of the apsu, the sweet water ocean.

[1] ANET, p. 96, n. 232. The conclusions drawn from this footnote by N. K. Sandars in his rendering of the GE in the form of a "straightforward narrative" are, as is his whole undertaking, a willful misrepresentation of the truth, unless one accepts the whisking away of the 1001 stumbling blocks and obscurities and the fabrication of a "Gilgamesh made easy" for a praiseworthy progress (*Gilgamesh Epic* [1960]. pp. 53, 113).

Although it is probable that the conception of one or more such "pipes" is the same as the Jewish one of the "channels," *shithim*, that went down to the tehom and were dug by God during the creation, this is not the place to deal broadly with this plot. In any case, Gilgamesh opening one or the other *rātu* comes close to David, who, when digging such a channel, found the Eben Shetiyyah. The relevant (and revealing) material has been assembled by D. Feuchtwang in his article, "Das Wasseropfer und die damit verbundenen Zeremonien," *Monatsschrift für Geschichte und Wissenschaft des Judentums* 54 (1910), pp. 535–52, 713–29; 55 (1911), pp. 43–63.

Of remarkable interest are pieces of information dealt with by Langdon (MAR 5, pp. 227–29) coming from Nicander, and Aelianus (*De natura animalium 6.51*), who in his turn refers to Sophocles,[2] and several poets whose works are lost. Aelianus—to whom, by the way, we are indebted for the only mention of our hero's name in Greek literature (*De natura animalium 12.21*: Gilgamos)—when dealing with a particularly fiendish small snake called *Dipsas* (literally "thirst"), tells the following:

> It is said that Prometheus stole fire, and the myth goes that Zeus was angered and bestowed upon those who laid information of the theft a drug to ward off old age. So they took it, as I am informed, and placed it upon an ass. The ass proceeded with the load on its back; and it was summer time, and the ass came thirsting to a spring in its need for a drink. Now the snake which was guarding the spring tried to prevent it and force it back, and the ass in torment gave it as the price of the loving-cup the drug that it happened to be carrying. And so there was an exchange of gifts: the ass got his drink and the snake sloughed his old age, receiving in addition, so the story goes, the ass's thirst. [The Sophocles fragment says that since then, snakes slough their old skin every year, kath'hekaston eniauton.]

Nicander, as quoted by Langdon, supplements the story by telling us of the date when this "exchange of gifts" took place, namely, on the occasion of a new distribution of the "Three Ways," reporting "that when Cronus' eldest son became master of Heaven, he divided up in his wisdom glorious governments among his brethren, and gave youth as a reward to short-lived men; so honouring them, because they disclosed the thief of fire, fools that they were! for they got no gain from their evil counsel."

---

[2] Frg. 362 (Pearson ed.) = frg. 335 *Tragicorum Graecorum Fragmenta*, ed. A. Nauck (1964), pp. 209f., from Kōphoi Saturoi.

Appendix 35

There are a few dim and blurred signals to be received from the regions of Styx flowing, as we have heard, in sight of the celestial South Pole. Photius[1] tells us about Hyllos, son of Herakles, who had a small horn growing out of the left side of his head, and how Epopeus[2] of Sikyon broke off this horn, after having killed Hyllos in a duel, fetched Styx water with this horn, and became king of the country. Why should he have procured this much dreaded water, if it did not enable him to become king?

Allegedly "late" are the legends claiming that Thetis made the child Achilles invulnerable by means of Styx water—his heel excepted, as we know. On the other hand it was fabled that Alexander was killed with water from the Styx, as Pausanias, who remained skeptical, reported (see also p. 201, n. 8). Thus, both of them were brought in touch with Stygian water, the one almost at the right moment, but only almost, and the other at a completely wrong time, far away from that unknown day in the year, where this fluid was supposed to make the drinker immortal, whereas it brought inevitable death on every other day.

Appendix 36

For related conceptions in Rome, see Festus (128M, BT [1965], p. 115): "*Manalem fontem* dici pro eo, quod aqua ex eo semper manet . . . *Manalem lapidem* putabant esse ostium Orci, per quod animae inferorum ad superos manarent, qui dicuntur manes." (Cf. F. Bömer, "Der sogenannte Lapis Manalis," ARW *33* [1936], p. 281; Kroll, RE *16* s.v. mundus, cols. 561f. To prevent one-sided conceptions from stealing into the picture, see also Festus 156M, p. 147: "Manes di ab auguribus vocabantur, quod eos per omnia manere credebant, eosque deos *superos atque inferos* dicebant.)

To this one should compare the rich material offered by F. M. Corn-

---

[1] *Bibliothèque*, ed. R. Henry (1962), vol. *2*, p. 56.
[2] M. Riemschneider (*Augengott und Heilige Hochzeit* [1953], p. 59) interprets the name: "der Hinaufschauer, der Hinaufwürfler."

ford ("The Eleusinian Mysteries," in *Festschrift Ridgeway* [1913], pp.
160ff.) about Greek underground structures, "*phrear,* the equivalent of
the Latin puteus." And about the "Curtius-Lake," Lacus Curtius—representing a *mundus*—which was to be found, according to Dion. Hal.
2.42, en mesō tēs Romaiōn agoras, i.e., right in the middle of the Forum
(see also Festus 49M, p. 42). Cornford explains (p. 162, note):

> The legend of Curtius, whose self-devotion stopped a flood, and who
> was honoured with *dona ac fruges* thrown into his *lakkos,* may throw
> light on the custom at Athens of throwing wheatmeal kneaded with
> honey into the cleft in the ground at the precinct of Ge Olympia
> where the water ran away after Deukalion's flood, *Paus. 1.18.7.*

The well, closed by a stone—here even by a veritable Roman general
and his horse—is not unfamiliar to us, meanwhile, after all that we heard
about Eben Shetiyyah, the well of the Ka'aba, etc. There are more
curious connections between wells and stones that ask for consideration
in future investigations, such as the following three items:

(1) The stone that was given by the Child to the Wise Men of the
East, according to a legend picked up by Marco Polo. "The Magi did
not understand the significance of the stone and cast it into a well. Then
straightaway there descended from Heaven a fire which 'they carried
into their own country and placed it in a rich and beautiful church.'"
L. Olschki[1] mentions also the Uigur version of this story, where "the
stone is detached by the Child from His crib and thrown into a well
because of its overwhelming weight which frustrated all human and
animal efforts to carry it away. A column of fire reaching the blue sky
is said to have risen from the well into which the stone had fallen and
to have kindled the fire worshiped by the Magi 'up to our days.'"

(2) The star of the Magi which fell into the well of Bethlehem,
according to Gervase of Tilbury,[2] after it had served its purpose to
guide the Wise Men to the "new way."

(3) The falling star that opened the abyss, according to Revelation
—a future event, for a change. Out of this well ascends smoke which
darkens the sun and air, and Franz Boll pointed aptly to the "smoke-

---

[1] "The Wise Men of the East in Oriental Traditions," in *Festschrift Popper*
(1951), p. 386.
[2] Sunt qui dicunt, stellam Magorum suo completo ministerio in puteum cecidisse
Bethlehemicum et illic eam intro videri autumant. See F. Liebrecht, *Des Gervasius
von Tilbury Otia Imperialia* (1856), pp. 1, 53.

barrel," south of Sagittarius and Scorpius: Ara, the Altar in the Galaxy,[3] and under this very Altar are the souls of the witnesses of God waiting for the last day (Rev. vi.9). According to Eratosthenes' catasterisms, at this Altar Zeus and his followers took their oath before attacking Kronos.[4]

The reader is likely to react unkindly claiming that there is no reason for whichever connection between legends about the Three Wise Men, Revelation, and the "Well of Gilgamesh." Yet, Franz Boll (*Offenbarung*, pp. 69ff.) has recognized in those strange locust demons of Revelation— they come out of the well of the abyss—who resemble horses with human heads, and have wings, and tails of scorpions, the Sagittarius-Centaur of Mesopotamian boundary stones, also to be found on the rectangular zodiac of Dendera. Revelation also states that they had wreaths as if of gold on their heads: the Egyptian Sagittarius wears a double crown, the Teukros tradition ascribes to the constellation "the royal face" (to prosōpon basilikon, Boll, *Sphaera*, pp. 181f.). In the Gilgamesh Epic Scorpion men watch the way to the other world; Virgil (*Aeneid* 6.286) makes it centaurs.

We must leave it at that: the chapter "Sagittarius and Saturn" would take us too far. We merely wanted to show that Gilgamesh's well and the opening of new ways are not "prehistoric drivel" that has nothing to do with our post-Greek, Christian civilization. It was with veritable awe that Boll stated (*Offenbarung*, p. 73, n. 4): "Von der Konstanz aller wesentlichen Charakteristiken in diesen Sternbildtypen macht sich der Fernerstehende schwer einen Begriff."

Although all this must remain posterior cura, we would like to mention the suggestion offered by Cornford, namely, "that one of these *phreata* (= wells) in Eleusis was closed at its mouth by the *agelastos petra*," i.e., the laughterless rock; Demeter was *agelastos* because of the loss of Persephone, and she was sitting upon this laughterless rock,

---

[3] Thymiatherion, or thyterion. Michael Scotus still made it "puteus sive sacrarius." See F. Boll, *Sphaera* (1903), p. 446; Boll, *Aus der Offenbarung Johannis* (1914), p. 75.

[4] It remains to be seen whether Ara has something to do with that enigmatical well of Gen. xxi.31, 33, called Beer-sheba, which is either "Well of Seven" or "Well of the Oath." The Septuaginta votes for the Oath," xxi.31: phrear horkismou; xxi.33: kai ephyteusen Abraam arouran epi tō phreati tou horkou kai epekalesato ekei to onoma Kyriou Theos aiōnios. (Compare also T. Nöldeke, "Sieben Brunnen," ARW 7 [1904], pp. 340–44.)

which Cornford ("The Eleusinian Mysteries," p. 161) proposes to take for "the double of the *anaklēthra* at Megara, which, as its name implies, was the place where Kore was 'called up.'" This might be, but it does not throw much light on the whole plot, whereas it seems important to recall how the "laughterless" state of the goddess was altered, namely, by the rather improper jokes of Baubo/Iambe. This very trait, now, occurs frequently within the scheme of world-ages. The Japanese sun goddess, Amaterasu, who, enraged by Susanowo's misdemeanor, had withdrawn into a rocky cave leaving the world in utter darkness, was caused to come out again only by the lascivious dances of "the ugly sky-female," Uzumue, dancing with the celestial jewel-tree upon her head, amidst the 800,000 gods assembled in the Milky Way, and producing fire afterward. Egyptian Ra, who had retired from a world which he did not like anymore, was "persuaded" by the same kind of jokes by Isis to take up again his duties ("And then the great god laughed at her"). The motif emerges again in the *Edda*, where Loke and a he-goat make the angry Skadi laugh, preventing her, thus, from avenging the murder of her father, Thiassi.[5] The story has also survived, although in dull disguise, in the Polynesian Marquesas Islands and, in excellent shape, with the Cherokee Indians; there the sex appeal is missing, admittedly, but the *agelastos* character is Mother Sun, desolate about the death of her only daughter: a true Demeter (her daughter resembling Eurydice: she had been brought back half of the way already, when the psychopompoi made a mistake that permitted her to return to Hades); the indecent dance is replaced by the concert of a juvenile orchestra.

We have heard (appendix #29) of an allegedly terrestrial *agelastos petra* with a river flowing at its foot, called Lethe. We also mentioned that Eleusis means "Advent," pointing to the circumstance that Demeter *arrived* there and that, before having borne Zeus, Demeter had the name of Rhea (Orph. frg. 145, Kern p. 188).

The moving of Rhea-Demeter to Eleusis is a huge and perplexing story, indeed, involving honeybees, a woodpecker—whose daughters were promoted to priestesses of Eleusinian Demeter—goats, and what else, and we are not likely to cover this event here and now. That we are up to a major change of residence can be taken from the parallel case of Amaterasu who, after having been caused by Uzumue's dance to

---

[5] See F. R. Schröder, *Skadi und die Götter Skandinaviens* (1941), pp. 19-25.

leave her cave, was respectfully guided into a "New Hall," as we hear in the Kogo-shui,[6] and "then Ama no Koyane no Mikoto and Futo-tama no Mikoto suspended an exalted Sun-rope around this Hall."[7]

Not being specialists in Eleusinian matters, topography, etc. (they remained secrets to the end), we do not feel entitled to deal earnestly with these items beyond raising some questions, such as which well was closed—if Cornford's suggestion is right—by the laughterless rock? Was it a former *lapis manalis?* What happened to the *agelastos petra* after Demeter had been moved to laugh? And how could this rock, closing a well connected with the underworld, be combined with the legends that hold Demeter responsible for the coming into being of the Stygian spring (Aelianus, *De natura animalium 10.40*), or for her having caused the waters of Styx to become black (O. Waser, Roscher 4.1572)? Demeter is supposed to have changed the color of the Stygian waters when she, on her search for Persephone, fleeing from Poseidon and changing into a mare, arrived at the Arcadian spring of Styx and perceived in the water her own mis-shape. And how, on the other hand, could the bringing into being of Styx and her sitting on the laughterless rock be combined with the Orphic claim, according to which Demeter "separated the double nourishment of the gods," splitting it up into Nectar and Ambrosia,[8] both of which come out of the "horn of Amaltheia," i.e., alpha Aurigae?

Considering the amount of testimonies for stones, shards, trees, plugs which close the one or the other well, abyss, whirlpool, or, by being pulled out or just removed announce major changes and great catastrophes, we might be expected to wrap up this whole parcel, from the "Holiest of Holies" replacing the Ark—its function, respectively, to cover the tehom—to Tahaki tearing up the tree of Tane-of-holy-waters, and to Alexander pulling out the pole-pin, or to mischievous Monkey who removed the basket. But apart from the fact that there are many more instances, unmentioned in this essay, which should also find their place in the said parcel, behind every tree, stone, and well lurks, as it

[6] K. Florenz, *Die historischen Quellen der Shinto-Religion* (1919), p. 423; see also pp. 37ff., 153–62; and *Nihongi*, trans. by W. Aston (1956), pp. 40–49.

[7] The question remains whether the "exalted Sun-rope" is the same as the "left rope"—being called thus because plaited from left to right—and the "rope whose root-ends are plaited together" by means of which Amaterasu was cut off from ever reentering that (laughterless) cave, according to Nihongi and Kojiki (see Florenz, *Quellen der Shinto-Religion*, p. 40, n. 22).

[8] Orph. frg. 189, Kern p. 216: Dēmētēr prōtē kai tas dittas trophas dieilen.

were, the danger of simplification and of ruthless identifying; to simplify, however, is the very danger that we most wish to avoid. In other words, we do not mean to make comparative mythology "easier," by procuring simple denominators upon which all these items could be brought; we think, on the contrary, that we are faced with an almost uncountable number of x's for which the fitting equations have to be worked out in long and cumbersome future investigations.

Appendix 37

A sidelight falls upon the notions connected with the stag by Horapollo's statement concerning the Egyptian writing of "A long space of time: A Stag's horns grow out each year. A picture of them means a long space of time."[1] Chairemon (hieroglyph no. 15, quoted by Tzetzes) made it shorter: "*eniautos: elaphos.*" Louis Keimer, stressing the absence of stags in Egypt, pointed to the Oryx (*Capra Nubiana*) as the appropriate "ersatz,"[2] whose head was, indeed, used for writing the word *rnp* = year, eventually in "the Lord of the Year," a well-known title of Ptah.[3] Rare as this modus of writing the word seems to have been—the *Wörterbuch der Aegyptischen Sprache* (eds. Erman and Grapow), vol. 2, pp. 429–33, does not even mention this variant—it is worth considering (as is every subject dealt with by Keimer), the more so as Chairemon[4] continues his list by offering as number 16: "*eniautos: phoinix,*" i.e., a different span of time, the much-discussed "Phoenix-period" (ca. 500 years). There are numerous Egyptian words for "the year," and the same goes for other ancient languages. Thus, we

[1] *The Hieroglyphs of Horapollo*, trans. by G. Boas (1950), p. 89 = Horap. 2.21.: "pōs polychronion. Elaphos kat'eniauton blastanei ta kerata, zōgraphoumenē de, polychronion sēmaiei."

[2] "Interpretation de plusieurs passages d'Horapollon," in *Suppl. 5 aux Annales du Service des Antiquités de l'Égypte* (1947), pp. 1–6. "Les Égyptiens avaient remarqué la resemblance existant entre les cornes d'un Bouquetin, caractérisées par de nombreux noeuds, et le signe . . . qui est originairement une branche de dattier" [this branch being the main part of the hieroglyph for "year"—*rnp*].

[3] M. Sandman Holmberg, *The God Ptah* (1946), pp. 22, 64f., 77, 178–80.

[4] F. J. Lauth, "Horapollon," SBAW (1876), p. 68. It remains a tragedy that only nineteen of Chairemon's explanations have been preserved by Tzetzes, who only stated that Chairemon had given "kai hetera myria."

propose to understand *eniautos* as the particular cycle belonging to the respective character under discussion: the mere word *eniautos* ("in itself," *en heauto;* Plato's *Cratylus* 410D) does not say more than just this. It seems unjustifiable to render the word as "*the* year" as is done regularly nowadays, for the simple reason that there is no such thing as *the* year; to begin with, there is the tropical year and sidereal year, neither of them being of the same length as the Sothic year. Actually, the methods of Maya, Chinese, and Indian time reckoning should teach us to take much greater care of the words we use. The Indians, for instance, reckoned with five different sorts of "year," among which one of 378 days, for which A. Weber did not have any explanation.[5] That number of days, however, represents the synodical revolution of Saturn. Nothing is gained by the violence with which the Ancient Egyptian astronomical system is forced into the presupposed primitive frame.

The *eniautos* of the Phoenix would be the said 500 (or 540) years; we do not know yet the stag's own timetable: his "year" *should* be either 378 days or 30 years, but there are many more possible periods to be considered than we dream of—Timaios told us as much. For the time being the only important point is to become fully aware of the plurality of "years," and to keep an eye open for more information about the particular "year of the stag" (or the Oryx), as well as for other *eniautoi,* especially those occurring in Greek myths which are, supposedly, so familiar to us, to mention only the assumed eight years of Apollo's indenture after having slain Python (Plutarch, *De defectu oraculorum,* ch. 21, 421C), or that "one eternal year (aidion eniauton)," said to be "8 years (oktō etē), that Cadmus served Ares (Apollod. 3.4.1; see also 2.5.11 with long note by Frazer).

Appendix 38

See RV *10.46.2,* ed. K. Geldner (1951); cf. V. Rydberg, *Teutonic Mythologie* (1907), p. 587. Geldner, vague as ever, spells it "der Gewässer Behausung." Agni, however, is a title, and the *Rigveda* stresses time and again that three Agnis already have gone away, "consumed"

[5] A. Weber, "Die Vedischen Nachrichten von den Naxatra," APAW *2* (1862), pp. 281–88, esp. pp. 286–87.

by the "sacrificial service." Agni, too, is not only coming from the confluence of the rivers like Gibil, but is also born in the "highest sky" (RV *6.8.2*): "Im höchsten Himmel geboren wachte Agni über die (Opfer) regeln als ihr Hüter. Der Klugsinnige mass den Luftraum aus." In fact, he has three birthplaces as a rule, in the "three worlds." (We have mentioned already that one of the Agnis had "seven mothers," like Heimdal.)

But wherever one of the Agnis is "found," he is a very busy surveyor. Says RV *6.7.6*: "Durch das Auge des Vaisvānara, durch das Wahrzeichen der Unsterblichkeit sind die Höhen des Himmels ausgemessen. Auf seinem Haupte (stehen) alle Welten, wie die Zweige sind seine sieben Arme (?) gewachsen." RV *6.7.1–2* calls the same Agni Vaisvānara, "head of the sky, leader[1] of the earth, born at the right time . . . the navel of the sacrifice." Stanza 5 of the same hymn states: "Vaisvānara! Diese deine hohen Anforderungen hat noch keiner angetastet, o Agni, der du im Schosse der beiden Eltern geboren, das *Wahrzeichen in der Reihenfolge der Tage* fandest." It is of another Agni, "just born," "the best path-finder," that RV *8.103.11* states: "Der bei (Sonnen-) aufgang die angebundenen Schätze erkundet." Whoever minds the "implex" is not going to think of the daily sunrise, if it is the sun at all: this is a conjecture of Geldner; we are up either to the heliacal rising of the "Agni-in-charge" at the vernal equinox, or at the rising day of Sirius. We wonder when the glorious day will finally arrive when the philologists begin to realize the purely cosmological significance of "sacrifices," and of "victims" chained to a "sacrificial post" or to a mountain.

The overwhelming amount of evidence on Agni and Soma ("lord of the world poles") as colures will have to be dealt with in the fitting frame, by means of an investigation of the so-called Shunashepa Hymns of the first Mandala of the *Rigveda*, Shunashepa being literally the same as Cynosoura, "Dog's Tail," i.e., Ursa Minor. In the present context we wish to point to only one more name of Agni—being himself a title— that is, Apām Napāt, a designation which belongs also to Iranian Tishtriya, Sirius. Usually it is translated into "child of the waters," but we

---

[1] Geldner's rendering of Sanskrit *aratí* into "Lenker" (Wagenlenker) has been contested by P. Thieme (*Untersuchungen zur Wortkunde und Auslegung des Rigveda* [1949], pp. 26–35). *Aratí* (fem.) from *ará*, the spoke, being the totality of spokes, according to Thieme, he translates RV *6.7.1*: "(den Agni), das Haupt des Himmels, den Speichenkranz der Erde," pointing also to *1.59.2*: "Agni ist das Haupt des Himmels, der Nabel der Erde. So ward er der Speichenkranz der beiden Welten."

cannot agree to this interpretation of napāt (whence also Neptunus) as "child." Not only does Boissacq allow only for nephews and nieces in connection with this radical, but we are always dealing with nephews in mythology, beginning with our own hero Amlethus; with Horus, nephew of Seth; with Kullervo, nephew of Untamo; with Reynard Fox, nephew of Isengrim; and so forth. What counts is a kind of "broken" relation, a subject deserving an extensive chapter, but since the understanding of the graphical sign that expresses best this "relation" ( ) comes from Mande tradition, West Sudan (where it marks circumcision, and the star of circumcision: Sirius), we postpone investigation of the whole complex.

Appendix 39

Excursus on Gilgamesh

There are many points from which to start new trips of exploration into the Gilgamesh Epic, once it is conceded that reasonable questions have to be asked. Among the many we single out two, without intending to "get at the bottom of the matter"; the first concerns the "ferryman," the second concerns "trees."

Face to face with the ferryman Urshanabi, a kind of personified *me* who was dragged away from the "confluence of the rivers" to check the proper measures of Uruk, it can hardly be taken for a farfetched idea that we ask for comparative "individuals" or "places" in other Mesopotamian texts. There is, indeed, no need for a frantic search: the Enuma elish offers us an equally decisive item from which depends the whole skeleton map, namely *Nibiru* (or *nēbēru*).

There are three passages of the so-called "Babylonian Genesis" that give—recognizable at first glance—details of the surveying of the new world as accomplished by Marduk/Jupiter. In Speiser's translation they read thus (ANET, pp. 67, 69):

4.141ff.   He crossed the heavens and surveyed the regions.
He squared Apsu's quarter, the abode of Nudimmud [Ea],
As the lord measured the dimensions of Apsu.
The Great Abode, its likeness, he fixed as Esharra
The Great Abode, Esharra, which he made as the firmament.
Anu, Enlil, and Ea he made occupy their places.

5.1–8   He constructed stations for the great gods,
Fixing their astral likenesses as *constellations*. [Heidel: The stars, their likeness(es), the *signs of the zodiac*, he set up][1]

He determined the year by designating the zones:
He set up three constellations for each of the twelve months.
After defining the days of the year (by means) of (heavenly) figures,
He founded the station of Nebiru to determine their (heavenly) bands,
That none might transgress or fall short.
Alongside he set up the stations of Enlil and Ea.

6.62f.   They raised high the head of Esagila equaling Apsu.
Having built a stage-tower as high as Apsu,
They sat in it an abode for Marduk, Enlil (and) Ea.

Leaving aside the specific charm of these passages—i.e., the circumstance that the *places* of Anu, Enlil, Ea in 4.146, and their *stations* in 5.8 are not the same—we concentrate on EE 5.6: "He founded the station of Nibiru to determine their (heavenly) bands" (Speiser), or: "He founded the station of Nibiru to make known their duties" (Heidel), or: "Er setzte ein den Nibirupunkt, um festzusetzen ihre Verknotung" (= *riksu;* Weidner, *Handbuch Babyl. Astr.,* p. 33), and on 5.8: "Alongside he set up the stations of Enlil and Ea" (Speiser), or: "He established the stations of Enlil and Ea together with it" (Heidel), or: "Den Enlilpunkt und den Eapunkt setzte er bei ihm fest" (Weidner). That means the position of the "Ways of Anu, Enlil, Ea" was a function of Nibiru; that only the setting up of the points, or stations, of

[1] The terminus is "Lumashi"-stars, and it is not yet certain which stars are meant. F. Kugler (*Sternkunde und Sterndienst in Babel.* [1907–13], vol. *1,* p. 259) voted for zodiacal *signs;* E. Weidner (*Reallexikon der Assyriologie* [1932], vol. *3,* p. 83) confined this signification to the 5th century B.C. and later, whereas O. Neugebauer (*The Exact Sciences in Antiquity* [1962], p. 140) stated that the zodiacal *signs* (instead of constellations) were not yet introduced at 418 B.C. There are texts which include among the Lumashi-stars Cygnus, Cepheus, Aquila, Orion, Sirius, Centaurus (A. Jeremias, HAOG, p. 200; P. Gössmann, *Planetarium Babylonicum,* 250), and this appears to rule out the zodiac. C. Bezold (Boll-Bezold, *Antike Beobachtungen farbiger Sterne* [1916], p. 149; see also Bezold, *Babylonisch-Assyrisches Glossar* [1926], p. 160) proposed to understand the Lumashi-stars as "Jupiter-stars"; this was accepted by B. Meissner (*Babylonien und Assyrien* [1932], vol. *2,* p. 408) but Weidner (RLA *3,* p. 80) claimed that Bezold had started from erroneous premises.

Enlil and Ea is mentioned suggests that Marduk/Jupiter claims the "Anu-ship" for himself.[2] The experts seem to be quite happy with the equation "Nibiru = Jupiter" (see below). But what is his "station," or point? Considering that upon this very station of Nibiru rests the whole tripartition of the universe during the age ruled by Marduk/ Jupiter, it is surprising how little the professionals care.

The plain meaning of *nibiru* is "ferry, ferryman, ford"—*mikis nibiri* is the toll one has to pay for crossing the river—from *eberu*, "to cross."[3] Alfred Jeremias insisted that Nibiru "in all star-texts of later times" indicated Canopus, taking this star for the provider of the meridian of the city of Babylon.[4] There have been other identifications (including even a comet!)—the summer solstice,[5] or the celestial North Pole;[6] the opinions and verdicts collected by Gössmann (*Planet.*, 311) show clearly that Nibiru remains an unknown factor for the time being.

This deplorable situation is not improved by means of the next occasion, when the ominous word is hurled at us anew, in EE 7.124ff., where fifty names are given to the new ruler, Marduk/Jupiter, among which is Nibiru.

Speiser translation:
> Nebiru shall hold the crossings of heaven and earth;
> Those who failed of crossing above and below,
> Ever of him shall inquire.
> Nebiru is the star which in the skies is brilliant.
> Verily, he governs their turnings, to him indeed they look
> Saying: "He who the midst of the Sea restlessly crosses,
> Let 'Crossing' be his name who controls its midst.
> May they uphold the course of the stars of heaven;
> May he shepherd all the gods like sheep."

[2] This dignity must have got lost after the first(?) flood (or by means of it?), otherwise Marduk could not ask reproachfully for the whereabouts of "Ninigi-*nangar*gid, the great carpenter of my Anu-ship" (Era Epic, tabl. *1*.155; Göss-mann, *Das Era-Epos* [1956], p. 98).

[3] Cf. C. Bezold, *Glossar*, p. 13f.; E. Ebeling, RLA *3*, p. 2f.; P. Jensen, *Kosmo-logie*, p. 128; E. Weidner, *Handbuch*, p. 26; P. Gössmann, *Planet.*, 311: "Nibiru ist eigentlich die 'Überfahrtsstelle.' Der 'Stern der Überfahrtsstelle' ist der Marduk-stern Jupiter, wenn er den Meridian überschreitet."

[4] HAOG, p. 134; Weidner (RLA *2*, p. 387): "Ob der Stern Marduk-Nebiru wirklich = Canopus, bleibt freilich ebenfalls unsicher." On p. 247, n. 2, Jeremias generalizes without much ado: "Kulminationspunkt der Sterne im Ortsmeridian."

[5] Weidner, *Handbuch* p. 33, but that was written at least thirty years earlier than his articles in RLA.

[6] Meissner, *Bab. und Assyr. 2*, p. 408.

Heidel translation:

> Nibiru shall be in control of the passages in heaven and on earth,
> For everyone above and below who cannot find the passage inquires of him.
> Nibiru is his star which they caused (?) to shine in the sky.
> He has taken position at the solstitial point (?), may they look upon him,
> Saying: "He who crosses the middle of the sea without resting,
> His name shall be Nibiru, who occupies the middle thereof;
> May he maintain the course of the stars in heaven;
> May he shepherd all the gods like sheep . . . "

Von Soden (ZA 47, p. 17):

> Nebiru soll die Übergänge von Himmel und Erde besetzt halten,
> denn droben und drunten fragt jeder, der den Durchgang nicht findet, immer wieder ihn.
> Nebiru ist sein Stern, den sie am Himmel sichtbar werden ließen;
> er fasste Posten am Wendepunkt, dann mögen sie auf ihn schauen
> und sagen: "Der die Mitte des Meeres (Tiamat) ohne Ruhe überschreitet,
> sein Name sei Nebiru, (denn) er nimmt die Mitte davon ein.
> Die Bahn der Sterne des Himmels sollen sie (unverändert) halten... "

How secure and unshakable the ground is upon which we walk, according to the inscrutable decree of the experts, may be guessed from the translation of lines 128–32 by Albrecht Götze[7] who starts from the conviction that *eberu* = "to bind, to enclose" which, combined with the "solution" that tam-tim means "struggle," apparently permits him to get rid of the "midst of Tiamat"

> Who enclosed (in his net) indeed amidst the struggle without loosening,
> May his name be "encloser," who seizes amidst (it).
> Of the stars of heaven may he uphold their courses
> May he shepherd the gods, all of them like sheep.

F. M. Th. Böhl[8] was at least perplexed enough to admit: "Der Passus gehört zu den sachlich schwierigsten der Tafel, ohne dass der ziemlich vollständig erhaltene Kommentar hierbei wesentliche Hilfe leistet." But he did not further the case by upholding opinions incompatible among themselves since based upon doubtful identifications. On the one hand, he claimed Nibiru to be the name given to "the planet and his hypsoma"; on the other hand, he took Nibiru for a star or constellation

---

[7] "Akkadian d/tamtum," in *Festschrift Deimel* (1935), pp. 185–91.
[8] "Die fünfzig Namen des Marduk," AfO *11* (1936), p. 210.

marking the point where Jupiter entered the "Way of Anu," to which he adds: "The time of observation is the night of the Vernal Equinox, when the Sun stands at the crosspoint of Equator and Ecliptic in the constellation Aries." He does not reveal from where he has this surprising knowledge; he seems to rely on the identification of "l-Iku" with Aries/Cetus, which is not the case: the Pegasus-square it is,[9] but the constellation is not mentioned in 7.124ff., so what? Apart from this, we do not know whether, in the time of the Enuma elish, Aries was taken for Jupiter's hypsoma; there seem to be reasons for recognizing —already at this time—Cancer (more precisely: Procyon) = Nangar = the Carpenter, as Jupiter's exaltation.[10] In the third place, if Böhl takes l-Iku for Aries ruling the vernal equinox, how could Jupiter *enter* there the "Way of Anu"?[11] The "Way of Anu" represents a band, accompanying the equator, reaching from 15 (or 17) degrees north of the equator to 15 (or 17) degrees south of it; the "Way of Enlil" runs parallel to that of Anu in the North, the "Way of Ea" in the South.[12] That, due to the precessional shifting of the crossroads of ecliptic and equator, the stars standing in these three "Ways" are not the same all the time, goes without saying.

[9] Böhl mentions this identification, with reference to Bezold and Schott, p. 211, n. 47.

[10] See E. Weidner, "Babylonische Hypsomatabilder," OLZ 22 (1910), cols. 14ff.; Weidner, *Gestirn-Darstellungen auf Babylonischen Tontafeln* (1967), pp. 9f., 134, n. 166, and plates V, VI (VAT 7847). A passage from the *Taittiriya Brahmana* (5.1.1) also has to be considered: "When Jupiter was first born, he defeated the nakshatra Pushya by his brilliance." P. Sengupta, who quoted the line in his introduction to Burgess' translation of the *Surya Siddhanta* (1935, p. xxxiv), misinterpreted it thoroughly by claiming it described "the discovery of Jupiter," and by adding, "the star group of Pushya (delta eta gamma Cancri) has no bright stars in it and the planet Jupiter was detected when it came near to this star group."

To the fully initiated expert who sternly points with outstretched finger to the circumstance that the nakshatra Pushya was formerly called Tishya (see, e.g., Scherer, *Gestirnnamen*, p. 150), and that means, Sirius, we can, for the time being, only answer that we are aware of this particular circumstance. Premature "solutions" are of no avail.

[11] To be sure, Böhl does not say so explicitly, his wording being as unprecise as possible. He claims that at the time of New Year (vernal equinox) "the orbit of Jupiter was observed particularly carefully." "Man beobachtete—so dürfen wir annehmen—wie er (wohl von der äusseren Ea-Sphäre her [sic!]) in den Anu-Bereich eintrat, diesen Bereich durchquerte (ebēru, itburu) und ihn dadurch gleichsam feierlich in Besitz nahm."

[12] For these much discussed "Ways," see van der Waerden, "The Thirty-Six Stars," JNES 8 (1949), p.16; Weidner, *Handbuch*, pp. 46–49; Meissner, *Bab. und Assyr. 2*, pp. 407f.; Bezold–Kopff–Boll, "Zenit- und Aquatorialgestirne," SHAW (1913); Schaumberger, *3. Erg.*, pp. 321–30.

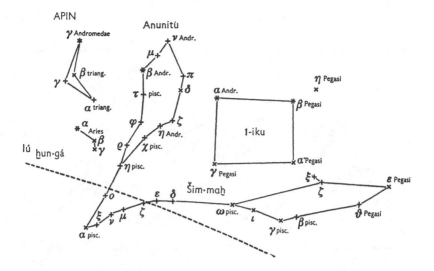

The Pegasus-square, called "l-Iku" (i.e., the standard field measure of the Sumerians), with the circumjacent constellations, as reconstructed out of Mesopotamian astronomical texts. The Fishes, then, must have had a larger extension than in our sphere.

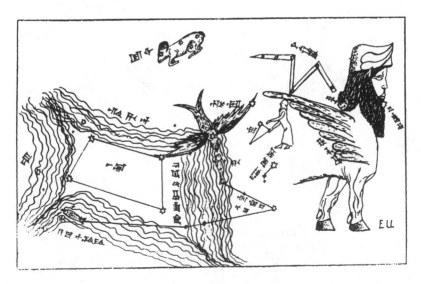

The same Babylonian constellation, according to A. Ungnad, who took "l-Iku" for the Paradise. The position of Ungnad's sketches is inverted with respect to the usual order of star maps.

The same square, correctly placed between the Pisces, as figured in the round (*above*) and the rectangular (*below*) zodiacs of Dendera (Roman Egypt).

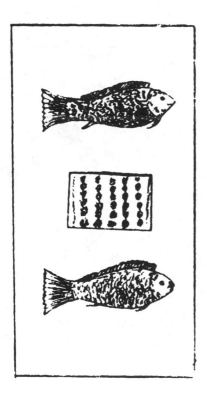

Everybody who is not inclined to think it is "obvious" or "in the very nature of things" to connect a "field" or a game board with two fishes, or with two lizards, or one turtle to two fishtails, is invited to compare the following items with the three previous illustrations.

A calabash from the Guinea Coast, Africa. The figure is really a hemisphere which is hard to reproduce, whence the open spaces.

Another calabash from the Guinea Coast.

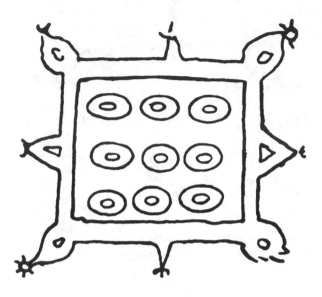

The zodiacal Pisces, as drawn by the Toba Batak of Sumatra.

This picture comes from the New World, and is described as "composite animal, head and body of a turtle, double tail of a fish."

But, as a matter of fact, "1-Iku," darkly hinted at by Böhl, does come into play, namely, in EE *6.62*, quoted above: "They raised high the head of *Esagila equaling Apsu.*" And concerning this Esagila (or Esagil) we hear in the ritual text of the New Year festival in Babylon[13] that the Urigallu-priest "shall go out to the Exalted Courtyard, turn to the north and bless the temple Esagil three times with the blessing: '*Iku-star, Esagil, image of heaven and earth.*'" "1-Iku," the Pegasus-square (= alpha beta gamma Pegasi, alpha Andromedae) is, indeed, of the utmost importance, 1-Iku representing the fundamental field measure,[14] and Ungnad (*Das wiedergefundene Paradies* [1923], p. 11) understood the constellation, enclosed by Pisces, for the "Paradise," the primordial field, so to speak. More important, Utnapishtim tells Gilgamesh (GE *11.57*) about his ark, which was, like the apsu, an exact cube: "One iku was its floor space."[15] (Before, *11.31*, Ea had ordered Utnapishtim: "Like the apsu thou shalt ceil her.") Remembering what we heard above: "Since the ark disappeared there was a stone in its place . . . which was called foundation stone," i.e., Eben Shetiyyah, that covered the abyss, this cubic ark, the floor space of which was one iku, cannot be without interest for us, the less so, when the gods "raised high the head of Esagila (= 1-Iku) equaling Apsu."

To be sure, this does not teach us where Marduk was supposed to be when he received the title Nibiru—it might have been decisive for the planet to rise heliacally together with 1-Iku, the celestial model of Esagil (representing the "foundation-stone-covering-the-apsu," maybe?), but when?[16] The heliacal rising of "1-Iku"—precisely, beta Pegasi—

[13] See Sachs translation, ANET, p. 232, l. 274f.

[14] About 3,600 square meters; see Heidel, GE, p. 82, n. 173.

[15] A. Schott translation: "Ein 'Feld' gross war seine Bodenfläche." Compare for details, Schott, "Zu meiner Übersetzung des Gilgamesch-Epos," ZA *42* (1934), pp. 37f., 40.

[16] One clue, at least (probably many more), to the situation is contained in the Cuneiform Tablet K 3476 dealing with the Babylonian New Year festival, translated and commented on by Heinrich Zimmern ("Zum babylonischen Neujahrsfest," BVSGW *58* [1906] *3*, pp. 127-36), which says that "Marduk lies with his feet within Ea" (lines 20-21: "(Das ist) Marduk [. . .] [der (?) mit (?)] seinen Füssen innerhalb (?) Eas liegt"). In a note, Zimmern proposes to understand this line as an "allusion to a constellation connected with Marduk (Auriga?) that reaches into a constellation connected with Ea (Aries?)." S. A. Pallis, not tending to astronomical notions, made it that "Marduk lies (?) before (?) Ea"; the unmistakable presence of the planet Venus in the second part of the sentence (kakkabuDIL. BAT) forced him to the concession: "perhaps it refers to certain astronomical conditions" (*The Babylonian Akītu Festival* [1926], p. 217). In 1926, sufficient literature about the "Three Ways" was available.

coincided with the winter solstice of 4000 B.C.; around 1000 B.C. it took place on January 25.[17] "1-Iku," the Pegasus-square, is called "the habitation of the deity Ea, the leader of the stars of Anu"[18] in the "Series ᵐᵘˡAPIN" (Plow-star, Triangulum), called by Weidner "a Babylonian compendium of astronomy."[19] According to van der Waerden ("The Thirty-Six Stars," p. 17) this series is a compilation "made about 700 B.C. or somewhat earlier[20]—in which material from different periods between −1400 and −700 was used": thus, 1-Iku as "leader" of the stars standing in the "Way of Anu" would rise in the end of January, quite a time away from the vernal equinox when the New Year's festival was held.

This is all very nice so far, and certainly not without highest interest, but do we know meanwhile what Nibiru, "ferry, ferryman, ford," was supposed to be? Even without worrying about Jupiter and his whereabouts? We know it not, and we feel tempted to say: "quod erat demonstrandum," namely, that the many verbose translations, eloquent articles, and books have not cleared up the decisive points of the cosmological system ruling the Enuma elish, the Gilgamesh Epic, the Era Epic and the other alleged "poems." Nibiru is only one case among many, but it is a rather significant model case for proving that no concrete problem is going to be solved as long as the experts of astronomy are too supercilious to touch "mythical" ideas—which are firmly believed to be plain nonsense, of course—as long as historians of religion swear to it that stars and planets were smuggled into originally "healthy" fertility cults and naïve fairy tales only "very late"—whence these unhealthy subjects should be neglected by principle—and as long as the philologists imagine that familiarity with grammar replaces that scientific knowledge which they lack, and dislike.

But even when the different specialists would condescend to renounce their common haughtiness, we do not think that there is much chance to arrive at a satisfying solution of concrete details, and the adequate understanding of the system as a whole, without taking into account comparable systems of other parts of the earth: Mesopotamia is by no means the only province of high culture where the astronomers worked

[17] See W. Hartner, "The Earliest History of the Constellations in the Near East," JNES *24* (1965), pp. 13, 15.

[18] Bezold-Kopff-Boll, p. 23.

[19] "Ein babylonisches Kompendium der Himmelskunde," AJSL *40* (1924), pp. 186–208.

[20] See also A. Schott, "Das Werden der babylonisch-assyrischen Positions-Astronomie und einige seiner Bedingungen," ZDMG *88* (1934), pp. 331, 333.

with a tripartition of the sphere—even apart from the notion allegedly most familiar to us, in reality most unknown—that of the "Ways" of Zeus, Poseidon, and Hades as given by Homer. The Indians have a very similar scheme of dividing the sky into Ways[21] (they even call them "ways"). And so have the Polynesians, who tell us many details about the stars belonging to the three zones (and by which planet they were "begotten"); but nobody has thought it worth listening to the greatest navigators our globe has ever seen; nor has any ethnologist of our progressive times thought it worth mentioning that the Polynesian megalithic "sanctuaries" (maraes) gained their imposing state of "holiness" (taboo) when the "Unu-boards" were present, these carved Unu-boards representing "the Pillars of Rumia," Rumia being comparable to the "Way of Anu," where Antares served as "pillar of entrance" (among the other "pillars": Aldebaran, Spica, Arcturus, Phaethon in Columba).

But now, is Nibiru as important as all that? We think so. Or, to say it the other way around: once this astronomical term, and two or three more, are reliably settled, one can begin in earnest to get wise to, and to translate, Mesopotamian "poetry."

II

The epics of Gilgamesh and Era offer too many trees for our modest demands. The several wooden individuals have, however, the one advantage that the expert's delight in uttering deep words about "the world-tree" wilts away.

There is, first, the mesh-tree, contained in the hero's name,[22] about the location of which Marduk asks stern questions of Era, followed by the cedar of Huwawa/Humbaba which was—as we have been taught by

---

[21] See W. Kirfel, *Die Kosmographie der Inder nach den Quellen dargestellt* (1920), pp. 140f. At first glance, it looks as if only the circle of lunar mansions was subdivided into these three ways, but the domains are extended far beyond the limits of the "inhabited world" in both directions, north and south, as are the Ways of Enlil and Ea.

[22] The identity of the tree is not settled. R. Labat (*Manuel d'Epigraphie Akkadienne* [1963], no. 314) proposes "*cèdre* (?micocoulier?) [*Celtis australis*, "gemeiner Zürgelbaum"—*Celtis occidentalis* is the American nettle tree] gisMEZ-MA-GAN-(NA) musskānu-mūrier (?micocoulier de Magan?)" (Cf. Labat, no. 296: "GIŠ, bois, arbre. Déterminatif précédant les noms d'arbres et d'objets en bois.") See also F. Delitzsch, *Assyrisches Handwörterbuch* (1896), p. 410 s.v. miskanu, musukanu, "ein Baum . . . , wechselt mit mis-ma-kan-na, d.i. MIS-Holz von Makan."

(Even this mes-wood from Magan cannot be dismissed as "not applicable" for the GE, because in the Sumerian myth "Gilgamesh and the Land of the Living" (Kramer, ANET, p. 49, *l.*111–15), when the hero is allegedly admonishing Enkidu

the specialists—felled by Gilgamesh and Enkidu. Yet, according to the "latest news" available to us,[23] Huwawa is "the guardian monster of the 'land of the cut cedar.' " Admittedly, also at an earlier occasion, Kramer stressed his opinion that "the far distant 'Land of the Living' was also the 'Land of the Felled Cedar,' "[24] but we have not yet found evidence of any thought, any consequence which should follow such alarming statements. But one cannot expect earnest thoughts to be wasted on Sumerian conceptions from a scholar who wrote about the fathers of hydraulic engineering (irrigation) that "to the Sumerian poets and priests the real sources of the Tigris and Euphrates in the mountains of Armenia were of little significance. They did not understand, as we do, that the volume of the waters of the two rivers depended upon 'feeding' from their tributaries, or that it was the melting winter snows which produced the annual overflow, or that the Tigris and Euphrates 'emptied' their swollen waters into the Persian Gulf. Indeed, their view was

---

not to shrink away from Humbaba, Gilgamesh utters the most enigmatical words: "Do thou help me [and] I will help thee, what can happen to us? After it had sunk, after it had sunk, After the *Magan-boat* had sunk, After the boat 'the might of Magilum' had sunk.")

See also F. Hommel, *Ethnologie und Geographie des Alten Orients* (1926), pp. 539, 783. According to Meissner, quoted by Weidner ("Gestirn-Darstellungen auf Babylonischen Tontafeln," SOAW *254* [1967], p. 18), ᵍⁱˢMES = mēsu is the *rowan*. As concerns the astrological system of connecting trees (and stones, and animals, etc.) with the zodiac, the tablets translated by Weidner put the mes-tree two times with Aquarius (pp. 18, 35), once with Aries (p. 31). Wood of the mēsu-tree and of the huluppu-tree occurs as building material for the chariot (*narkabtu*) of Ningirsu, in the Gudea Cylinder A VII, 16–18 (cf. A. Salonen, *Prozessionswagen* [1946], p. 6; Salonen, *Die Landfahrzeuge des Alten Mesopotamien* [1951], pp. 111f.).

This tree is also part of the name of MES.LAM.TA.E₃.A, taken for the oldest name known of the god Nergal (see J. Böllenrücher, *Gebete und Hymnen an Nergal* [1904], p. 7) and the name of the one of the Gemini, MES.LAM.TA.E.A., means "who comes forth from MES.LAM." MES.LAM was the name given to Nergal's sanctuary in Kutha, and means "the luxuriantly growing MES-tree," according to Gössmann (*Das Era-Epos*, p. 67), who continues with respect to the name MES.LAM.TA.E.A.: "Später diente der Name in erster Linie als Bezeichnung eines der beiden Zwillinge (*Planetarium Babylonicum*, 271), bezw. als Tummelplatz philologischer Spielereien. Auf Grund solcher Philologeme wurde der Name auf Marduk und Gilgamesh übertragen (Tallqvist, 374)." It is not in the best scientific style to dispose of difficult formulae by declaring them philological pastimes. Since MES.LAM appears to be a "fixed" topos, we can hardly expect that "to come forth from MES.LAM" has been a monopoly of Nergal-Mars. But see below, p. 449.

[23] S. N. Kramer, *The Sumerians* (1963), p. 277.

[24] In *Gilgamesh et sa légende*, ed. by P. Garelli (1958), p. 64. In "Gilgamesh and the Land of the Living," JCS *1* (1947), p. 4, he had styled it more modestly: "the far distant Land of the Living (also known as cedar land)."

just the opposite; it was the Persian Gulf which was responsible for the waters of the Tigris and Euphrates and for their all-important overflow. Mythologically expressed, it was Enki who filled the Tigris and Euphrates with sparkling water, and who, by riding the sea, makes its waters and those of the Tigris and Euphrates, turbulent and violent . . . In short, as the Sumerians saw it, it was not the rivers that 'fed' the sea, . . . but rather the sea that 'fed' the rivers."[25]

Apart from the mes-tree and the unexplained cedar of Huwawa/ Humbaba—whether it was felled by our heroes or not—the Gilgamesh Epic confronts us with the *huluppu*-tree, taken for a willow by Labat (nos. 371, 589), for an oak by the *Assyrian Dictionary* (vol. 6, pp. 55f.), for a kind of Persea by Salonen (*Landfahrzeuge*, pp. 111f.)—all the identifications decently equipped with a question mark. This specimen crosses our way in the Sumerian version of the Gilgamesh Epic, part of which was incorporated as Tablet XII in the Akkadian Epic; the Sumerian text has been translated by C. J. Gadd,[26] and by S. N. Kramer.[27] We quote the summary given by Kramer in his first translation (1938, p. 12) for the simple reason that it is shorter than the one offered in JAOS 64 (1944), pp. 19–21. In the meantime, this text had been given a different name, i.e., "Gilgamesh, Enkidu, and the Nether World." The first half of the first sentence is, of course, no quotation, and it is not likely to be subscribed to by the author.

On occasion of a new distribution of the "Three Ways,"[28] "on that day," it happened that "a huluppu tree (very likely a willow) which had been planted on the bank of the Euphrates and nourished by its waters was uprooted by the South Wind and carried off by the Euphrates. A goddess wandering along the bank seized the floating tree,

---

[25] "Dilmun, the Land of the Living," BASOR *96* (1944), p. 28; we pass in silence the identification of this "Land of the Living" with Dilmun, as claimed in this article, and as upheld in all later publications.

[26] "Epic of Gilgamesh, Tablet XII," in RA *30* (1933), pp. 129–43.

[27] "Gilgamesh and the Huluppu-Tree," *Assyriological Studies 10* (1938). Cf. Kramer, *Sumerian Mythology* (1944), pp. 33–37, and *From the Tablets of Sumer* (1956), pp. 222–26.

[28] As concerns the beginning of this text, one shock after the other receives the reader who studies eagerly the various "translations": it is hard to believe that they are meant to render the same Sumerian original. Out of the first lines Kramer (*Sumerian Mythology*, pp. 30ff.) built up the Sumerian creation story which he took (and takes?) for unknown; in JAOS *64*, p. 19, he stressed again: "The first thirteen lines of this passage contain some of our basic data for the analysis of the Sumerian concept of the creation of the universe." Of the following lines 14–25 he constructed a dragon-fight. By means of hitherto unpublished pieces, Kramer claimed, in 1958 (*Gilgamesh et sa légende*, p. 66), that "the first seven lines of the

and at the word of Anu and Enlil she brought it to Inanna's [i.e., Ishtar's] garden in Uruk. Inanna tended the tree carefully and lovingly, hoping to have made of its wood a throne and bed for herself. After ten years had passed and the tree had matured, Inanna, to her chagrin, found herself unable to realize her hopes. For in the meantime a dragon had set up its nest at the base of the tree, the Zu-bird had placed his young in its crown, and in its midst the demoness Lilith had built her house. But Gilgamesh, informed of Inanna's distress, rushed to her aid. Making light of his weighty armor, the giant slew the dragon with his huge bronze ax, seven talents and seven minas in weight. Thereupon the Zu-bird fled with his young to the mountain, while Lilith, terror-stricken, tore down her house and escaped to the desert. After Gilgamesh had uprooted the liberated tree, his followers, the men of Uruk, cut down its trunk and gave part of it to Inanna for her throne and bed. Of the remainder—i.e., root and crown—"Gilgamesh makes for himself the pukku and mikku, two wooden objects of magic significance."

---

poem can now be completely restored." He added, however: "Unfortunately, the *meaning of the passage* is by no means certain and the mythological implications are rather obscure, as is obvious from the following tentative translation:

> The days of creation, the distant days of creation,
> The nights of creation, the far-off nights of creation,
> The years of creation, the distant years of creation,—
> After in (?) days of yore everything needful had been brought into existence,
> After in (?) days of yore everything needful had been commanded,
> After in the shrines(?) of the land bread(?) had been tasted(?)
> After in the ovens of the land, bread(?) had been baked(?)."

Nobody is likely to contradict the stated uncertainty of the meaning; it would be advisable to mind the utterance of Margarete Riemschneider (*Augengott und Heilige Hochzeit* [1953], p. 190): "So lange sie sinnlos sind, stimmen unsere Übersetzungen nicht." The objections raised by stern expert reviewers (T. Jacobsen, "Sumerian Mythology," JNES 5 [1946], pp. 128–52; M. Witzel, "Zur sumerischen Mythologie," Or. 17 [1948], pp. 393–415) remain throughout within the usual frame of specialists on grammar and "religion," and it is hard to decide who carries off the laurels in this race of arbitrary interpretations. The remarkable point of the new "distribution" seems to be that Ereshkigal belongs henceforward to the "nether world." (In 1938 Kramer translated line 12: "After Ereshkigal had been presented (?) as a gift (?) to (?) the netherworld"; in his *Sumerian Mythology*, after having "discovered" the dragon-fight, he made it: "After Ereshkigal had been carried off into Kur as its prize." Witzel (Or. 17, p. 402) rendered the line: "Als (der) Ereshkigal mit der Unterwelt Geschenk 'aufgewartet' worden war.") Since we do not know yet which star or constellation Ereskigal was meant to represent, this does not tell us more than that the (unknown) asterism had "entered" the Way of Ea, i.e., that it had sunk below the 15th (or 17th) degree of southern latitude.

(It goes without saying that there is no whiff of "magic significance" to be found in the text.) Here, the summary of 1938 comes to its end, and we continue with JAOS *64*, p. 20: "Follows a passage of twelve lines which describes Gilgamesh's activity in Erech with this pukku and mikku, with this 'drum' and 'drumstick' [see below]. Despite the fact that the text is in perfect condition, it is still impossible to penetrate its meaning. It is not improbable, however, that it describes in some detail the overbearing and tyrannical acts which, according to the first tablet of the Epic of Gilgamesh, brought woe to the inhabitants of Erech, and which, again according to the Babylonian epic only, led to the creation of Enkidu."

According to this verdict, Kramer does not even try to translate literally the lines 24–35 which are allegedly in "perfect condition." Gadd (RA *30*, p. 131) renders the passage as follows:

22. He makes its root into his pukku   [^{gis}RIM (ellag)]
23. Its top he makes into his mikku   [^{gis}E.AG]
24. He says "ellag," except (?) "ellag" let him not speak
25. Saying . . .   except (?) . . . let him not speak
26. The men of his city   say "ellag"
27. He viewed his little company which did not . . .
28. ? ? his lament they make   [a-geštin-nu a-geštin-nu]
29. He that had a mother, (she) brought bread for her son
30. He that had a wife, (she) poured out water for her "brother"
31. The Wine (?) was taken away   [^{d}geštin-an-na]
32. (In) his place where the pukku was set he draws a circle
33. The pukku he raised before him and went into the house
34. In the morning his place where the circle was drawn he viewed
35. The adults (?)   do not . . .
36. (But) at the crying of a little girl . . .

Kramer continues: "When the story becomes intelligible once again, it continues with the statement that 'because of the outcry of the young maidens,' the pukku and mikku fell into the nether world. Gilgamesh put in his hand as well as his foot to retrieve them, but was unable to reach them. And so he seats himself at the gate of the nether world and laments:

O my pukku, O my mikku.
My pukku whose lustiness was irresistible,
My mikku whose pulsations could not be drowned out.[29]

---

[29] In *From the Tablets of Sumer* (1956), p. 224, Kramer translated: "My pukku with lustiness irresistible, My mikku with dance-rhythm unrivaled."

(With the following line, representing line 1 of Tablet XII, the Akkadian translation sets in):[30]

> In those days when verily my pukku was with me in the house of the carpenter,[31]
> (When) verily the wife of the carpenter was with me like the mother who gave birth to me
> (When) verily the daughter of the carpenter was with me like my younger sister,
> My pukku, who will bring it up from the nether world,
> My mikku, who will bring it up from the "face" of the netherworld?

His servant Enkidu, his constant follower and companion, thereupon volunteers to descend to the nether world and bring them up for him . . . Hearing his servant's generous offer, Gilgamesh warns him of a number of the nether world tabus which he is to guard against . . . But Enkidu heeds not the instructions of his master and commits all those very acts against which Gilgamesh has warned him. And so he is seized by Kur and is unable to reascend to the earth."

We can do, here, without the following description of the goings-on in the "underworld," a description which is common to the Sumerian myth of the huluppu-tree, and to Tablet XII of the Akkadian Epic. Kramer (JAOS 64, p. 23) closes his inquiry on the Sumerian sources of the Gilgamesh Epic: "In conclusion, a comparison of the text of the 'twelfth' tablet of the Epic of Gilgamesh with that of our Sumerian poem Gilgamesh, Enkidu, and the Nether World, proves beyond all doubt what has long been suspected, that is, that the 'twelfth' tablet is an inorganic appendage attached to the Babylonian epic whose first eleven tablets constitute a reasonably well-integrated poetic unit." We do wish neither to consent nor to disagree; we do not like those "beyond all doubt's," and similar verdicts, considering how frightfully little we know of the Epic. (If there is something that is, really, "beyond all doubt," it is only this, that the eleven tablets of the Epic do not "con-

---

[30] Thus, directly after Urshanabi's checking of the measures of Uruk (*11*.307), there follows as catchline line 308 = *12*.1: "In those days, when . . ."

[31] A. Heidel (p. 95) translates lines 1–3: "O that today I had left the pukku in the house of the carpenter! O that I had left it with the wife of the carpenter, who was to me like the mother who bore me! O that I left it with the daughter of the carpenter, who was to me (like) my younger sister." In a footnote he explains: "Had Gilgamesh left his pukku and his mikku in the house of the carpenter, they would have been safe and would not have fallen into the underworld." He adds: "The translation of the first three lines is somewhat tentative." Of only the first three lines?

stitute a reasonably well-integrated poetic unit," *not* the translated Epic.)

Of course, it would be to our great advantage were we to know more about the objects "pukku" and "mikku," that have withstood the honest efforts of several scholars, first among whom is Sidney Smith ("b/pukk/qqu and mekku," RA *30* [1933], pp. 153–68). Nets have been proposed, wind instruments (pipes and horns), and Margarete Riemschneider (*Augengott*, pp. 50f.) voted for a particular trap, the very same rather uncanny trap which is known to us from the Pyramid Texts (representing the "palace" of Upper Egypt). Most interpreters have accepted Landsberger's first proposition "drum" and "drumstick";[32] there is nothing to say against this solution *per se*, as long as the significance of celestial drums is recognized (see chapter VIII, "Shamans and Smiths"), and under the condition that comparable celestial drums are properly investigated—e.g., those of the Chinese sphere. For the time being there is no cogent reason to stick to "drum" and "drumstick," the less so as Landsberger dropped his earlier notion—about which he states explicitly that he never substantiated it—for the sake of "hoop" and "driving stick."[33] In the present situation, however, we know nothing of the function of pukku and mikku, and this fact should prevent idle speculations.

No less lamentable than the loss of these objects is the circumstance that we do not know more about Inanna's unwelcome subtenants in her huluppu-tree, about Lilith, and about the dragon at the root; that he corresponds to Nidhöggr of the *Edda* does not enlighten us concerning his identity. The Zu-bird, at least, is known to us: the planet Mars it is,[34] but we do not yet risk drawing specific conclusions from this identifica-

[32] Marius Schneider votes for "drum" (Rahmentrommel) and "harp" or "lyre," in his article "Pukku und Mikku. Ein Beitrag zum Aufbau und zum System der Zahlenmystik des Gilgamesh-Epos," *Antaios 9* (1967), pp. 280f.

[33] "Einige unerkannt gebliebene oder verkannte Nomina des Akkadischen," WZKM *56* (1960), pp. 124–26. It is advisable to take into consideration that Landsberger does not recognize the occurrence of the word pukku in GE *1*.II.22, because Schott and Schmoekel brought the alleged drum pukku into the first tablet of their translations without hesitating. At first glance, it might seem irrelevant whether or not pukku occurs in the first tablet. A little concentrated thinking will correct this impression: pukku having been made from the wood of the felled huluppu-tree, the whole timetable of the Epic, particularly the appropriate allocation of the 12th tablet and the Sumerian poem of the huluppu-tree, might hinge upon the valid answer to this very question: whether or not pukku does make its appearance in the first tablet.

[34] See Gössmann, *Planetarium Babylonicum*, 195: ᵐᵘˡ ᵈIMDUGUDᵐᵘˢᵉⁿ.

tion to the "nest" or "house" of the planet that was taken away from him.

The deadlock is hardly to be overcome by Mesopotamian texts alone, and this goes for the huluppu-tree, the mes-tree, Huwawa's cedar, and that tree in the Era Epic of which Era announces (Tablet 4.123–26, Gössmann, pp. 30f.; Langdon, MAR 5, p. 144): "Irkalla will I shake and the heavens shall tremble. The brilliancy of Jupiter [^{II}ŠUL.PA.E$_3$] will I cause to fall and the stars will I suppress.[35] The root of the tree will I tear up and its sprout will not thrive."

In case we wished to go on this errand in the future we should start from two Indian *nakshatras* (lunar mansions) and the legends connected with them: *mūla* (or *mūra*), "the root," also called "the tearer out of the root" (see also appendices #4 and #30), and even "Yama's two un-fasteners," i.e., the Sting of Scorpius,[36] (lambda upsilon Scorpii)—in Babylonian astronomy ^{mul}ŠAR.UR and ^{mul}ŠAR.GAZ, the weapons of Marduk in the "battle" against Tiamat; and the *nakshatra* containing Antares (alpha Scorpii) which bears the names "the oldest," or "who slays the oldest"[37]—in Tahiti: "parent pillar of the world."

From India we should turn to the hero Tahaki of Tuamotuan texts,[38] already mentioned, because he represents the almost "professional" avenger of his father. Right from the beginning of events, Tahaki's mother laments that the hero is destined to die in a faraway country;

[35] Langdon (MAR 5, pp. 144f.) points to the prophecy against Babylon and its king, in Isa. XIII, XIV, "clearly reminiscent of this passage . . . 'I will make the heavens to tremble and the earth shall be shaken out of her place.' So prophesied the Hebrew writer, and even more obvious is his borrowing from the Irra myth when he compares the king of Babylon to Hêlêl: 'How art thou fallen from Heaven, O Hêlêl, son of morning!' In the cuneiform text of the Irra myth Marduk is called Shulpae, the name of Jupiter in the early morning, and there can be little doubt that Hêlêl is a transcription of a Babylonian title of Marduk-Jupiter, *elil*, 'the shining one.'"

[36] The Indians claim that exactly opposite to *mūla* was Betelgeuse, ruled by "Rudra-the-destroying-archer," whereas the Coptic list of lunar mansions (Kircher, *Oedipus Aegyptiacus 2* [1653], pt. 2, p. 246) calls the Sting of Scorpius (al-Sha'ula) "Soleka statio translationis caniculae in coelum . . . unde et Siôt vocatur, statio venationis," which is of the utmost importance since it elucidates the role of a "sea-star" common to Sirius and the Scorpion-goddess.

[37] A. Weber ("Die Vedischen Nachrichten von den Naxatra," APAW 2 [1862], pp. 291f.) renders Jyesthaghnī: "die ältesten (Geschwister) tödtend" which re-minds us, *nolens volens*, of Mercer, who translates Pyramid text 399 ab: "It is N. who judges with him whose name is hidden (on) this day of slaying the eldest (gods), and N. is lord of offerings, who knots the cord."

[38] J. F. Stimson, *The Legends of Maui and Tahaki*, Bull. BPB Mus. (1933), pp. 50–77.

and again and again throughout the unfolding of the legend, Tahaki sings: "I go to the night-realm of Kiho, the last bourne of repose." When still a child, his cousin, with whom he plays diving for pearls, kills and dismembers him; but his foster brother saves the vital parts (unlike the case of Osiris), from which his mother revives him again. He sets out with this brother to free his father from the "goblin myriads" (see above, p. 175). When reaching the home of his grand-parents, he wins the love of Hapai, daughter of Tane, the Deus Faber. When Hapai tells her father about the young man, he answers: "If he is really Tahaki go and say to him: 'Tane-of-ancient-waters told me that if you can pass before his face you must be Tahaki; if you can sit upon his four-legged stool, you must be Tahaki; if you can pull up his sacred tree by the roots then you are surely Tahaki.' Then Tahaki went to Tane-of-ancient-waters and stood beside him; and immediately he passed before his face; he sat upon his high four-legged stool—and it broke to pieces under him; then Tahaki pulled up his sacred tree by the roots—and Tahaki looked down and saw the entrance to Havaiki beneath.[39] Then Tahaki and Tane-of-ancient-waters chanted a song about the death of Tahaki."[40] Nonetheless, with Tane's consent, the pair

[39] Compare Handy, on Marquesas (*Bull. BPB Mus. 69*, p. 132): "When Vaka-Uhi had reached a certain spot in the sea, he could see Havaiki down at the bottom of the ocean." We seem to be still circling the spot beneath the whirlpool, de-scribed by Adam of Bremen, and by the Cherokee (see pp. 106f.).

[40] Stimson, p. 73. The antiphony of the chant does not allow for a summary: there are "First Voice," "Second Voice," "Chorus," "Refrains" sung by Tane, and Tahaki gets some lines in between, also. Noteworthy is the mentioning of a "way-opener," but we do not know who he is, the Polynesians being more prone to "titles" and kenningar than other mythographers. Sung by (a) Tahaki, (b) the first voice, (c) the second voice, we hear (a) "It was Puga-ariki-tahi-"; (b) "The first Puga-ariki who came at last"; (c) "To *Fare-kura*-templed abode of the vener-ated learning of the gods—there in the spirit-world where thou dwellest." This *Fare-kura* (*fare* = house, *kura* = red, or purple; Maori: *Whare kura;* Samoa: *Fale ula*, etc.) was, according to the "Lore of the Whare-wananga" of New Zealand, a temple "at Te Hono-i-wairua . . . in the spot where the teaching of the Whare-wananga originated" (i.e., remarks S. Smith, p. 82, "where man was first taught the doctrines brought down from Heaven by Tane"). The Te Hono-i-wairua (the gathering place of the spirits) was in Hawaiki, the so-called "primordial home" of the Polynesians, and the sage states (Smith, p. 101): "Whakaahu, a star (*Castor*, in the constellation Gemini) was appointed (or set up) at Te Hono-i-wairua in Hawaiki . . . whilst Puanga (Rigel of Orion) was fixed at the east of Rarohenga (Hades)." Later he explains (p. 113) that "those spirits which by their evil con-duct on this earth . . . left the temple [Whare kura] by the Takeke-roa (or long rapid, descent) to Rarohenga, or Hades," while the others ascended slowly to the "realm of Io the Supreme God," i.e., the same as Kiho-tumu, Kiho-the-All-Source, of Tuamotu.

still lived together "many months until a certain day when trouble arose between them . . . So Tahaki went far far away to a distant land hoping that he might be killed there. And the land where Tahaki was slain at last was known as Harbor-of-refreshing-rain."

After an extended excursion into Mexico and the "broken tree," the symbol of *Tamoanchan*, "the house of descending," where the gods were hurled down for having plucked the forbidden flowers, the broken tree being claimed to be the Milky Way (W. Krickeberg, "Der mittel-amerikanische Ballspielplatz und seine religiöse Symbolik," *Paideuma 3* [1944-49], p. 132), we should return once more to the storehouse of magnificent survivals, Finland, particularly to the many variants of the "cutting of the large oak" (K. Krohn, FFC *52* [1924], pp. 183-99). This was by no means an easy task to accomplish, but the oak had made trouble right from the start. When (in the second rune of the *Kalevala*) Sämpsä Pellervöinen had sowed trees, it was the oak alone that would not grow until four or five lovely maidens from the water, and a hero from the ocean, had cleared the ground with fire and planted an acorn in the ashes; and once it had started, the growth of the tree could not be stopped:

> *And the summit rose to heaven*
> *And its leaves in air expanded,*
> *In their course the clouds it hindered,*
> *And the driving clouds impeded,*
> *And it hid the shining sunlight,*
> *And the gleaming of the moonlight.*
>
> *Then the aged Vainamoinen,*
> *Pondered deeply and reflected,*
> *"Is there none to fell the oak-tree,*
> *And o'erthrow the tree majestic?*
> *Sad is now the life of mortals,*
> *And for fish to swim is dismal,*
> *Since the air is void of sunlight,*
> *And the gleaming of the moonlight."*

"One sought above in the sky, below in the lap of the earth," as we are informed by variants, but then Vainamoinen asked his divine mother for help.

Then a man arose from ocean
From the waves a hero started,
Not the hugest of the hugest,
Not the smallest of the smallest.
As a man's thumb was his stature;
Lofty as the span of woman.

The "puny man from the ocean," whose "hair reached down to his heels, the beard to his knees," announces, "I have come to fell the oak tree/And to splinter it to fragments." And so he does. In several variants the oak is said to have fallen over the Northland River, so as to form the bridge into the abode of the dead. Holmberg (quoted by Lauri Honko, "Finnen," *Wb. Myth.*, p. 369) took the oak for the Milky Way.

Considering that the same puny character was alone able to kill the huge ox—we might call it "bull" quietly—whose mere sight chased all heroes to the highest trees, we can hardly overlook the possibility that we are up to some kind of "grandson" of hairy Enkidu, and the oak would be a faint reflection of the cedar. Whereas an Esthonian variant sounds—although suffering from atrophy—more like the story of the huluppu-tree. A damsel plants the acorn—it is typical that Krohn (p. 187) calls the versions of Russian Karelia "disfigured," where this acorn is called "taivon tähti," i.e., sky-star—the growing tree endangers the sky, trying to "tear the celestial luminaries, or to darken them." The maiden, therefore, asks her brother to cut off the tree. Out of its wood presents are made for the relatives of the bridegroom, and for the virgin herself a chest is fabricated.

Since we do not mean to undertake the expedition into comparative tree-lore here and now, we have to leave it at that. That mythical "trees" are not of terrestrial provenance, and that we cannot cope with the different specific tree individuals under the heading "the world tree"—not *although*, but *because* they are "cosmic" trees—could have been expected by everybody who has spent time and thought on the tree of the Cross; on Yggdrasil (and Ashvatta); on the "Saltwater-tree" of the Cuna Indians; on Zeus' oak, part of which was built into Argo; on the fig tree at the vortex which saved Odysseus; on the laurel which did not yet mark the omphalos of Delphi, when Apollo slew Python ("nondum laurus erat," Ovid)—it had to be brought from Tempe after Apollo's indenture of eight great years; on Uller's yew-tree (belonging to Sirius) by means of whose juice Hamlet's father was murdered; on—apart from the mentioned Mesopotamian tree individuals—the "dark kishkannu-tree" growing in Eridu, where no mortal is ever admitted; on the tamarisk at Be'ersheba in Genesis XXI; on the heather tree that "enfolded and embraced the chest with its growth and concealed it within its trunk," the "chest" being the coffin of Osiris (Plutarch, *De Iside et Osiride*, ch. 14–15, 356E–357A); and on the king of the country who "cut off the portion that enfolded the chest, and used it as pillar to support the roof of his house," until Isis carried off this "pillar." Those who prefer to overlook these items (and very many more) might recall the many times that we hear of much sighing and crying over trees cut down, sawed in two, and the like[41]—after all, our very Yima-Jamshid was sawed in two, by Azhi Dahak—as Tammuz "the lord of the great tree, overcome by the rage of his enemies," and the numerous comparisons of Mesopotamian temples with trees (cf. M. Witzel, *Texte zum Studium Sumerischer Tempel und Kultzentren* [1932], pp. 37f.; Witzel, *Tammuz-Liturgien und Verwandtes* [1935], pp. 108f.).

It would be an imposition to expect the reader to listen to such endless rambling on without telling him the aim that we hope to attain, sooner or later, by digging into these trees and posts: we do want to know which "New Way" it was that has been "opened" by Gilgamesh who was "wood" from the mes-tree, and we wish to find out the chronological sequence of the celestial events as told in the Enuma elish, the Gilgamesh Epic and the Era Epic. The irrelevancy of the scholarly quest for "poets" (and who cribbed from whom) has been understood, meanwhile: it is the celestial phenomena that move and change, and not the "mythopoetic fantasy" or the "doctrines" of poets and pontiffs. We have to find out, therefore, who came first as ruler of "the underworld," Nergal or Gilgamesh, or whether these two should really be the same, which we doubt for the time being. Yet, we have already heard (pp. 437f., n. 22) that Nergal's name MES.LAM.TA.E.A. was given to Gilgamesh. As Lambert states (*La Légende de Gilgamesh*, pp. 39f.): "After his life on earth Gilgamesh became king of the under-

---

[41] See R. Eisler, *Orphisch-Dionysische Mysterien-Gedanken in der christlichen Antike* (1925; repr. 1966), pp. 246, 248. Compare also the "epitheton" of Ugaritic Baal, 'alíyn, and its possible derivation from Hebrew 'allôn ('êlôn), Oak, Therebynth, holy tree, and allânati as name of the fourth month, i.e., the month of Tammuz. (H. Birkeland, *Norsk Tidskrift for Sprogvidenskap 9* [1938], pp. 338–45; W. Robertson Smith, *The Religion of the Semites* [1957], p. 196, n. 4.)

world, a Babylonian Osiris. A formal statement of this is given in a late religious text: 'Meslamtaea is Gilgamesh, Gilgamesh is Nergal, who resides in the underworld.' This comes from one of the texts which explain the functions of deities by equating them with other gods or goddesses, a very significant type of exposition."

This "significant type of exposition" is, in fact, the technique of the Old Norse skalds, and we have some perfect kenningar from Mesopotamia, such as "Ninurta is the Marduk of strength," "Nergal is the Marduk of battle," "Nabu is the Marduk of business,"[42] "Enzak is the Nabu of Tilmun."[43] Now, the passage quoted by Lambert says: "^{d}gilgameš ^{d}nergal (u.gur) āšib (dúr) ersetim^{tim}." In the text (quoted above) that addresses Gilgamesh as "supreme king, judge of the Anunnaki . . . you stand in the underworld and give the final verdict," it is again *ersetu*, and according to GE *12.56* it is *ersetu* that has seized Enkidu. Thus, that line might try to tell us "Gilgamesh is the Nergal of Ersetu," whereas Nergal's own "underworld" is *Arallu* (Aralu). Says Albright:[44] "Eridu is employed as a name of the apsū, just as Kutu (Kutha), the city of Nergal, is a common name of Aralu." Thus, it would be the very confidence in the custom of giving many names to the same topos —and in "synonyms" in general—which enforces, so to speak, distorted translations. It is a matter of course that the final decision will rest with those who know Sumerian and Akkadian, in the future: spontaneous angry refusals should not be accepted. Taught by bad experience with the Egyptian dictionary (*Aegyptisches Wörterbuch*) that renders thirty-seven Egyptian special termini with the one word *Himmel*, we suspect the Assyriologists to handle their "underworld" accordingly— and their "heaven," of course. The authors of the *Assyrian Dictionary* do try to be as specific as possible, admittedly, so they deliver several particular significations of *ersetu* (vol. *4*, pp. 308–13): "(1) the earth (in cosmic sense); (2) the nether world; (3) land, territory, district, quarter of a city, area; (4) earth (in concrete sense), soil, ground, dry land"; but translations being a function of the expectations of the translator, the categories are bound to look fundamentally different, once several of them are expected to represent sections of the sphere.

[42] Jeremias, HAOG, p. 190; see also Meissner, *Babylonien und Assyrien 2*, p. 133; Witzel, *Tammuz-Liturgien*, pp. 470f.

[43] D. O. Edzard, "Die Mythologie der Sumerer und Akkader," *Wörterbuch der Mythologie 1*, p. 130.

[44] "The Mouth of the Rivers," AJSL *35* (1919), p. 165; see also K. Tallqvist, *Sumerisch-Akkadische Namen der Totenwelt* (1934), p. 35.

But where does the proportion, Gilgamesh belongs to Ersetu, Nergal to Arallu, lead us to? This is not yet to be made out properly; too many riddles lurk behind every word. About the mes-tree, Marduk knew to tell (in the Era Epic) that it *"had* its roots in the wide sea, in the depth of *Arallu,* and its top attained High Heaven," asking Era reproachfully "Because of this work which thou, o hero, didst command to be done, *where* is the mes-tree, flesh of the gods, adornment of kings?" (S. Langdon, *Semitic Mythology* [1931], p. 140). Concerning the Mashu mountain (Mashu = twin) watched by the Scorpion-men, the GE says: "Whose peaks reach to the vault of heaven (And) whose breast reach to the nether world below," this "nether world" being *Arallu.* We knew all the time, certainly, that we were up to Scorpius (probably with a part of Sagittarius), but the huge constellation offers sufficient space for more than one way of descending. It is for this reason particularly that we hope for a better understanding from the Indian lunar mansions (1) lambda upsilon Scorpii, alias "the root," alias "the tearer out of the root," alias "Yama's two unfasteners," and (2) Antares, "the eldest," alias "who slays the eldest": in the sense of Precession, the sting of the Scorpion antecedes Antares.

If we knew the precise "extension" of the Scorpion-goddess (Ishara tamtim, Egyptian Selket) we should be better off. And this is the reason: GE Tablet 7, col. 4, 10f., dealing with Enkidu's alleged sick-bed hallucinations, makes Enkidu prophesy to that "harlot"—in the texts of Boghazköi it is she who has the name Siduri—who had lured him into the city: "[On account of thee (?)] the wife, the mother of seven, shall be forsaken." (Speiser: "[On thy account] shall be forsaken the wife (though) a mother of seven." Ebeling, AOTAT, p. 105: "[Um deinet-willen soll] verlassen werden die Mutter der sieben, die Hauptgattin.") This "mother of seven" should be Ishara tamtim, the Scorpion-goddess whose seven sons are notorious with her[45]—it is preposterous, anyhow, to associate one or the other righteous housewife in Uruk or elsewhere; but whenever well-bred scholars meet a "harlot" they accept it as their duty to discover moral lectures in the text surrounding her, very touchy they are! The first part of the line, however, is not in existence, and it is, again, their expectation that urges the philologists to supply "[On account of thee (?)]." Here, for a change, Freud would come in handy, but for the sake of the translators, not for the text. The readable part of

[45] Meissner, *Babylonien und Assyrien 2,* p. 26; Edzard, Wb. Myth., p. 90.

the line states nothing else but that "the wife, the mother of seven shall be forsaken." But since we do not know yet the whole extension of the Lady Ishara tamtim who was going to be forsaken, we still do not know the position of Gilgamesh's "new way"—to *ersetu,* as we assume, or by way of *ersetu. Ersetu might* have replaced Ishara tamtim, because we learn right in the beginning of the Era Epic (Tablet *1.*28–29, Göss-mann, p. 8) that Anu begets "the Sevengods" (ᴵᴵSIBIᵗᴵ) on *Ersetu,* trans-lated "the Earth," as companions for Era. The one who doubts that "begetting" is done up there might begin to ponder over the Hurrian texts, where MAR.GID.DA, the Big Dipper (alias the Seven Rishis), begets twins on "the Earth."[46] It is evident that we are still far away from the first among the proposed goals, but we prefer to confess to this state of things rather than fall into the bottomless pit of speculation —the very many inviting pits, respectively.

[46] The Big Dipper does it on the order of Ea. See H. Otten, *Mythen vom Gotte Kumarbi. Neue Fragmente* (1950)`, pp. 7f.

BIBLIOGRAPHY

BIBLIOGRAPHY

Ackerman, Phyllis. "The Dawn of Religion," in *Forgotten Religions*, V. Fermi, ed. New York, 1950.

Aelianus. *De natura animalium*, A. F. Scholfield, trans. 1958, 1959, 3 vols. LCL.

Afrikanistiche Studien. Diedrich Westermann zum 80. Geburtstag gewidmet, J. Lukas, ed. Berlin, 1955.

Albright, William Foxwell. "The Mouth of the Rivers," AJSL, vol. *35* (1919), pp. 161–95.

———. "Notes on Assyrian Lexicography," RA, vol. *16* (1919), pp. 173–94.

———. "The Goddess of Life and Wisdom," AJSL, vol. *36* (1919–20), pp. 258–94.

———. "Gilgamesh and Engidu," JAOS, vol. *40* (1920), pp. 307–35.

Albright, William Foxwell, and Dumont, P. E. "A Parallel Between Indian and Babylonian Sacrificial Ritual," JAOS, vol. *54* (1934), pp. 107–28.

Alexander, Hartley Burr. *North American Mythology*. MAR, vol. *10* (1916).

———. *Latin American Mythology*. MAR, vol. *11* (1920).

Alföldi, Andreas. "Smith as a Title of Dignity" (Hungarian), *Magyar Nyelv*, vol. *28* (1932), pp. 205–20.

Allen, Richard Hinkley. *Star Names: Their Lore and Meaning*. New York, 1963 (1st ed. 1899, under the title *Star Names and Their Meanings*).

Altorientalische Texte zum Alten Testament. In Verbindung mit Erich Ebeling, Hermann Ranke, Nikolaus Rhodokanakis herausgegeben von Hugo Gressmann, 2nd ed. Berlin-Leipzig, 1926.

Amiet, Pierre. *La Glyptique Mesopotamienne Archaique*. Paris, 1961.

Ancient Near Eastern Texts Relating to the Old Testament, James B. Pritchard, ed. (2nd ed. corrected and enlarged). Princeton, 1955.

Anderson, Andrew Runni. *Alexander's Gate, Gog and Magog, and the Inclosed Nations*. Mediaeval Academy of American Publ. 12. Cambridge, Mass., 1932.

Anderson, R. B. *See Edda.*

Anklesaria, B. T. *Zand-Akâsîh: Iranian or Greater Bundahishn*. Transliteration and translation in English. Bombay, 1956.

Apollodorus. *The Library*, and *Epitome*, Sir James George Frazer, trans. 1954, 1956. 2 vols. First printed 1921, LCL.

Apollonius Rhodius. *Argonautica*, R. C. Seaton, trans. 1955. First printed 1921, LCL.

Aratos. *Phaenomena*, G. R. Mair, trans. 1955. First printed 1921, LCL.

———. *Sternbilder und Wetterzeichen*. Übersetzt und eingeleitet von Albert Schott, mit Anmerkungen von Robert Böker. München, 1958.

———. *See also* Maass.

Aristotle. Aristotelis *Opera* edidit Academia Regia Borussica. Aristoteles Graece ex Recognitione Immanuelis Bekkeri. Darmstadt, 1960. 2 vols. 1st ed. Berlin, 1831.

———. *Metaphysics*. A revised text with introduction and commentary by W. D. Ross. Oxford, 1953. 2 vols. 1st ed. 1924.

———. *Meteorologica*, H. D. P. Lee, trans. 1962, LCL.

The Assyrian Dictionary of the Oriental Institute of the University of Chicago, I. J. Gelb, T. Jacobsen, B. Landsberger, and A. L. Oppenheim, eds. Chicago, 1964–.

Atharva Veda Sanhita, William Dwight Whitney, trans. Delhi, 1962. 2 vols. Reprint from the Cambridge, Mass. ed. 1905. *Harvard Oriental Series 7–8*.

———. *Hymns of the Atharva-Veda*, Maurice Bloomfield, trans. Delhi, 1964. Reprint from the Oxford University Press ed. 1897. SBE 42.

Athenaeus. *Deipnosophistae*, Charles Burton Gulick, trans. 1951–1957. 7 vols. First printed 1927–1941, LCL.

Avenol, Louis. *See* Wou Tch'eng Ngen.

Baravalle, Hermann von. *Die Erscheinungen am Sternenhimmel*. Stuttgart, 1962.

Barb, Alfons A. "Der Heilige und die Schlangen," MAGW, vol. *82* (1953), pp. 1–21.

———. "St. Zacharias the Prophet and Martyr," *Journal of the Warburg and Courtauld Institutes*, vol. *11* (1948), pp. 35–67.

Barthel, Thomas S. "Einige Ordnungsprinzipien im Aztekischen Pantheon: Zur Analyse der Sahagunschen Götterlisten," *Paideuma*, vol. *10* (1964), pp. 77–101.

Bartholomae, C. *Altiranisches Wörterbuch*. Strassburg, 1904.

Bastian, Adolf. *Die Heilige Sage der Polynesier*. Leipzig, 1881.

Baudissin, W. Graf. *Adonis und Esmun*. Leipzig, 1911.

Baumann, Hermann. *Das Doppelte Geschlecht: Ethnologische Studien zur Bisexualität in Ritus und Mythos*. Berlin, 1955.

Baumgartner, W. "Untersuchungen zu den akkadischen Bauausdrücken," ZA, vol. *36* (1925).

Beckwith, Marta. *Hawaiian Mythology*. New Haven-London, 1940.

––––––. *The Kumulipo: A Hawaiian Creation Chant*. Translated and edited with a commentary. Chicago, 1951.

Benveniste, E. "La Légende de Kombabos," in *Mélanges Syriens offerts à R. Dussaud* (1939), pp. 249–58.

Berger, E. H. *Mythische Kosmographie der Griechen*. Leipzig, 1904. Supplementary volume to Roscher's *Lexikon*.

Bernhardi, Anna. "Vier Könige: Ein Beitrag zur Geschichte der Spiele," BA, vol. *19* (1936), pp. 148–80.

Bertholet, Alfred. "Der Schutzengel Persiens," *Festschrift Pavry* (1933), pp. 34–40.

Best, Elsdon. *The Astronomical Knowledge of the Maori*. Wellington, 1955. Dominion Museum Monograph 3. Reprint from the 1922 ed.

Bezold, Carl. *Babylonisch-Assyrisches Glossar*. Heidelberg, 1926.

––––––. "Zenit- und Aequatorialgestirne am babylonischen Fixsternhimmel," mit astronomischen Beiträgen von August Kopff und Zusätzen von Franz Boll. SHAW, vol. *11* (1913).

Bhagavata Purana: The Srimad-Bhagavatam of Krishna-Dwaipayana Vyasa, J. M. Sanyal, trans. (English), 2nd ed. Calcutta, n.d. Preface, 1952.

Al-Biruni. *The Chronology of Ancient Nations*, Edward C. Sachau, trans. London, 1879.

––––––. *Alberuni's India*, Edward C. Sachau, trans. Delhi, 1964. 2 vols. Reprint from London ed. 1888.

Boas Anniversary Volume. Anthropological Papers Written in Honor of Franz Boas. New York, 1906.

Boas, Franz. *Indianische Sagen von der Nord-Pacifischen Küste Amerikas*. Berlin, 1895.

Boas, George. *The Hieroglyphs of Horapollo*. New York, 1950. Bollingen Series 28.

Böhl, F. M. T. "Die fünfzig Namen des Marduk," AfO, vol. *11* (1936), pp. 191–218.

––––––. "Zum babylonischen Ursprung des Labyrinths," *Festschrift Deimel* (1935), pp. 6–23.

Böker, Robert. *See* Aratos.

Böllenrücher, Josef. *Gebete und Hymnen an Nergal*. Leipziger Semitistische Studien *1*.6 (1904).

Bömer, Franz. "Der sogenannte Lapis Manalis," ARW, vol. *33* (1936), pp. 270–81.

Boll, Franz. *Sphaera: Neue Griechische Texte und Untersuchungen zur Geschichte der Sternbilder*. Leipzig, 1903.

––––––. *Aus der Offenbarung Johannis: Hellenistische Studien zum Weltbild der Apokalypse*. Leipzig-Berlin, 1914. Stoicheia 1.

————. "Antike Beobachtungen farbiger Sterne," mit einem Beiträge von Carl Bezold. ABAW *30*.1 (1916).

Bouché-Leclerq, A. *L'Astrologie Grecque*. Paris, 1899. Reprint Brussels, 1963.

Breasted, J. H. *Development of Religion and Thought in Ancient Egypt*. New York, 1959. 1st ed. 1912.

Brennand, W. *Hindu Astronomy*. London, 1896.

The Brihadāranyaka Upanishad. With the commentary of Shankaracarya. Swāmi Mādhavānanda, trans. 4th ed. Calcutta, 1965.

The Brihad-Devata. Attributed to Saunaka. A. A. Macdonell, ed. and trans. Delhi, 1965. Reprint of the 1904 edition, *Harvard Oriental Series* 5–6.

Brodeur, Arthur Gilchrist. *See Edda*.

Bromwich, Rachel. *Trioedd Ynys Prydein: The Welsh Triads*. Edited and with an introduction, translation, and commentary. Cardiff, 1961.

Brugsch, Karl Heinrich. *Thesaurus Inscriptionum Aegyptiacarum*. Altägyptische Inschriften. Graz, 1968. 1st ed. Leipzig, 1883–1891.

————. *Die Aegyptologie*. Leipzig, 1891.

————. *Religion und Mythologie der alten Aegypter*, nach den Denkmälern bearbeitet. Leipzig, 1891.

Buck, Sir Peter. *Mangaian Society*. Bulletin BPB Mus. 122. Honolulu, 1934.

Budge, Sir E. A. W. *The Life and Exploits of Alexander the Great*. Translated from several Ethiopic versions. London, 1896.

————. *The Book of the Dead*. An English translation of the chapters, hymns, etc., of the Theban Recension, 2nd ed., rev. and enlarged, 7th impression. London, 1956.

Buffière, Félix. *Les Mythes d'Homère et la Pensée Grecque*. Paris, 1956.

Bundahishn. See Anklesaria.

Burgess, E. *See Sūrya Siddhānta*.

Burrows, Eric, S.J. "Some Cosmological Patterns in Babylonian Religion," in *The Labyrinth* (London-New York, 1935), pp. 45–70.

————. "The Constellation of the Wagon and Recent Archeology," in *Festschrift Deimel* (1935), pp. 34–40.

Casanova, Paul. "De quelques Légendes astronomiques Arabes, considérées dans leurs rapports avec la Mythologie Egyptienne," BIFAO, vol. 2 (1902), pp. 1–39.

Censorinus. *Censorini De Die Natali Liber*, recensuit et emendavit Otto Jahn. Amsterdam, 1964. Reprint of the 1845 ed., Berlin.

Charpentier, Jarl. *Kleine Beiträge zur indoiranischen Mythologie*. Uppsala, 1911. Uppsala Univ. Arsskrift, fil/hist. Vetensk. 2.

Chimalpahin Quauhtlehuanitzin, Domingo de San Anton Muñon. *Memorial Breve acerca de la Fundación de la Ciudad de Culhuacan* . . .

Aztekischer Text mit deutscher Übersetzung von W. Lehmann und G. Kutscher Stuttgart, 1958. Quellenwerke zur Alten Geschichte Amerikas 7.

Christensen, Arthur. *Le premier Homme et le premier Roi dans l'Histoire légendaire des Iraniens.* Archives d'Études Orientales, vol. *14*, pt. 1 (1917), pt. 2 (1934).

———. *Les Kayanides.* Kgl. Danske Vidensk. Selskab. Hist/filol. Medd., vol. *19*. Copenhagen, 1932.

Chwolson, D. *Die Ssabier und der Ssabismus.* St. Petersburg, 1856. 2 vols.

Cicero. *De divinatione*, William Armistead Falconer, trans. 1964. First printed 1923, LCL.

———. *De natura deorum*, H. Rackham, trans. 1956. First printed 1933, LCL.

Claudian, Maurice Platnauer, trans. 1956. 2 vols. First printed 1922, LCL.

Cleasby, Richard. *An Icelandic-English Dictionary.* Revised, enlarged, and completed by Gudbrand Vigfusson, 2nd ed. with a supplement by W. A. Craigie. Oxford, 1962. Reprint.

Clemens Alexandrinus. *Stromata*, Bks. I–VI, O. Stählin, ed. 3rd ed. by L. Früchtel. Berlin, 1960.

———. *Die Teppiche (Stromateis).* Deutscher Text nach der Übersetzung von Franz Overbeck. Basel, 1936.

Collitz, Hermann. "Konig Yima und Saturn," in *Festschrift Pavry* (1933), pp. 86–108.

Comparetti, Domenico. *The Traditional Poetry of the Finns*, J. M. Anderton, trans. London-New York-Bombay, 1898.

Cook, A. B. "The European Sky-God," *Folk-Lore*, vol. *15* (1904), pp. 264–315.

Cornford, F. M. "The *Aparchai* and the Eleusinian Mysteries," *Festschrift Ridgeway* (1913), pp. 153–66.

———. *See also* Plato.

Creuzer, Friedrich. *Symbolik und Mythologie der Alten Völker, besonders der Griechen*, 3rd ed. Leipzig-Darmstadt, 1837–42.

Culin, S. "Chess and Playing Cards," AR U.S. National Museum for 1896 (Washington, 1898), pp. 667–842.

Cumont, Franz. *After Life in Roman Paganism.* New York, 1959.

———. "Adonis et Sirius," Extrait des *Mélanges Glotz*, vol. *1* (1932), pp. 257–64.

Curwen, E. Cecil. "Querns," *Antiquity*, vol. *11* (1937), pp. 135–51.

Dähnhardt, Oskar. *Natursagen.* Leipzig-Berlin, 1907–12. 4 vols.

Dahse, Johannes. "Ein zweites Goldland Salomos: Vorstudien zur Geschichte Westafrikas," ZfE, vol. *43* (1911), pp. 1–79.

Daressy, Georges. "L'Égypte Céleste," BIFAO, vol. *12* (1916), pp. 1–34.

Deimel, Anton. *Pantheon Babylonicum: nomina deorum e textibus cuneiformibus excerpta et ordine alph. ordinata.* Rome, 1914.

Delitzsch, Friedrich. *Assyrisches Handwörterbuch.* Leipzig-Baltimore-London, 1896.

Deonna, Waldemar. *Un divertissement de la table "à cloche-pied."* Brussels, 1959. Collection Latomus 40.

Deonna, Waldemar, and Renard, M. *Croyances et Superstitions de la Table dans la Rome antique.* Brussels, 1961. Collection Latomus 46.

Diakonoff, I. M. "Review article on the GE translations of F. M. T. Böhl and P. L. Matous," *Bibliotheca Orientalis,* vol. *18* (1961), pp. 61–67.

Diels, Hermann. "Das Aphlaston der antiken Schiffe," *Zeitschrift der Vereins für Volkskunde in Berlin* (1915), pp. 61–80.

―――. *Die Fragmente der Vorsokratiker* (Griechisch und Deutsch), 6th ed. Herausgegeben von Walther Kranz. Berlin, 1951. 3 vols.

―――. *See also Doxographi Graeci.*

Dieterich, Albrecht. *Nekyia: Beiträge zur Erklärung der neuentdeckten Petrusapokalypse.* Leipzig, 1893.

―――. *Eine Mithrasliturgie.* Leipzig-Berlin, 1923.

Dieterlen, Germaine. "The Mande Creation Story," *Africa,* vol. *27* (1957), pp. 124–38.

―――. *See also Griaule, Marcel.*

Dittrich, Ernst. "Gibt es astronomische Fixpunkte in der älteren babylonischen Chronologie?" OLZ, vol. *15* (1912), cols. 104–07.

Dixon, Roland B. *Oceanic Mythology.* MAR, vol. *9* (1916).

Dowson, John. *A Classical Dictionary of Hindu Mythology and Religion, Geography, History and Literature,* 8th ed. London, 1953.

Doxographi Graeci Collegit recensuit prolegomenis indicibusque instruxit H. Diels, 2nd ed. Berlin, 1929.

Dreyer, J. L. E. *A History of Astronomy from Thales to Kepler* (formerly titled *History of the Planetary Systems from Thales to Kepler*). Revised with a foreword by W. H. Stahl. New York, 1953.

Duemichen, J. "Die Bauurkunden der Tempelanlage von Edfu," *Zeitschrift für Aegyptische Sprache und Altertumskunde,* vol. *9* (1871), pp. 25–32.

Dumézil, Georges. *Le Festin d'Immortalité: Étude de Mythologie Comparée Indo-Européenne.* Paris, 1924. Annales du Musée Guimet. Bibliothèque d'Études, vol. *34.*

―――. *La Saga de Hadingus (Saxo Grammaticus I, 5–8). Du Mythe au Roman.* Paris, 1953. Bibl. École Hautes Études, Section Sciences Relig., vol. *66.*

Dupuis, Charles. *Origine de tous les Cultes et toutes les Religions.* Paris, 1795.

Ebeling, Erich. *Tod und Leben nach den Vorstellungen der Babylonier.* Berlin-Leipzig, 1931.

Edda. *Eddalieder,* Finnur Jónsson, ed. Halle, 1888–90. Altnordische Textbibliothek 2–3.

————. *Die Lieder der sogenannten Älteren Edda, nebst einem Anhang: Die mythischen und heroischen Erzählungen der Snorre Edda.* Übersetzt und erläutert von Hugo Gering. Leipzig-Vienna, n.d. Preface, 1892.

————. Nach der Übersetzung von Karl Simrock neu bearbeitet von Hans Kuhn. Leipzig: Reclam, n.d.

————. *The Prose Edda by Snorri Sturluson.* Translated from the Icelandic with an introduction by Arthur Gilchrist Brodeur. New York-London, 1929.

————. Übertragen von Felix Genzmer. *Thule,* vol. *1* (1928).

————. *The Younger Edda, also called Snorre's Edda, or the Prose Edda.* An English version with introduction and notes by Rasmus B. Anderson. Chicago-London, 1880.

————. *Die Jüngere Edda.* Übertragen von Gustav Neckel und Felix Niedner. *Thule,* vol. *20* (1942).

Edzard, D. O. "Die Mythologie der Sumerer und Akkader," in *Wörterbuch der Mythologie,* vol. *1,* pp. 19–139.

Eisenmenger, J. A. *Entdecktes Judenthum.* Königsberg, 1711. 2 vols.

Eisler, Robert. *Weltenmantel und Himmelszelt.* München, 1910.

————. *Orpheus the Fisher.* London, 1921.

————. *Orphisch-Dionysische Mysterien-Gedanken in der christlichen Antike.* Leipzig-Berlin, 1925. Vorträge Bibliothek Warburg, vol. *2.*

————. *The Royal Art of Astrology.* London, 1946.

Elwin, Verrier. *The Agaria.* Oxford, 1941.

Emerson, Nathaniel B. *Unwritten Literature of Hawaii.* BAE Bulletin 38. Washington, 1909.

Emory, Kenneth P. *Tuamotuan Religious Structures and Ceremonies.* Bulletin BPB Mus. 191. Honolulu, 1947.

Encyclopaedia of Religion and Ethics, James Hastings, ed. Edinburgh-New York, 1955. 13 vols. 1st impression 1908.

Eratosthenes. *Catasterismorum Reliquiae,* C. Robert, ed. Berlin, 1878. Reprint 1963.

Erman-Grapow. See *Wörterbuch der Aegyptischen Sprache.*

Essays and Studies presented to William Ridgeway. Cambridge University Press, 1913.

Falkenstein, A. "Die Anunna in der sumerischen Überlieferung," *Festschrift Landsberger* (1965), pp. 127–40.

Falkenstein, A., and Soden, W. von. *Sumerische und Akkadische Hymnen und Gebete*. Zürich, 1953.

Fausbøll, V. *Indian Mythology According to the Mahabharata*. London, 1902.

Feng-Shen-Yen-I. Die Metamorphosen der Götter. Historisch-mythologischer Roman aus dem Chinesischen. Übersetzung der Kapitel 1–46 von Wilhelm Grube. Durch eine Inhaltsangabe der Kapitel 47–100 ergänzt, eingeleitet und herausgegeben von Herbert Mueller. Leiden, 1912.

Festschrift Boas. See Boas Anniversary Volume.

――― *Deimel. See Miscellanea Orientalia*.

――― *Georg Jacob zum 70. Geburtstag, 26. Mai 1932*. T. Menzel, ed. Leipzig, 1932.

――― *Landsberger. See Studies in Honor of Benno Landsberger*.

――― *Pavry. See Oriental Studies*.

――― *Popper. See Semitic and Oriental Studies*.

――― *Ridgeway. See Essays and Studies*.

―――, *Publication d'Hommage offerte au P. W. Schmidt*. W. Koppers, ed. Vienna, 1928.

――― *Westermann. See Afrikanistische Studien*.

Festugière, A. J. *See Proclus*.

Festus. *Sexti Pompei Festi De verborum significatu quae supersunt cum Pauli Epitome*, W. M. Lindsay, ed. Hildesheim, 1965. Reprint from Leipzig ed. 1913. BT.

Feuchtwang, D. "Das Wasseropfer und die damit verbundenen Zeremonien," *Monatsschrift für Geschichte und Wissenschaft des Judentums*, vol. *54* (1910), pp. 535–52, 713–28; vol. *55* (1911), pp. 43–63.

Filliozat, Jean. "L'Inde et les Échanges scientifiques dans l'Antiquité," *Cahiers d'Histoire Mondiale*, vol. *1* (1953), pp. 353–67.

Firdausi. *The Shahnama of Firdausi*, Arthur George Warner and Edward Warner, trans. London, 1905–09. Vols. 1–4.

―――. *Das Königsbuch*, Deutsch von H. Kanus-Credé. Lieferung 1, Buch 1–5. Glückstadt, 1967.

Florenz, Karl. *Der historischen Quellen der Shinto-Religion*. Göttingen-Leipzig, 1919. Quellen der Religionsgeschichte.

Forbes, R. J. *Studies in Ancient Technology*, vol. 3. Leiden, 1955.

Fornander, Abraham. *An Account of the Polynesian Race, its Origin and Migrations*, vol. *1*. London, 1878.

―――. *Fornander Collection of Hawaiian Antiquities and Folk-Lore*. With translations edited and illustrated with notes by Thomas G. Thrum. Mem. BPB Mus. 4–6. Honolulu, 1916–20.

Frankfort, Henry. *Cylinder Seals: A Documentary Essay on the Art and Religion of the Ancient Near East.* London, 1939.

Frazer, Sir James George. *Pausanias' Description of Greece.* Translated with a commentary. London. 6 vols.

——. *The Dying God* (= *The Golden Bough*, Part 3), 3rd ed. London, 1963.

——: *Folk-Lore in the Old Testament.* London, 1918. 3 vols.

——. *Myths of the Origin of Fire.* London, 1930.

——. *See also* Apollodorus.

Friedrich, J. "Die hethitischen Bruchstücke des Gilgameš-Epos," ZA, vol. *39* (1929–30), pp. 1–82.

Frobenius, Leo. *Das Zeitalter des Sonnengottes.* Berlin, 1904.

——. "Das Archiv für Folkloristik," *Paideuma*, vol. *1* (1938–40), pp. 1–18.

——. *The Childhood of Man.* New York, 1960.

Gadd, C. J. "Epic of Gilgamesh, Tablet XII," RA, vol. *30* (1933), pp. 129–43.

Ganay, Solange de. *See* Zahan.

Gardiner, Sir Allan. *Egyptian Grammar*, 3rd ed. Oxford-London, 1957.

Geminos. *Gemini Elementa Astronomiae.* Ad codicum fidem recensuit Germanice interpretatione et commentariis instruxit Carolus Manitius. Leipzig, 1898. BT.

Genzmer, Felix. *See Edda.*

Gering, Hugo. *Kommentar zu den Liedern der Edda*, nach dem Todes des Verf. hrsg. von B. Sijmons. Halle, 1927. Germanische Handbibliothek 7.III.1.

——. *See also Edda.*

Gervasius von Tilbury. *Otia Imperialia.* In einer Auswahl neu herausgegeben und mit Anmerkungen begleitet von Felix Liebrecht. Hannover, 1856.

Gibson, G. E. "The Vedic Nakshatras and the Zodiac," *Festschrift Popper* (1951), pp. 149–65.

Gifford, E. W. *Tongan Myths and Tales.* Bulletin BPB Mus. 8. Honolulu, 1924.

Das Gilgamesch-Epos. Neu übersetzt von Arthur Ungnad und gemeinverständlich erklärt von Hugo Gressmann. Göttingen, 1911.

Das Gilgamesch Epos. Eingeführt, rhythmisch übertragen und mit Anmerkungen versehen von Hartmut Schmökel. Stuttgart, 1966.

Das Gilgamesch-Epos. Neu übersetzt und mit Anmerkungen versehen von Albert Schott. Durchgesehen und ergänzt von Wolfram von Sodern. Stuttgart, 1958.

The Epic of Gilgamesh. An English Version with an introduction by N. K. Sandars. Harmondsworth, 1966. 1st ed. 1960.

Gilgamesh et sa légende. Études recueillies par Paul Garelli à l'occasion de la VIIe Rencontre Assyriologique Internationale Paris 1958, in *Cahiers du Groupe François-Thureau-Dangin,* vol. *1.* Paris, 1960.

Gilgamesh. See also Heidel, A.; Kramer, S. N.; Speiser, E. A.

Gill, W. W. *Myths and Songs from the South Pacific.* London, 1876.

Ginsberg, H. C. (trans.) "Ugaritic Myths, Epics and Legends," ANET, pp. 129–55.

Ginzberg, Louis. *The Legends of the Jews.* Translated from the German ms. by Henrietta Szold. Philadelphia, 1954. 7 vols.

Gladwin, Harold S. *Men Out of Asia.* New York, 1947.

Gössmann, P. F. *Planetarium Babylonicum* (= Deimel, *Sum. Lex. 4.2*). Rome, 1950.

———. *Das Era-Epos.* Würzburg, 1956.

Götze, Albrecht. "Akkadian d/tamtum," *Festschrift Deimel* (1935), pp. 185–91.

Gollancz, I. *Hamlet in Iceland.* London, 1898. Northern Library, vol. *3.*

Gordon, Cyrus H. *Ugaritic Literature.* Rome, 1949.

Granet, Marcel. *Danses et Légendes de la Chine Ancienne.* Paris, 1959. 1st ed. 1926. Annales du Musée Guimet. Bibliothèque d'Études 64.

———. *Chinese Civilization,* K. E. Innes and M. R. Brailsford, trans. New York, 1961.

Grassmann, Hermann. *Wörterbuch zum Rig-Veda,* 3rd ed. Wiesbaden, 1955.

Grégoire, Henri, Goossens, R. and Matthieu, M. *Asklepios, Apollon Smintheus et Rudra.* Brussels, 1949. Mém. Acad. R. de Belgique. Classe des Lettres 45.1.

Grey, Sir George. *Polynesian Mythology,* Christchurch-London, 1956. 1st Eng. ed. 1855.

Griaule, Marcel. "Le Rôle du Silure Clarias Senegalensis dans la Procréation au Soudan Français," *Festschrift Westermann* (1955), pp. 299–311.

———. *Conversations with Ogotemmêli: An Introduction to Dogon Religious Ideas.* With an introduction by Germaine Dieterlen. Oxford, 1965 (= *Dieu d'Eau: Entretiens avec Ogotemmeli.* Paris, 1948).

———. "Symbolisme des Tambours Soudanais," *Mélanges historiques offerts à Paul-Marie Masson,* vol. *1* (1955), pp. 79–86.

Griaule, Marcel, and Dieterlen, Germaine. *Signes Graphiques Soudanais.* Paris, 1951. Actualités Scientifiques et Industrielles 1158.

Grimm, Jacob. *Geschichte der Deutschen Sprache,* 2nd ed. Leipzig, 1853.

————. *Deutsche Mythologie.* Tübingen, 1953. Reprint of the 4th ed. 1876. 1st ed. 1835.

————. *Reinhard Fuchs.* Berlin, 1854.

————. *Teutonic Mythology.* Translated from the 4th ed. with notes and appendix by J. S. Stallybrass. London, 1883. Reprint 1966.

Grimm, Wilhelm. *Die Deutsche Heldensage.* Unter Hinzufügung der Nachträge von Karl Müllenhoff und Oskar Jänicke. Darmstadt, 1957. Reprint from 3rd ed. 1889.

Gruenwedel, Albert. *Altbuddhistische Kultstätten in Chinesisch Turkestan.* Berlin, 1912.

Gruppe, O. *Griechische Mythologie und Religionsgeschichte.* München, 1906. Handbuch der Klassischen Altertumswissenschaft, vol. 5, Abt. 2, pts. 1–2.

Guérin, J. M. G. *Astronomie Indienne.* Paris, 1847.

Gundel, Wilhelm. *De Stellarum Appellatione et Religione Romana.* Giessen, 1907. Religionsgeschichtliche Versuche und Vorarbeiten 2.

————. *Dekane und Dekansternbilder.* Glückstadt, 1936. Stud. Bibl. Warburg 19.

————. *Neue Astrologische Texte des Hermes Trismegistos,* ABAW phil/hist. Kl. N. F. *12.* München, 1936.

Guthrie, W. K. C. *Orpheus and Greek Religion,* 2nd ed. London, 1952.

Haavio, Martti. *Vainamoinen, Eternal Sage.* FFC, vol. *144* (1952).

————. *Der Etanamythos in Finnland.* FFC, vol. *154* (1955).

Hagar, Stansbury. *The Peruvian Asterisms and Their Relation to the Ritual.* 14. Internationaler Amerikanisten-Kongress, 1904 (Stuttgart, 1906), pp. 593–602.

————. "Cherokee Star-Lore," *Festschrift Boas* (1906), pp. 354–66.

Hahn, J. G. von. *Griechische und Albanesische Märchen,* 2nd ed. München-Berlin, 1918. 2 vols.

————. *Sagwissenschaftliche Studien.* Jena, 1876.

Hako, M. *Das Wiesel in der europäischen Volksüberlieferung.* FFC, vol. *167* (1956).

Hamel, J. G. van. "The Game of the Gods," *Arkiv för Nordisk Filologi,* vol. *50* (1934), pp. 218–42.

Harris, J. Rendel. *Boanerges.* Cambridge, 1913.

Harrison, Jane Ellen. *Themis: A Study of the Social Origins of Greek Religions.* New York, 1962. 1st ed. 1912.

Hartmann, R. "Ergeneqon," *Festschrift Georg Jacob* (1932), pp. 68–79.

Hartner, Willy. "The Pseudoplanetary Nodes of the Moon's Orbit in Hindu and Islamic Iconographies," *Ars Islamica,* vol. *5* (1938), pt. 1. Reprinted in Willy Hartner, *Oriens-Occidens* (1968), pp. 349–404.

————. *Le Problème de la Planète Kaïd*. Paris, 1955. Conférences du Palais de la Découverte, sér. D, no. 36. Reprinted in *Oriens-Occidens*, pp. 268–86.

————. "Zur astrologischen Symbolik des 'Wade Cup,'" *Festschrift Kühnel* (1959), pp. 234–43. Reprinted in *Oriens-Occidens*, pp. 405–14.

————. "The Earliest History of the Constellations in the Near East, and the Motif of the Lion-Bull Combat," JNES, vol. *24* (1965), nos. 1, 2. Reprinted in *Oriens-Occidens*, pp. 227–59.

————. *Oriens-Occidens: Ausgewahlte Schriften zur Wissenschafts- und Kulturgeschichte*. Festschrift zum 60. Geburtstag. Hildesheim, 1968. Collectanea 3.

Harva, Uno. *See* Holmberg.

Heidel, Alexander. *The Gilgamesh Epic and Old Testament Parallels*. Chicago, 1963. 1st ed. 1946.

————. *The Babylonian Genesis*. Chicago, 1963. 1st ed. 1942.

Heine-Geldern, Robert von. "Weltbild und Bauform in Südostasien," *Wiener Beiträge zur Kunst- und Kulturgeschichte Asiens*, vol. *4* (1930), pp. 28–78.

Heinzel, Richard. *Über das Gedicht vom König Orendel*. Vienna, 1892. SOAW, phil.-hist. Kl. *126.1*.

Held, G. J. *The Mahabharata*. London, 1935.

Henry, Teuira. *Ancient Tahiti: Based on material recorded by J. M. Orsmond*. Bulletin BPB Mus. 48. Honolulu, 1928.

Herodotus, A. D. Godley, trans. 1956–57. 4 vols. First printed 1920–25, LCL.

Herrmann, Paul. *Die Heldensagen des Saxo Grammaticus*, pt. 2. Erläuterungen zu den ersten neun Büchern. Leipzig, 1922.

————. *Deutsche Mythologie*. Leipzig, 1898.

Herzfeld, Ernst. *Zoroaster and His World*. Princeton, 1947.

Hesiod, Hugh G. Evelyn-White, trans. 1964. First printed 1914, LCL.

Higgins, Godfrey. *Anacalypsis: An attempt to draw aside the Veil of the Saitic Isis, or an Inquiry into the Origins of Languages, Nations and Religions*. New York, 1927. Reprint.

Hinke, W. J. *A New Boundary Stone of Nebuchadnezzar I from Nippur*. Philadelphia, 1907. Babyl. Exped. U. of Pennsylvania, ser. D, no. 4.

Hinze, Oscar Marcel. "Studien zum Verständnis der archaischen Astronomie," *Symbolon, Jahrbuch für Symbolforschung*, vol. *5* (1966), pp. 162–219.

Höfler, Otto. "Cangrande von Verona und das Hundesymbol der Langobarden," in *Brauch und Sitte, Festschrift Eugen Fehrle* (1940), pp. 101–37.

Holma, Harri. *Die Namen der Körperteile im Assyrisch-Babylonischen.* Helsinki, 1911. Ann. Acad. Sci. Fenn. ser. B.7.1.

Holmberg, Uno. *Der Baum des Lebens.* Helsinki, 1922. Ann. Acad. Sci. Fenn. ser. B.*16*.3.

———. "Finno-Ugric and Siberian Mythology," MAR, vol. *4* (1927).

———. *Die religiösen Vorstellungen der altaischen Völker.* Übers. von E. Kunze. FFC, vol. *125* (1938).

Homer. *Homeri Ilias,* W. Dindorf and C. Hentze, eds. Leipzig, 1928–30. BT.

———. *Homeri Odyssea,* W. Dindorf and C. Hentze, eds. Leipzig, 1930. BT.

———. *The Iliad,* W. H. D. Rouse, trans. New York, 1964. 1st ed. Edinburgh, 1938.

———. *The Odyssey,* W. H. D. Rouse, trans. New York, 1961. 1st ed. Edinburgh, 1937.

Hommel, Fritz. "Das 'Reis' des Gilgamis," OLZ, vol. *12* (1909), pp. 473–77.

———. *Die Schwur-Göttin Esch-Ghanna und ihr Kreis.* München, 1912 (= Anhang zu S. A. B. Mercer. *The Oath in Babylonian and Assyrian Literature*).

———. *Ethnologie und Geographie des Alten Orients.* München, 1926. Handbuch der Altertumswissenschaft, Abt. 3, pt. 1, vol. *1*.

Honko, Lauri. "Finnen," in *Wb. Myth.* Stuttgart: Klett. Pt. 6, n.d.; pt. 7, 1965.

Horwitz, Hugo T. "Die Drehbewegung in ihrer Bedeutung für die Entwicklung der materiellen Kultur," *Anthropos,* vol. *28* (1933), pp. 721–75; vol. *29* (1934), pp. 99–126.

Hüsing, G. *Beiträge zur Kyrossage.* Leipzig, 1906.

———. *Die einheimischen Quellen zur Geschichte Elams.* Leipzig, 1916.

Hyde, Thomas. *Veterorum Persarum et Parthorum Religionis Historia,* 2nd ed. Oxford, 1760.

Hyginus. *Hygini Fabulae,* recensuit H. J. Rose. Leiden, 1963. Reprint of 1933 ed.

Ideler, Ludwig. *Historische Untersuchungen über die astronomischen Beobachtungen der Alten.* Berlin, 1806.

———. *Untersuchungen über den Ursprung und die Bedeutung der Sternnamen.* Berlin, 1809. (LXXII pp. Einleitung; pp. 1–372: Zakaria Ben Mahmud El-Kazwini Gestirnbeschreibung deutsch, mit Erläuterungen die Sternnamen betreffend; pp. 373ff.: El-Kazwini Gestirnbeschreibung arabisch.)

Jablonski, Paul Ernst. *Pantheon Aegyptiorum sive de Diis eorum Commentarius.* Frankfurt/Oder, 1750–52. 3 parts.
Jacobi, Hermann G. *Mahabharata: Inhaltsangabe, Index und Konkordanz der Calcuttaer und Bombayer Ausgaben.* Bonn, 1903.
Jacobsen, Thorkild. "Parerga Sumerologica," JNES, vol. *2* (1943), pp. 117–21.
———. "Sumerian Mythology: A Review Article," JNES, vol. *5* (1946), pp. 128–52.
———. *The Sumerian King List.* Chicago, 1964. Assyriological Studies 11.
Jastrow, Morris (Jr.). "Sun and Saturn," RA, vol. 7 (1909), pp. 163–78.
Jensen, Peter. *Die Kosmologie der Babylonier.* Strassburg, 1890.
———. *Assyrisch-Babylonische Mythen und Epen.* Berlin, 1909. Keilinschriftliche Bibliothek, vol. *6*, pt. 1.
Jeremias, Alfred. *Das Alte Testament im Lichte des Alten Orients,* 3rd ed. rev. Leipzig, 1916.
———. *Handbuch der Altorientalischen Geisteskultur,* 2nd ed. rev. Berlin-Leipzig, 1929.
Jiriczek, Otto L. "Hamlet in Iran," *Zschr. des Vereins für Volkskunde,* vol. *10* (1900), pp. 353–64.
Jung, C. G., and Kerényi, K. *Essays on a Science of Mythology: The Myths of the Divine Child and the Divine Maiden.* New York, 1949.
Justi, Ferdinand. *Iranisches Namenbuch.* Marburg, 1895.
Kalevala: The Land of the Heroes, W. F. Kirby, trans. London, 1956. Everyman's Library, vols. *259–60.* 1st ed. 1907.
Kampers, Franz. *Mittelalterliche Sagen vom Paradiese und vom Holze des Kreuzes Christi.* Köln, 1897.
———. *Alexander der Grosse und die Idee des Weltimperiums in Prophetie und Sage.* Freiburg, 1901.
———. *Vom Werdegange der abendländischen Kaisermystik.* Leipzig-Berlin, 1924.
Karsten, G. *Maass und Gewicht in alten und neuen Systemen.* Berlin, 1871.
Kautzsch, E. (ed.) *Die Apokryphen und Pseudoepigraphen des Alten Testaments.* Tübingen, 1900. 2 vols.
Keeler, Clyde S. *Secrets of the Cuna Earthmother.* New York, 1960.
Kees, Hermann. *Der Götterglaube im Alten Aegypten,* 2nd ed. Berlin, 1956.
———. *Der Opfertanz des Aegyptischen Königs.* Leipzig, 1912.
Keimer, Louis. *Interprétation de plusieurs passages d'Horapollon.* Cairo, 1947. Supplement vol. *5* of Annales du Service des Antiquités de l'Égypte.

Keith, A. Berriedale. *Indian Mythology*. MAR, vol. *6* (1917).

Kennedy, E. S. "The Sasanian Astronomical Handbook Zîj-i Shâh, and the Astrological Doctrine of 'Transit' (Mamarr)," JAOS, vol. *78* (1958), pp. 246–62.

Kepler, Johannes. *Opera Omnia*, C. Frisch, ed. Frankfurt-Erlangen, vol. *2* (1859); vol. *4* (1863).

———. "De Trigono Igneo" (=chapters 2–11 of *De Stella Nova in Pede Serpentarii* [Prague, 1606]), in *Gesammelte Werke*, Max Caspar, ed. München, 1938. Vol. *1*, pp. 165–208.

———. *Mysterium Cosmographicum*. Editio altera, Franz Hammer, ed. München, 1963. *Gesammelte Werke*, vol. *8*.

Kerényi, Karl. "Zum Urkind-Mythologem," *Paideuma*, vol. *1* (1938–40), pp. 241–78.

Kern, Otto. See *Orphicorum Fragmenta*.

Keynes, Lord John Maynard. "Newton the Man," in *The Royal Society. Newton Tercentenary Celebrations, 15–19 July 1946* (Cambridge, 1947), pp. 27–34.

Kircher, Athanasius, S.J. *Oedipus Aegyptiacus*. Rome, 1653.

———. *Mundus Subterraneus*. Amsterdam, 1665.

Kirfel, Willibald. *Die Kosmographie der Inder nach den Quellen dargestellt*. Bonn-Leipzig, 1920.

Kleomedes. *Die Kreisbewegung der Gestirne*, A. Czwalina, trans. and commentator. Leipzig, 1927. Ostwalds Klassiker der Naturwissenschaften, vol. *220*.

Klibansky, Raymond, Panofsky, E., and Saxl, F. *Saturn and Melancholy*. London, 1964.

Knapp, Martin. *Antiskia: Ein Beitrag zum Wissen um die Präzession im Altertum*. Basel, 1927.

———. *Pentagramma Veneris*. Basel, 1934.

Krämer, Augustin. *Die Samoa-Inseln*. Stuttgart, 1902. 2 vols.

Kramer, Samuel Noah. *Gilgamesh and the Huluppu-Tree*. Chicago, 1938. Assyriological Studies 10.

———. *Sumerian Mythology*. Philadelphia, 1944.

———. "Dilmun, the Land of the Living," BASOR, vol. *96* (1944), pp. 18–28.

———. "The Epic of Gilgamesh and Its Sumerian Sources," JAOS, vol. *64* (1944), pp. 7–23.

———. "Gilgamesh and the Land of the Living," JCS, vol. *1* (1947), pp. 3–46.

———. *Enmerkar and the Lord of Aratta*. Philadelphia, 1952. U. of Pennsylvania Mus. Mono.

———. *From the Tablets of Sumer*. Indian Hills, Colorado, 1956.

———. *The Sumerians: Their History, Culture, and Character*. Chicago, 1963.

———. (trans.) "Sumerian Myths and Epic Tales," ANET, pp. 37–59.

Krappe, Alexander H. "Apollon Smintheus and the Teutonic Mysing," ARW, vol. *33* (1936), pp. 40–56.

Krause, Ernst. *Tuisko-Land*. Glogau, 1891.

Krickeberg, Walter. *Indianermärchen aus Nordamerika*. Jena, 1924.

———. *Märchen der Azteken und Inkaperuaner, Maya und Muisca*. Jena, 1928.

———. "Mexikanisch-peruanische Parallelen," *Festschrift P. W. Schmidt* (1928), pp. 378–93.

———. "Der mittelamerikanische Ballspielplatz und seine religiöse Symbolik," *Paideuma*, vol. *3* (1944–49), pp. 118–90.

Kritzinger, Hans H. *Der Stern der Weisen: Astronomisch-kritische Studie*. Gütersloh, 1911.

Krohn, Kaarle. "Lappische Beiträge zur germanischen Mythologie," FUF, vol. *6* (1906), pp. 155–80.

———. "Windgott und Windzauber," FUF, vol. 7 (1907), pp. 173ff.

———. *Magische Ursprungsrunen der Finnen*. FFC, vol. *52* (1924).

———. *Kalevalastudien 1. Einleitung*. FFC, vol. *53* (1924).

———. *Kalevalastudien 3. Ilmarinen*. FFC, vol. *71* (1927).

———. *Kalevalastudien 4. Sampo*. FFC, vol. *72* (1927).

———. *Kalevalastudien 5. Vainamoinen*. FFC, vol. *75* (1928).

———. *Kalevalastudien 6. Kullervo*. FFC, vol.*76* (1928).

Kugler, Franz Xaver, S.J. *Sternkunde und Sterndienst in Babel*, Münster i. W., 1907–13.

———. *Drittes Erganzungsheft zum 1.u.2. Buch*, von Johannes Schaumberger. Münster, 1935.

———. *Sibyllinischer Sternkampf und Phaethon in naturgeschichtlicher Beleuchtung*. Münster, 1927.

Kuhn, Adalbert. *Die Herabkunft des Feuers und des Göttertranks*, 2nd ed. Gütersloh, 1886.

Kunitzsch, Paul. *Arabische Sternnamen in Europa*. Wiesbaden, 1959.

Labat, René. *Manuel d'Epigraphie Akkadienne*, 4th ed. Paris, 1963.

The Labyrinth: Further Studies in the Relation Between Myth and Ritual in the Ancient World, Samuel Henry Hooke, ed. London-New York, 1935.

Lagercrantz, Sture. "The Milky Way in Africa," *Ethnos* (1952), pp. 64–72.

Landsberger, Benno. *Die Fauna des Alten Mesopotamien nach der 14. Tafel der Serie HAR-RA = HUBULLU*. Unter Mitwirkung von J.

Krumbiegel, Leipzig, 1934. Abhandlungen d. Sächsischen Akad. d. Wiss. phil-hist. Kl. *42.6.*

———. "Einige unbekannt gebliebene oder verkannte Nomina des Akkadischen," WZKM, vol. *56* (1960), pp. 109–29.

Langdon, Stephen. *Semitic Mythology.* MAR, vol. *5* (1931).

———. *The Legend of Etana and the Eagle, or the Epical Poem "The City They Hated."* Paris, 1932.

Lauth, Franz Joseph. "Horapollon," SBAW (1876), pp. 57–115.

———. *Zodiaques de Denderah.* München, 1865.

Lehmann-Haupt, C. F. *Babyloniens Kulturmission einst und jetzt.* Leipzig, 1903.

Lehmann-Nitsche, Robert. "Das Sternbild des Kinnbackens in Vorderasien und Sudamerika," offprint from *Zweiter Tagungsbericht der Gesellschaft für Völkerkunde.* Leipzig, 1936. 14 pp.

Lessmann, Heinrich. *Die Kyrossage in Europa.* Programm Stadtische Realschule. Charlottenburg, 1906.

Lewy, Hildegard. "Origin and Significance of the Mâgên Dâwîd," *Archiv Orientalni,* vol. *18,* pt. 3 (1950), pp. 330–65.

Lewy, Hildegard and Julius. "The Origin of the Week and the Oldest Westasiatic Calendar," HUCA, vol. *17* (1942–43), pp. 1–152.

———. "The God Nusku," *Orientalia,* vol. *17* (1948), pp. 146–59.

Liebrecht, Felix. *Zur Volkskunde.* Alte und neue Aufsätze. Heilbronn, 1879.

———. *See also* Gervasius von Tilbury.

Littmann, Enno. "Sternensagen und Astrologisches aus Nordabessinien," ARW, vol. *11* (1908), pp. 298–319.

Lockyer, Sir J. Norman. *The Dawn of Astronomy.* Cambridge, Mass., 1964. 1st ed. 1894.

Loeb Classical Library, founded by James Loeb. London: Heinemann; Cambridge, Mass.: Harvard University Press.

Lommel, Hermann. *Die Yäsht's des Awesta, übersetzt und eingeleitet.* Göttingen-Leipzig, 1927. Quellen der Religionsgeschichte.

———. "Kavy Uçan," in *Mélanges Linguistiques offerts à C. Bally* (Geneva, 1939), pp. 209–14.

Loomis, Roger Sherman (ed.) *Arthurian Literature in the Middle Ages.* Oxford, 1959.

———. *The Grail: From Celtic Myth to a Christian Symbol.* Cardiff-New York, 1963.

L'Orange, H. P. *Studies on the Iconography of Cosmic Kingship in the Ancient World.* Inst. Sammenlignende Kulturforskning, ser. A.23. Oslo-London-Wiesbaden-Cambridge, Mass., 1953.

Loth, J. *Les Mabinogion du Livre Rouge de Hergest.* Ed. revue, augmentée. Paris, 1913. 2 vols.

Lucian. "De Dea Syria," A. M. Harmon, trans., in vol. 4 of *Lucian*. LCL.

———. "De Saltatione," "Astrologia," A. M. Harmon, trans., in vol. 5 of *Lucian*. LCL.

———. "Saturnalia," K. Kilburn, trans., in vol. 6 of *Lucian*. LCL.

Lüders, Heinrich. *Das Würfelspiel im Alten Indien*. Göttingen, 1907. Abhdl. Ges. Wiss. New Series 9.2.

———. "Die Sage von Rishyashringa," in *Philologica Indica: Ausgewählte Kleine Schriften von Heinrich Lüders* (1940), pp. 1–42.

———. "Zur Sage von Rishyashringa," in *Philologica Indica*, pp. 47–73.

———. *Varuna*. Aus dem Nachlass herausgegeben von Ludwig Alsdorf. Göttingen, vol. 1, 1951; vol. 2, 1959.

Luedtke, W. "Die Verehrung Tschingis-Chans bei den Ordos-Mongolen. Nach dem Berichte G. M. Potanins aus dem Russischen ubersetzt und erläutert," ARW, vol. 25 (1917), pp. 83–129.

Ludendorff, Hans. "Zur astronomischen Deutung der Maya-Inschriften," SPAW, phys.–math. Kl. (1936), pp. 65–88.

Lycophron, A. W. Mair, trans. 1955. First printing 1921, LCL.

Lydus. *Joannis Laurentii Lydi Liber de Mensibus*, R. Wuensch, ed. Stuttgart, 1967. 1st ed. 1898. BT.

Lynam, Edward. *The Carta Marina of Olaus Magnus: Venice 1539 and Rome 1572*. Jenkintown, Pa., 1949. Tall Tree Library Publ. 12.

Maass, Alfred. "Sternkunde und Sterndeuterei im Malaiischen Archipel," Tijdschrift Ind. Taal-, Land-, en Volkenkunde, vol. 64 (1924–25), pp. 1–172, 347–459; vol. 66 (1926), pp. 618–70.

Maass, Ernst. *Commentariorum in Aratum Reliquiae*. Berlin, 1898. Reprint 1958.

The Mabinogion: A New Translation by Gwyn Jones and Thomas Jones. London, 1949. Everyman's Library, vol. 97.

Macdonell, Arthur Anthony. *A Practical Sanskrit Dictionary*. Oxford, 1929.

McGuire, J. D. *A Study of the Primitive Methods of Drilling*. AR U.S. Nat. Mus. for 1894 (Washington, 1896), pp. 623–756.

Macrobius. *Conviviorum Saturnaliorum Septem Libri* (Latin and French). Traduction nouvelle par Henri Bornecque (I–III) and François Richard (IV–VII). Paris, n.d. Classiques Garnier.

———. *Commentary on the Dream of Scipio*. Translated with introduction and notes by William Harris Stahl. New York, 1952. Records of Civilization, Sources and Studies, vol. 48.

The Mahabharata of Krishna-Dwaipayana Vyasa. Translated into English prose from the original Sanskrit Text by Pratap Chandra Roy, 2nd ed. Calcutta, n.d. 12 vols.

Makemson, Maud W. *The Morning Star Rises: An Account of Polynesian Astronomy*. New Haven-London, 1941.

Manilius. *The Five Books of M. Manilius: Containing a System of the Ancient Astronomy and Astrology. Together with the Philosophy of the Stoicks, done into English Verse with Notes.* London, 1697. Republished in 1953 by the National Astrological Library, Washington, D.C.

————. *M. Manili Astronomicon,* recensuit F. Jacob. Berlin, 1846.

Mannhardt, Wilhelm. *Wald- und Feldkulte.* Vol. *1: Der Baumkultus der Germanen und ihrer Nachbarstämme* (1875); Vol. *2: Antike Wald- und Feldkulte aus nordeuropäischer Überlieferung erläutert* (1877). Berlin.

Mansikka, V. J. *Über russische Zauberformeln mit Berücksichtigung der Blut- und Verrenkungssagen.* Helsinki, 1909. Ann. Acad. Sci. Fenn. ser. B.1.

Marsham, Johannes. *Canon chronicus Aegypticus, Ebraicus, Graecus.* London, 1672.

Massignon, Louis, and others. "Les sept dormants d'Ephèse (Ahl-Al-Kahf) en Islam et en Chrétienté. Recueil documentaire et iconographique," *Revue des Études Islamiques,* Paris, 1954 (1955), pp. 61–112; 1955 (1956), pp. 93–106; 1957, pp. 1–11; 1958, pp. 1–10; 1959, pp. 1–8; 1960, pp. 107–13; 1961, pp. 1–18; 1963, pp. 1–5.

Matthews, W. H. *Mazes and Labyrinths: A General Account of Their History and Developments.* New York-Toronto-Bombay, 1922.

Mayer, Maximilian. *Die Giganten und Titanen in der antiken Sage und Kunst.* Berlin, 1887.

Mayrhofer, Manfred. *Kurzgefasstes Etymologisches Wörterbuch des Altindischen.* Heidelberg, 1956–.

Meier, Gerhard. "Ein Kommentar zu einer Selbstprädikation des Marduk aus Assur," ZA, vol. 47 (1942), pp. 241–46.

Meissner, Bruno. *Alexander und Gilgamos.* Habilitationsschrift. Leipzig, 1894.

————. *Babylonien und Assyrien,* vol. 2. Heidelberg, 1925.

————. *Beiträge zum Assyrischen Wörterbuch.* Chicago, 1932. Assyriological Studies, vol. *1,* pt. 1.

Mette, Hans Joachim. *Sphairopoiia: Untersuchungen zur Kosmologie des Krates von Pergamon.* München, 1936.

Michatz, Paul. *Die Götterlisten der Serie An ilu A-nu-um.* Doctoral Dissertation. Breslau, 1909.

Miscellanea Orientalia dedicata A. Deimel. An. Or., vol. *12.* Rome, 1935.

Monod, T. "Le ciel austral et l'orientation (autour d'un article de Louis Massignon)," Bull. Institut Français d'Afrique Noire *25,* ser. B (1963), pp. 415–26.

Mooney, James. "Myths of the Cherokee," 19th ARBAE 1897–98 (Washington, 1900), pp. 3–548.

Moritz, L. A. *Grain Mills and Flour in Classical Antiquity*. Oxford, 1958.

Movers, Franz Karl. *Die Phönizier*. Aalen, 1967. Reprint of the 1841–56 ed. 3 vols.

Mowinckel, Sigmund. *Die Sternnamen im Alten Testament*. Offprint from Norsk Teologisk Tidsskrift *29* (1928).

Much, R. "Der germanische Himmelsgott," in *Abhandlungen zur germanischen Philologie. Festgabe für R. Heinzel* (Halle, 1898), pp. 189–278.

Müller, J. G. *Geschichte der Amerikanischen Urreligionen*. Basel, 1855.

Müller, W. Max. *Egyptian Mythology*. MAR, vol. *12* (1918).

Mus, Paul. *Barabudur*. Hanoi, 1935.

Musurillo, Herbert, S.J. "The Mediaeval Hymn, Alma Redemptoris," *Classical Journal*, vol. *52* (1957), pp. 171–74.

Nauck, August. See *Tragicorum Graecorum Fragmenta*.

Neckel. *See Edda*.

Needham, Joseph. *Science and Civilization in China*. With the Research Assistance of Wang Lin. Vol. *3: Mathematics and the Sciences of the Heavens and the Earth* (1959); vol. *4.1: Physics* (1962); vol. *4.2: Mechanical Engineering* (1965). Cambridge University Press.

Negelein, Julius von. "Zum kosmologischen System in der ältesten indischen Literatur," OLZ, vol. *29* (1926), cols. 903–07.

Neugebauer, Otto. *The Exact Sciences in Antiquity*, 2nd ed. New York, 1962.

Neugebauer, Otto, and Parker, R. A. *Egyptian Astronomical Texts*. Brown University Bicentennial Publications. Vol. *1: The Early Decans* (1960); vol. *2: The Ramesside Star Clocks* (1964). London.

Nihongi: Chronicles of Japan from the earliest times to A.D. *697*. Translated from the original Chinese and Japanese by W. G. Aston. London, 1956. Reprint.

Nöldeke, Theodor. "Sieben Brunnen," ARW, vol. 7 (1904), pp. 340–44.

Nonnos. *Dionysiaca*, W. H. D. Rouse, trans. 1955–56. First printing 1940. 3 vols. LCL.

Normann, Friedrich. *Mythen der Sterne*. Gotha-Stuttgart, 1924.

Numelin, Ragnar. *Les Migrations Humaines*, traduit par Victor Forbin. Paris, 1939.

Numuzawa, F. K. *Die Weltanfänge in der Japanischen Mythologie*. Paris-Lucerne, 1946.

Nyberg, H. S. *Die Religionen des Alten Iran*. Deutsch von H. H. Schaeder. Osnabrück, 1966. Reprint of the 1938 ed.

Oberhuber, Karl. *Der numinose Begriff ME im Sumerischen*. Innsbruck, 1963. Innsbrucker Beiträge zur Kulturwissenschaft. Sonderheft 17.

O'Curry, Eugene. *On the Manners and Customs of the Ancient Irish:*

A Series of Lectures. Edited with an introduction, appendices, etc., by W. K. Sullivan. London-Edinburgh-Dublin-New York, 1873. 3 vols.

Ohlmarks, A. *Heimdalls Horn und Odins Auge.* Lund-Copenhagen, 1937.

————. *Stellt die mythische Bifröst den Regenbogen oder die Milch-strasse dar?* Medd. Lunds Astron. Observ. ser. II, no. 110 (1941).

Ohrt, F. *The Spark in the Water: An Early Christian Legend—A Finnish Magic Sound.* FFC, vol. 65 (1926).

Olcott, William Tyler. *Star Lore of All Ages.* New York-London, 1911.

Olrik, Axel. *The Heroic Legends of Denmark,* L. M. Hollander, trans. New York-London, 1919.

————. *Ragnarök: Die Sagen vom Weltuntergang.* Übers. von W. Ranisch. Berlin-Leipzig, 1922.

Olschki, Leonardo. "The Wise Men of the East in Oriental Traditions," *Festschrift Popper* (1951), pp. 375–95.

Onians, R. D. *The Origins of European Thought About the Body, the Mind, the Soul, the World, Time, and Fate,* 2nd ed. Cambridge University Press, 1954.

Oppenheim, A. L. "Akkadian pul(u)h(t)u and melammu," JAOS, vol. 63 (1943), pp. 31–34.

————. "Mesopotamian Mythology," Or., vol. *16* (1947), pp. 207–38; vol. *17* (1948), pp. 17–58; vol. *19* (1950), pp. 129–58.

Oriental Studies in Honour of C. R. Pavry. Oxford, 1933.

Orphicorum Fragmenta. Otto Kern, ed. 2nd ed. Berlin, 1963.

Orphei Hymni. W. Quandt, ed. 2nd ed. Berlin, 1962.

Otten, H. *Mythen vom Gotte Kumarbi: Neue Fragmente.* Berlin, 1950. Veröffentlichungen des Instituts für Orientforschung.

Ovid. *Fasti,* Sir James George Frazer, trans. 1951. LCL.

————. *Metamorphoses,* Frank Justus Miller, trans. Vol. *1* (1956); vol. *2* (1958). First printing 1916, LCL.

Pallis, S. A. *The Babylonian Akîtu-Festival.* Copenhagen, 1926. K. Danske Videnskab. Sels. Hist/phil. Medd. *12.1.*

Pâques, Viviana. *L'Arbre Cosmique dans la Pensée populaire et dans la Vie quotidienne du Nord-Ouest Africain.* Paris, 1964. Université de Paris. Travaux et Mémoires de l'Institut d'Ethnologie 70.

Pascal, Blaise. *Oeuvres Completes.* Paris, 1963.

————. *Pensées,* W. F. Trotter, trans. New York, 1941 (Modern Library).

Pausanias. *Description of Greece,* W. H. S. Jones, trans. 1954–55. 4 vols. First printing 1918–35, LCL.

Pechuel-Loesche, E. *Volkskunde von Loango.* Stuttgart, 1907.

Penzer, N. M. *The Ocean of Story, being C. H. Tawney's Translation of Somadeva's Katha Sarit Sagara.* Edited with an introduction, explanatory notes and terminal essay by N. M. Penzer. London, 1924.

Petronius. *The Satyricon,* William Arrowsmith, trans. New York, 1960. New American Library.

Photius. *Bibliothèque.* Texte établi et traduit par René Henry. Paris, 1959–67 [to be continued].

Pindar, Sir John Sandys, trans. 1957. First printing 1915, LCL.

Pingree, David. "Astronomy and Astrology in India and Iran," ISIS, vol. *54* (1963), pp. 229–46.

Pischel, R. and Geldner, K. F. *Vedische Studien.* Stuttgart, 1889–1901. 3 vols.

Plato. *Platonis Opera,* recognovit brevique adnotatione instruxit Ioannes Burnet. Oxford, 1957–59. 5 vols. First published 1900–07. Oxford Classical Texts.

———. *Plato's Cosmology: The Timaeus of Plato translated with a running commentary,* by Francis Macdonald Cornford. New York, 1957. The Library of Liberal Arts, vol. *101.*

———. *The Republic of Plato.* Translated with introduction and notes by Francis Macdonald Cornford. New York–London, 1952. 1st ed. 1941.

———. *Plato's Phaedo: The Phaedo of Plato translated with introduction, notes and appendices,* by R. S. Bluck. New York, 1955. The Library of Liberal Arts, vol. *110.*

———. *The Works of Plato,* selected and edited by Irwin Edman (from the third edition of the Jowett translation). New York, 1956. The Modern Library.

Plutarch. *De Iside et Osiride. De defectu oraculorum,* Frank Cole Babbitt, trans. 1957. First printing 1936, LCL. Moralia, vol. *5.*

———. *De facie quae in orbe lunae apparet,* Harold Cherniss and William C. Helmbold, trans. 1957. LCL. Moralia, vol. *12.*

Pogo, Alexander. "The Astronomical Ceiling Decoration in the Tomb of Senmut (18th Dyn.)," ISIS, vol. *14* (1931), pp. 301–25.

———. "Zum Problem der Identifikation der nördlichen Sternbilder der alten Aegypter," ISIS, vol. *16* (1931), pp. 102–14.

Pokorny, Julius. "Ein neun-monatiges Jahr im Keltischen," OLZ, vol. *21* (1918), cols. 130–33.

Popol Vuh: The Sacred Book of the Ancient Quiché Maya. English version by Delia Goetz and Sylvanus G. Morley from the Spanish translation by Adrián Recinos. London-Edinburgh, 1951.

Potanin, G. M. *See* Luedtke, W.

Preisendanz, Karl (ed. and trans.). *Papyri Graecae Magicae: Die Griechischen Zauberpapyri.* Leipzig–Berlin, 1928. 2 vols.

Preller, L. *Griechische Mythologie*, 5th ed. Bearbeitet von Carl Robert. Berlin-Zürich, 1964.

Proclus. *Procli Diadochi in Platonis Timaeum Commentaria*, E. Diehl, ed. Leipzig, 1903–06. 3 vols. BT. [To be continued.]

———. *Commentaire sur le Timée*. Traduction et notes par A. J. Festugière. Paris, 1966–68. 4 vols.

Pseudo-Callisthenes. *Historia Alexandri Magni*. Vol. *1*: Recensio vetusta, W. Kroll, ed. Berlin, 1958. Reprint from the 1926 ed.

Ptolemy. *Claudii Ptolemaei opera quae exstant omnia*, vols. *1* and *2*: *Syntaxis Mathematica*, J. L. Heiberg, ed. Leipzig, 1898–1903. BT.

———. Ptolemaeus. *Handbuch der Astronomie* [Almagest]. Deutsche Übersetzung und erläuternde Anmerkungen von K. Manitius. Vorwort und Berichtigungen von Otto Neugebauer. Leipzig, 1963. 2 vols. BT.

———. *Tetrabiblos*, F. E. Robbins, trans. 1956. First printing 1940, LCL.

The Pyramid Texts in translation and commentary. Samuel A. B. Mercer, ed. New York-London-Toronto, 1952. 4 vols.

Rabuse, Georg. *Der kosmische Aufbau der Jenseitsreiche Dantes*. Graz-Köln, 1958.

Realencyclopaedie der Classischen Altertumswissenschaft (*Pauly's Realencyclopaedie*). Neu bearbeitet und unter Mitwirkung zahlreicher Fachgenossen herausgegeben von Georg Wissowa. Stuttgart, 1893–.

Reallexikon der Assyriologie. Unter Mitwirkung zahlreicher Fachgelehrter hrsg. von Erich Ebeling und Bruno Meissner. Vol. *3* (1959ff.) Ernst Weidner. Berlin-Leipzig, 1932–.

Reuter, Otto Sigfrid. *Germanische Himmelskunde: Untersuchungen zur Geschichte des Geistes*. München, 1934.

Riemschneider, Margarete. *Augengott und Heilige Hochzeit*. Leipzig, 1953. (= Riemschneider. Fragen zur Vorgeschichtlichen Religion, vol. 1.)

Der Rig-Veda. Aus dem Sanskrit ins Deutsche übersetzt und mit einem laufenden Kommentar versehen von Karl Friedrich Geldner. Cambridge, Mass., 1951. 3 vols. Vol. *4*: Namen- und Sachregister . . . von J. Nobel, 1957. *Harvard Oriental Series* 33–36.

Roeder, Günther. *Urkunden zur Religion des Alten Aegypten*. Jena, 1915. Religiöse Stimmen der Völker.

———. *Altaegyptische Erzählungen und Märchen*. Jena, 1927.

Roennow, Karsten. "Zur Erklärung des Pravargya, des Agnicayana und des Sautrāmanī," *Le Monde Oriental* (1929), pp. 113–73.

Rohde, Erwin. "Zum griechischen Roman," Rh. Mus., vol. *48* (1893), pp. 110–40.

Rooth, Anna B. *The Raven and the Carcass: An Investigation of a Motif in the Deluge Myth in Europe, Asia and North America*. FFC, vol. *186* (1962).

Roscher, W. H. D. (ed.) *Ausführliches Lexikon der griechischen und römischen Mythologie.* Leipzig, 1884–1937. Reprint Hildesheim, 1965. 10 vols.

Roscher, W. H. "Die Legende vom Tode des Grossen Pan," *Fleckeisens Jahrbücher für Classische Philologie,* vol. *38* (1892), pp. 465–77.

———. *Omphalos.* Leipzig, 1913. Abhdl. Sächs. Ges. Wiss. phil/hist. Kl. *29.9.*

———. *Der Omphalosgedanke bei verschiedenen Völkern, besonders den semitischen.* BVSGW, vol. *70.2* (1918).

Roth, Walter E. *An Inquiry into the Animism and Folk-lore of the Guiana Indians.* 30th ARBAE 1908–09 (Washington, 1915), pp. 103–386.

Rydberg, Viktor. *Teutonic Mythology.* Authorized translation from the Swedish by Rasmus B. Anderson. London-Copenhagen-Stockholm-Berlin-New York, 1907.

Sachs, A. (trans.) "Akkadian Rituals," ANET, pp. 331–45.

Sahagún. *Einige Kapitel aus dem Geschichtswerk des Fray Bernardino de Sahagún.* Aus dem Aztekischen übersetzt von Eduard Seler. Stuttgart, 1927.

Sahagún, Fray Bernardino de. *General History of the Things of New Spain. Florentine Codex.* Translated from the Aztec into English with notes and illustrations by Arthur J. O. Anderson and Charles E. Dibble. Santa Fe, New Mexico, 1953. Book 7: *The Sun, Moon, and Stars, and the Binding of the Years.* With an appendix consisting of the first five chapters of Book VII from the *Memoriales con escolios.*

Salonen, Armas. *Die Wasserfahrzeuge in Babylonien nach sumerisch-akkadischen Quellen.* Helsinki, 1936. Studia Orientalia Societas Orientalis Fennica *8.4.*

———. *Nautica Babyloniaca: Eine lexikalische und kulturgeschichtliche Untersuchung.* Helsinki, 1942. Studia Orientalia Societas Orientalis Fennica *11.1.*

———. *Die Landfahrzeuge des Alten Mesopotamien.* Ann. Acad. Sci. Fenn. ser. B.*72.2* (1951).

Sandars, N. K. *See The Epic of Gilgamesh.*

Sandman Holmberg, Maj. *The God Ptah.* Lund, 1946.

Santillana, Giorgio de. *The Origins of Scientific Thought: From Anaximander to Proclus 600 B.C. to 500 A.D.* Chicago-London, 1961.

———. *Prologue to Parmenides.* Princeton, N.J., 1964 (Lectures in Memory of Louise Taft Semple). Reprinted in G. de Santillana, *Reflections on Men and Ideas* (1968), pp. 82–119.

———. *Reflections on Men and Ideas.* Cambridge, Mass., 1968.

———. "The Role of Art in the Scientific Renaissance," in *Critical*

Problems in the History of Science, Marshall Clagett, ed. Madison, Wis., 1959, pp. 33–65. Reprinted in *Reflections on Men and Ideas* (1968), pp. 137–66.

Santillana, G. de, and Pitts, W. "Philolaos in Limbo, or, What happened to the Pythagoreans?" ISIS, vol. *42* (1951), pp. 112–20. Reprinted in *Reflections on Men and Ideas*, pp. 190–201.

Saussure, Léopold de. *Les Origines de l'Astronomie Chinoise*. Paris, 1930. Reproduction photomécanique posthume d'articles parus dans le *T'oung Pao*.

Saxo Grammaticus. *The First Nine Books of the Danish History of Saxo Grammaticus*. Translated by Oliver Elton with some considerations by Frederick York Powell. London, 1894. Folk-Lore Society Public. 33.

———. *Saxonis Gesta Danorum*, primum a C. Knabe et P. Herrmann recensita recognoverunt et ediderunt J. Olrik and H. Raeder. Vol. *1: Textum continens* (1931); vol. *2: Indicem verborum conficiendum curavit Franz Blatt* (1957). Copenhagen.

Schaeder, Hans Heinrich. "Der iranische Zeitgott und sein Mythos," ZDMG, vol. *95* (1941), pp. 268–99.

Schaumberger. *See* Kugler.

Scheftelowitz, J. *Die Zeit als Schicksalsgottheit in der indischen und iranischen Religion*. Stuttgart, 1929.

Scherer, Anton. *Gestirnnamen bei den indogermanischen Völkern*. Heidelberg, 1953.

Schirren, C. *Die Wandersagen der Neuseeländer und der Mauimythos*. Riga, 1856.

Schlegel, Gustave. *L'Uranographie Chinoise*. Leiden, 1875. Reprint Taipei, 1967.

Schmidt, Leopold. "Pelops und die Haselhexe," *Laos*, vol. *1* (1951), pp. 67–78.

Schmidt, Pater Wilhelm. *Der Ursprung der Gottesidee*, vol. II: *Die asiatischen Hirtenvölker*. Münster i. W., 1954.

Schmökel, Hartmut. *See Das Gilgamesch Epos*.

Schneider, Marius. "Pukku und Mikku: Ein Beitrag zum Aufbau und zum System der Zahlenmystik des Gilgamesch-Epos," *Antaios*, vol. *9* (1967), pp. 262–83.

Schott, Albert. "Das Werden der babylonisch-assyrischen Positions-Astronomie und einige seiner Bedingungen," ZDMG, vol. *88* (1934), pp. 302–37.

———. "Zu meiner Übersetzung des Gilgamesch-Epos," ZA, vol. *42* (1934), pp. 92–143.

———. *See also* Aratos; *Das Gilgamesch-Epos*.

Schröder, Franz Rolf. *Altgermanische Kulturprobleme*. Berlin, 1929.

———. *Skadi und die Götter Skandinaviens.* Tübingen, 1941.
Scriptores Rerum Mythicarum Latini Tres Romae Nuper Reperti. Edidit ac scholiis illustravit Georgius Henricus Bode. Hildesheim, 1968. 2 vols. 1st ed. Celle, 1834.
Sède, Gérard de. *Les Templiers sont parmi nous.* Paris, 1962.
Seler, Eduard. *Codex Vaticanus Nr. 3773 (Codex Vaticanus B).* Berlin, 1902.
———. *Gesammelte Abhandlungen zur Amerikanischen Sprach- und Altertumskunde.* Graz, 1960–61. 5 vols. Reprint of the 1902–23 ed., Berlin.
———. See also Sahagún.
Semitic and Oriental Studies presented to William Popper. Berkeley, 1951.
Servii Grammatici qui feruntur in Vergilii Carmina Commentarii. Rec. G. Thilo and H. Hagen. Hildesheim, 1961. Reprint of the 1884 ed., Leipzig.
Setälä, E. N. "Kullervo-Hamlet," FUF, vol. 3 (1903), pp. 61–97; vol. 7 (1907), pp. 188–264; vol. 10 (1910), pp. 44–127.
The Shatapatha-Brāhmana according to the Text of the Mādhyandina School, Julius Eggeling, trans. Delhi, 1963. Reprint from the 1882–1900 ed. 5 vols. SBE 12, 26, 41, 43, 44.
Sicard, Harald von. *Ngoma Lungundu: Eine afrikanische Bundeslade.* Uppsala, 1952. Studia Ethnographica Upsaliensia 3.
Sieg, Emil. *Die Sagenstoffe des Rigveda und die indische Itihāsatradition.* Stuttgart, 1902.
Simeon, Remi. *Dictionnaire de la Langue Nahuatl.* Graz, 1964. Reprint of the 1885 ed.
Simrock, Karl. *Der ungenähte Rock oder König Orendel, wie er den grauen Rock gen Trier brachte.* Gedicht des 12. Jahrhunderts, übersetzt. Stuttgart, 1845.
———. *Handbuch der Deutschen Mythologie,* 3rd ed. Bonn, 1869.
———. *Die Quellen des Shakespeare in Novellen, Märchen und Sagen.* Bonn, 1870.
———. See also Edda.
Singer, Charles, Holmyard, E. J., and Hall, A. R. (eds.) *A History of Technology,* vol. 1. Oxford, 1954.
Smith, Sidney. "b/pukk/qqu and mekku," RA, vol. 30 (1933), pp. 153–68.
Smith, S. Percy (ed. and trans.) *The Lore of the Whare-wānanga.* New Plymouth, 1913. Memoirs Polynesian Society 3.
Smith, W. Robertson. *The Religion of the Semites.* New York, 1957. Reprint.
Snorri Sturluson. See Edda.

Soden, Wolfram von. "Licht und Finsternis in der sumerischen und babylonisch-assyrischen Religion," *Studium Generale*, vol. *13* (1960), pp. 647–53.

————. "Neue Bruchstucke zur sechsten und siebenten Tafel des Weltschöpfungsepos Enuma elish," ZA, vol. *47* (1942), pp. 1–26.

————. "Zu einigen altbabylonischen Dichtungen," Or., vol. *26* (1957), pp. 306–20.

Sörensen, S. *An Index to the Names in the Mahabharata*. With short explanations and a concordance to the Bombay and Calcutta editions and P. C. Roy's translation. Delhi, 1963. 1st ed. London, 1904.

Speck, Frank G., and Moses, Jesse. *The Celestial Bear Comes Down to Earth: The Bear Sacrifice Ceremony of the Munsee-Mohican in Canada, as related by Nekatcit*. Reading, Pennsylvania, 1945. Reading Public Museum and Art Gallery, Scientific Public. 7.

Speiser, E. A. (trans.) "Akkadian Myths and Epics," ANET (1966), pp. 60–119.

Stegemann, Viktor. *Astrologie und Universalgeschichte: Studien und Interpretationen zu den Dionysiaka des Nonnos von Panopolis*. Stoicheia, vol. *9* (1930).

Stein, Sir Aurel. *Innermost Asia: Detailed Report of Explorations in Central Asia, Kan-Su and Eastern Iran*, vol. *3*. Oxford, 1928.

Stimson, J. Frank. *Tuamotuan Religion*. Bulletin BPB Mus. 103. Honolulu, 1933.

————. *The Cult of Kiho-Tumu*. Bulletin BPB Mus. 111. Honolulu, 1934.

————. *The Legends of Maui and Tahaki*. Bulletin BPB Mus. 127. Honolulu, 1934.

Stokes, Whitley. "The Prose Tales in the Rennes Dindsenchas," *Revue Celtique*, vol. *15* (1894), pp. 272–336, 418–84; vol. *16* (1895), pp. 31–83, 135–67, 269–307.

Storck, John, and Teague, W. D. *Flour for Man's Bread: A History of Milling*. Minneapolis-London, 1952.

Ström, Ake V. "Indogermanisches in der Volüspa," *Numen*, vol. *14* (1967), pp. 167–208.

Stucken, Eduard. *Astralmythen*. Leipzig, 1896–1907.

————. *Der Ursprung des Alphabets und die Mondstationen*. Leipzig, 1913.

Studies in honor of Benno Landsberger on his 75th Birthday. Chicago, 1965. Assyriological Studies 16.

Stutterheim, W. *Râma-Legenden und Râma-Reliefs in Indonesien*. München, 1925.

Sūrya Siddhānta. Translation of the *Sūrya Siddhānta* with notes and an appendix by Rev. Ebenezer Burgess. Reprinted from the 1860 ed., P.

Gangooly, ed. With an introduction by P. Sengupta. U. of Calcutta, 1935.

Taittiriya Sanhita: The Veda of the Black Yajus School, entitled Taittiriya Sanhita. Translated from the original Sanskrit prose and verse by Arthur Berriedale Keith. Delhi, 1967. Reprinted from the 1914 ed., *Harvard Oriental Series* 18–19.

Tallqvist, Knut. *Sumerisch-akkadische Namen der Totenwelt.* Helsinki, 1934. Studia Orientalia Soc. Orient. Fenn. 5.4.

———. *Akkadische Götterepitheta.* Helsinki, 1938. Studia Orientalia Soc. Orient. Fenn. 7.

Taqizadeh, S. H. *Old Iranian Calendars.* London, 1938.

Taylor, A. D. *A Commentary on Plato's Timaeus.* Oxford, 1928.

Testa, Monsignor Domenico. *Il Zodiaco di Dendera Illustrato.* Genova, 1822.

Theophrastus. *Enquiry into Plants,* Sir Arthur Hort, trans. 1958–59. 2 vols. First printing 1916, LCL.

Thieme, P. *Untersuchungen zur Wortkunde und Auslegung des Rigveda.* Halle, 1949.

Thureau-Dangin, F. *Rituels Accadiens.* Paris, 1921.

Thurneysen, Rudolf. *Die irische Helden- und Königssage bis zum 17. Jahrhundert.* Halle, 1921.

Tilak, Bal Gangadhar. *The Orion, or Researches into the Antiquity of the Vedas.* Bombay, 1893.

Tragicorum Graecorum Fragmenta, A. Nauck, ed. Hildesheim, 1964. Reprint. BT.

Tregear, Edward. *The Maori-Polynesian Comparative Dictionary.* Wellington, 1891.

Ungnad, Arthur. *Das wiedergefundene Paradies.* Breslau, 1923. Kulturfragen 3.

———. *See also Das Gilgamesch-Epos.*

Usener, Hermann. *Götternamen: Versuch einer Lehre von der religiösen Begriffsbildung,* 3rd ed. Frankfurt, 1948.

Vajda, Laszlo. "Zur phaseologischen Stellung des Schamanismus," *Ural-Altaische Jahrbücher,* vol. 31 (1959), pp. 456–85.

Varāhamihira. *The Brihad Sanhita,* H. Kern, trans. JRAS, vol. 5 (1871), pp. 45–90, 231–88.

Vatican Mythographers. *See Scriptores Rerum Mythicarum Latini.*

Virgil, H. Rushton Fairclough, trans. Rev. ed. 2 vols. 1954–56. LCL.

Vishnu Purana. A System of Hindu Mythology and Tradition, translated from the original Sanskrit and illustrated by notes by H. H. Wilson, 3rd ed. Calcutta, 1961. 1st ed. London, 1840.

Vries, Jan de. *Altnordisches Etymologisches Wörterbuch.* Leiden, 1961.

Waerden, B. L. van der. "The Thirty-Six Stars" (Babylonian Astronomy 2), JNES, vol. 8 (1949), pp. 6–26.

———. Die Anfänge der Astronomie (Erwachende Wissenschaft 2). Groningen, n.d.

Wainwright, G. A. "A Pair of Constellations," in Studies Presented to F. L. Griffith (1932), pp. 373–83.

Walde-Hofmann. Lateinisches Etymologisches Wörterbuch, von A. Walde. 3rd rev. ed. by J. B. Hofmann. Heidelberg, 1938–56.

Weber, Albrecht. "Die Vedischen Nachrichten von den Naxatra (Mondstationen)," pt. 1: APAW 1860, pp. 283–332; pt. 2: APAW 1862, pp. 267–399.

———. "Miszellen aus dem indogermanischen Familienleben," in Festgruss an Rudolf von Roth (1893), pp. 135–38.

Weidner, Ernst F. Handbuch der Babylonischen Astronomie. Leipzig, 1915. Assyriologische Bibliothek 23.1.

———. "Babylonische Hypsomatabilder," OLZ, vol. 22 (1919), pp. 10–16.

———. "Ein babylonisches Kompendium der Himmelskunde," AJSL, vol. 40 (1924), pp. 186–208.

———. "Gestirn-Darstellungen auf Babylonischen Tontafeln," SOAW, vol. 254 (1967), no. 2.

———. See also Reallexikon der Assyriologie.

Weinreich, Otto. "Zum Tode des Grossen Pan," ARW, vol. 13 (1910), pp. 467–73.

Werner, Edward T. C. Myths and Legends of China. London-Calcutta, 1927.

———. A Dictionary of Chinese Mythology. New York, 1961. 1st ed. Shanghai, 1932.

Werner, Helmut. "Die Verstirnung des Osiris-Mythos," IAfE, vol. 16 (1952), pp. 147–62.

Wheeler, Post. The Sacred Scriptures of the Japanese. New York, 1952.

White, Lynn, Jr. Medieval Technology and Social Change. Oxford, 1962.

Wiedemann, Alfred. Herodots Zweites Buch, mit sachlichen Erläuterungen. Leipzig, 1890.

Wildhaber, Robert. Das Sündenregister auf der Kuhhaut. FFC, vol. 163 (1955).

Williamson, Robert W. Religious and Cosmic Beliefs of Central Polynesia. Cambridge, 1924. 2 vols.

Witzel, Maurus. "Texte zum Studium Sumerischer Tempel und Kultzentren," An. Or., vol. 4 (1932).

———. "Tammuz-Liturgien und Verwandtes," An. Or., vol. 10 (1935).

————. "Zur sumerischen Mythologie," Or., vol. *17* (1948), pp. 393–415.

Wörterbuch der Aegyptischen Sprache. Im Auftrag der Deutschen Akademien hrsg. von Adolf Erman und Hermann Grapow. 2nd reprint. Berlin, 1957. 6 vols.

Wörterbuch der Mythologie. Herausgegeben von H. W. Haussig. Stuttgart, n.d. Pt. 1 ca. 1960.

Wohleb, L. "Die altrömische und hethitische evocatio," ARW, vol. *25* (1927), pp. 206–09.

Wou Tch'eng Ngen. *Si Yeou ki, ou le Voyage en Occident.* Traduit du Chinois par Louis Avenol. Paris, 1957. 2 vols.

Zaehner, R. C. *Zurvan: A Zoroastrian Dilemma.* Oxford, 1955.

Zahan, D., and de Ganay, S. "Études sur la cosmologie des Dogon et des Bambara du Soudan Français," *Africa,* vol. *21* (1951), pp. 13–23.

Zenker, Rudolf. *Boeve-Amlethus: Das Altfranzösische Epos von Boeve de Hamtone und der Ursprung der Hamletsage.* Berlin, 1905. Literarhistorische Forschungen 32.

Zerries, Otto. "Sternbilder als Ausdruck jägerischer Geisteshaltung in Südamerika," *Paideuma,* vol. *5* (1951), pp. 220–35.

Zimmern, Heinrich. "Zum babylonischen Neujahrhfest," BVSGW, vol. *58* (1906), pp. 126–56; vol. *70* (1918), pt. 3, 52 pp.

————. "Die sieben Weisen Babyloniens," ZA, vol. *35* (1923), pp. 151–54.

————. "Zur Herstellung der grossen babylonischen Götterliste An-(ilu) Anum," BVSGW, vol. *63* (1911), pt. 4.

Zinzow, Adolf. *Die Hamletsage an und mit verwandten Sagen erläutert.* Halle, 1877.

The Zohar. Translated by Harry Sperling and Maurice Simon. London, 1956. 5 vols. 1st ed. 1933.

INDEX

INDEX